AF389408

Light-Controlled Modulation and Analysis of Neuronal Functions

Light-Controlled Modulation and Analysis of Neuronal Functions

Editors

Piotr Bregestovski
Carlo Matera

MDPI • Basel • Beijing • Wuhan • Barcelona • Belgrade • Manchester • Tokyo • Cluj • Tianjin

Editors
Piotr Bregestovski
Institut National de la Santé
et de la Recherche Médicale,
Institut de Neurosciences des
Systèmes
Aix-Marseille University
Marseille
France

Carlo Matera
Department of
Pharmaceutical Sciences
University of Milan
Milano
Italy

Editorial Office
MDPI
St. Alban-Anlage 66
4052 Basel, Switzerland

This is a reprint of articles from the Special Issue published online in the open access journal *International Journal of Molecular Sciences* (ISSN 1422-0067) (available at: www.mdpi.com/journal/ijms/special_issues/light_modulation).

For citation purposes, cite each article independently as indicated on the article page online and as indicated below:

LastName, A.A.; LastName, B.B.; LastName, C.C. Article Title. *Journal Name* **Year**, *Volume Number*, Page Range.

ISBN 978-3-0365-6563-7 (Hbk)
ISBN 978-3-0365-6562-0 (PDF)

Contents

International Journal of
Molecular Sciences

Editorial

Light-Controlled Modulation and Analysis of Neuronal Functions

Carlo Matera [1,*] and Piotr Bregestovski [2,3,4,*]

[1] Department of Pharmaceutical Sciences, University of Milan, 20133 Milan, Italy
[2] Institut National de la Santé et de la Recherche Médicale, Institut de Neurosciences des Systèmes, Aix-Marseille University, 13005 Marseille, France
[3] Institute of Neurosciences, Kazan State Medical University, 420111 Kazan, Russia
[4] Department of Normal Physiology, Kazan State Medical University, 420111 Kazan, Russia
* Correspondence: carlo.matera@unimi.it (C.M.); piotr.bregestovski@univ-amu.fr (P.B.)

Citation: Matera, C.; Bregestovski, P. Light-Controlled Modulation and Analysis of Neuronal Functions. *Int. J. Mol. Sci.* **2022**, *23*, 12921. https://doi.org/10.3390/ijms232112921

Received: 20 October 2022
Accepted: 21 October 2022
Published: 26 October 2022

Publisher's Note: MDPI stays neutral with regard to jurisdictional claims in published maps and institutional affiliations.

Light is an extraordinary tool allowing us to read out and control neuronal functions thanks to its unique properties: it has a great degree of bioorthogonality and is minimally invasive; it can be precisely delivered with high spatial and temporal precision; and it can be used simultaneously or consequently at multiple wavelengths and locations. In the last 15 years, light-based methods have revolutionized the way we analyse and control biological systems. It turns out that with the help of light it is possible to study the functions of cells and cell ensembles [1–3], regulate the functions of deoxyribonucleic acid (DNA) [4], perform the photocontrol of peptide conformations [5], modulate the activity of voltage-gated and receptor-operated channels [6–8], and measure the concentrations of ions [9,10] and other cellular components [11,12]. It has become possible to control the behaviour of organisms [13,14], as well as to explore new ways of treating certain diseases [15–17]. Three novel approaches have been emerging more than others: optogenetics, optosensorics, and photopharmacology (optopharmacology).

Optogenetics is an elegant method that combines optical technology and genetic engineering to control the functions of biological systems (cells, tissues, organs, organisms) genetically modified to express photosensitive proteins [2]. By stimulating with light, this method provides high-spatiotemporal and high-specificity resolutions, in contrast to conventional pharmacological or electrical stimulation [18,19]. Optogenetics provides a route to study synaptic circuits [20] and underlying movement diseases [21,22] and has become an effective technology to revolutionize brain research for the therapy of vision [23–25], cardiovascular [26], and neurodegenerative disorders [27,28]. It has also been extended to other biomedical fields [29,30].

Optosensorics is the direction of research exploring the possibility of non-invasively monitoring intracellular ion concentrations and the activity of enzymes, lipids, and other cellular components using specific optical sensors. Most of the sensors developed in the past years are genetically encoded tools. These probes possess fluorophore groups capable of changing fluorescence when interacting with certain ions or molecules [31]. The use of modern optical and fluorescent technologies in combination with molecular genetic techniques has provided the possibility for the biosensoric monitoring of reactive oxidative species [32,33], cAMP [34], glucose [35,36], glutamate [37,38], pyruvate [39], lactate [40], different ions [41,42], and neuronal activity [43].

In contrast to the above-mentioned methods, photopharmacology relies on synthetic photoswitchable ligands that are externally supplied and usually does not require genetic manipulation. Photopharmacological agents are obtained via the incorporation of a molecular photoswitch (e.g., azobenzene and its derivatives) into the structure of a biologically active compound, so that light can be used as an external signal to "switch" the ligand conformation and hence its activity between two different states (ideally "off" and "on") [44,45]. This is ultimately performed to achieve control over biological systems with a high degree of temporal and spatial precision for investigational and therapeutic purposes. Potential applications of photopharmacology include neurobiology [46–48], cancer [49,50], microbial infections [51], pain [52,53], and blindness [54,55], among others.

In this Special Issue, several papers focus on neurobiological aspects, using light as a tool for experimental analysis. In the paper published by Gerasimov et al., the authors describe an optogenetic approach to stimulate astrocytes with the aim to modulate neuronal activity [56]. They applied light stimulation to astrocytes expressing a version of ChR2 (ionotropic opsin) or Opto-α1AR (metabotropic opsin). Optimal optogenetic stimulation parameters were determined using patch-clamp recordings of hippocampal pyramidal neurons' spontaneous activity in brain slices as a readout. They observed that the activation of the astrocytic Opto-α1AR, but not ChR2, results in an increase in the fEPSP slope in hippocampal neurons. The authors conclude that Opto-α1AR expressed in hippocampal astrocytes provides an opportunity to modulate the long-term synaptic plasticity optogenetically and suggest that this approach may potentially be used to normalize the synaptic transmission and plasticity defects in a variety of neuropathological conditions, including models of Alzheimer's disease and other neurodegenerative disorders.

In another work, Abd El-Aziz et al. used an optogenetic system to study the activity of the epithelial Na^+ channel (ENaC) [57]. The activity of ENaC is strongly dependent on the membrane phospholipid phosphatidylinositol 4,5-bisphosphate (PIP2). PIP2 binds two distinct cationic clusters within the N termini of β- and γ-ENaC subunits (βN1 and γN2). The purpose of this study was to determine whether each independent PIP2–ENaC interaction site is sufficient to abolish the response of ENaC to changes in PIP2 levels. The authors had previously determined the affinities of these sites using short synthetic peptides. In this paper, they describe their role in sensitizing ENaC to changes in PIP2 levels in the cellular system. For this purpose, they compared the effects of PIP2 depletion and recovery on ENaC channel activity and intracellular Na^+ levels $[Na^+]_i$. They tested the effects on ENaC activity with mutations to the PIP2 binding sites using the optogenetic system CIBN/CRY2-OCRL to selectively deplete PIP2. Whole cell patch-clamp measurements showed a complete lack of response to PIP2 depletion and recovery in ENaC with mutations to βN1 or γN2 or both sites compared to wild-type ENaC. These results suggest that the βN1 and γN2 sites on ENaC are each necessary to permit maximal ENaC activity in the presence of PIP2.

Lilja et al. presented the results obtained with a novel optogenetic tool for the investigation of neuroplasticity [58]. The activation of tropomyosin receptor kinase B (TrkB), the receptor of brain-derived neurotrophic factor, plays a key role in induced juvenile-like plasticity (iPlasticity), which allows the restructuring of neural networks in adulthood. In this work, they evaluated the utility of a new, highly sensitive, optically activatable tropomyosin receptor kinase B receptor, named OptoTrkB (E281A). OptoTrkB (E281A) was successfully transduced in parvalbumin-positive (PV+) interneurons and in alpha calcium/calmodulin-dependent protein kinase type II positive (CKII+) pyramidal neurons specifically. Light stimulation through transparent skulls or even through a high-opacity barrier (intact skull and fur) promoted the phosphorylation of ERK and CREB, downstream signals of TrkB, in the neurons expressing optoTrkB (E281A) at a certain level. Their results indicate that this highly sensitive optoTrkB (E281A) receptor can be activated using wireless optogenetic methods and can thus be used for a broad range of behavioural studies. Overall, these findings show that the highly sensitive optoTrkB (E281A) can be used in iPlasticity studies of both inhibitory and excitatory neurons, with flexible stimulation protocols in behavioural studies.

The application of novel optogenetic/chemogenetic tools and advanced in vitro models, including those based on iPSC-derived cells, organoids, or utilizing 3D brain-on-chip platforms, are of great importance for the development of new therapeutic options and the assessment of aberrant neurogenesis in Parkinson's-type neurodegeneration. In their review article, Salmina et al. discuss current approaches to assess neurogenesis and prospects in the application of optogenetic protocols to restore neurogenesis in patients with Parkinson's disease [59].

In their paper, Ivashkina et al. present a methodological development and experimental application of a genetically engineered optosensoric tool that allows the imaging of activity-induced neuronal c-fos expression [60]. The authors show a novel light-controlled approach for the long-term analysis of calcium activity in the cortical neurons that were specifically tagged through c-fos expression during a particular cognitive episode. In addition, using in vivo two-photon imaging of Fos-GFP PtA neurons, they report specific changes in neuronal activity in this cortical area during both the acquisition and retrieval of associative fear memory. Taken together, these results suggest that Fos-Cre-GCaMP mice are suitable for the investigation of calcium activity in the neurons which were specifically activated during a particular learning episode.

In the work by Zhilyakov et al., the authors reported the results from their investigation on the relationship between the nicotine-induced autoregulation of acetylcholine (ACh) release and the changes in the concentration of presynaptic calcium levels [61]. For this purpose, they used a pharmacological approach, electrophysiological techniques, and a method for the optical registration of changes in calcium levels in the motor nerve endings. The authors found that an agonist of nicotinic receptors (at a concentration not significantly affecting the state of the postsynaptic membrane) leads to a decrease in the amount of released ACh quanta. This effect is accompanied not by a decrease but by an increase in calcium ion entry into the motor nerve terminal. These results suggest that the nicotinic cholinergic receptors responsible for the mechanism of ACh release autoregulation are the receptors of neuronal type. Activation of these receptors leads to the upregulation of the Cav1 type of VGCCs, resulting in the enhancement of Ca^{2+} entry into the nerve ending. This study adds new elements to the understanding of the cholinergic system functioning and the envisioning of novel potential approaches for the treatment of diseases associated with cholinergic dysfunctions.

In the field of optosensorics, Ponomareva et al. described the effectiveness of the genetically encoded biosensor, ClopHensor, expressed in transgenic mice for the estimation of $[H^+]_i$ and $[Cl^-]_i$ concentrations in brain slices [62]. They performed simultaneous monitoring of $[H^+]_i$ and $[Cl^-]_i$ under different experimental conditions, including changing external concentrations of ions (Ca^{2+}, Cl^-, K^+, Na^+) and the synaptic stimulation of Shaffer's collaterals of hippocampal slices. The results obtained illuminate the different pathways regulating Cl^- and pH equilibrium in neurons and demonstrate that ClopHensor, expressed in transgenic mice, represents an efficient tool for the non-invasive monitoring of intracellular Cl^- and H^+ ions.

In their work, Sotskov et al. performed in vivo imaging of neuronal activity in the CA1 field of the mouse hippocampus using genetically encoded green calcium indicators, including the novel NCaMP7 and FGCaMP7, designed specifically for in vivo calcium imaging [63]. The purpose of this study was to investigate the dynamics of the initial place field formation in the mouse hippocampus. Their data show that neuronal activity recorded with genetically encoded calcium sensors revealed fast behaviour-dependent plasticity in the mouse hippocampus, resulting in the rapid formation of place fields and population activity that allowed the reconstruction of the geometry of the navigated maze. Taken together, these results reveal the fast emergence and tuning dynamics of place cell codes and demonstrate the applicability of novel calcium indicators NCaMP7 and FGCaMP7 in the light-controlled analysis of neural functions in behaving mice.

In the field of photopharmacology, Matera et al. reported a novel photoswitchable ligand that enables reversible spatiotemporal control of dopaminergic transmission [64]. They demonstrated that this new photoswitch, named azodopa, activates D_1-like receptors in vitro in a light-dependent manner. Moreover, azodopa enables reversibly photocontrolling zebrafish motility on a timescale of seconds and allows separating the retinal component of dopaminergic neurotransmission. Finally, they proved that azodopa increases the overall neural activity in the cortex of anesthetized mice and displays illumination-dependent activity in individual neurons. Azodopa is the first photoswitchable dopamine agonist with demonstrated efficacy in wild-type animals and opens the way forward to remotely controlling dopaminergic neurotransmission for investigational and therapeutic purposes.

Finally, the review article by Nin-Hill et al. discusses the application of structure-based computational methods, such as homology modelling, molecular docking, molecular dynamics, and enhanced sampling techniques, to photoswitchable ligands targeting voltage- and ligand-gated ion channels [65]. Notably, the examples presented by the authors show how the integration of computational modelling with experimental data can greatly facilitate photoswitchable ligand design and optimization and provide structural insights to understand the observed light-regulated effects. They conclude by stating that the latest advances in structural biology will further support computer-assisted approaches in photopharmacology.

Overall, this Special Issue of the *International Journal of Molecular Sciences* contains original contributions reporting recent advances in the fields of optogenetics, optosensorics, and photopharmacology, as well as review papers discussing key achievements and prospects in the field. Therefore, we expect that these articles will be of interest to many scientists working with light-based biological methods and will inspire further investigations in relevant research areas.

Author Contributions: Conceptualization, C.M. and P.B.; writing—original draft preparation, C.M. and P.B.; writing—review and editing, C.M. and P.B. All authors have read and agreed to the published version of the manuscript.

Funding: This research was funded by the Russian Scientific Foundation (project no. 18-15-00313 to P.B.).

Conflicts of Interest: The authors declare no conflict of interest.

References

1. Deisseroth, K. Optogenetics. *Nat. Methods* **2011**, *8*, 26–29. [CrossRef] [PubMed]
2. Deisseroth, K. Optogenetics: 10 years of microbial opsins in neuroscience. *Nat. Neurosci.* **2015**, *18*, 1213–1225. [CrossRef] [PubMed]
3. Kim, C.K.; Adhikari, A.; Deisseroth, K. Integration of optogenetics with complementary methodologies in systems neuroscience. *Nat. Rev. Neurosci.* **2017**, *18*, 222–235. [CrossRef] [PubMed]
4. Kashida, H.; Liang, X.; Asanuma, H. Rational design of functional DNA with a non-ribose acyclic scaffold. *Curr. Organ. Chem.* **2009**, *13*, 1065–1084. [CrossRef]
5. Szymański, W.; Beierle, J.M.; Kistemaker, H.A.; VelemNa, W.A.; Feringa, B.L. Reversible photocontrol of biological systems by the incorporation of molecular photoswitches. *Chem. Rev.* **2013**, *113*, 6114–6178. [CrossRef]
6. Gorostiza, P.; Isacoff, E.Y. Optical switches for remote and noninvasive control of cell signaling. *Science* **2008**, *322*, 395–399. [CrossRef]
7. Banghart, M.R.; Mourot, A.; Fortin, D.L.; Yao, J.Z.; Kramer, R.H.; Trauner, D. Photochromic blockers of voltage-gated potassium channels. *Angew. Chem. Int. Ed. Engl.* **2009**, *48*, 9097–9101. [CrossRef]
8. Bregestovski, P.; Maleeva, G.; Gorostiza, P. Light-induced regulation of ligand-gated channel activity. *Br. J. Pharmacol.* **2018**, *175*, 1892–1902. [CrossRef]
9. Bregestovski, P.; Waseem, T.; Mukhtarov, M. Genetically encoded optical sensors for monitoring of intracellular chloride and chloride-selective channel activity. *Front. Mol. Neurosci.* **2009**, *2*, 15. [CrossRef]
10. Suzuki, J.; Kanemaru, K.; Iino, M. Genetically encoded fluorescent indicators for organellar calcium imaging. *Biophys. J.* **2016**, *111*, 1119–1131. [CrossRef]
11. Imamura, H.; Nhat, K.P.H.; Togawa, H.; Saito, K.; Iino, R.; Kato-Yamada, Y.; Nagai, T.; Noji, H. Visualization of ATP levels inside single living cells with fluorescence resonance energy transfer-based genetically encoded indicators. *Proc. Natl. Acad. Sci. USA* **2009**, *106*, 15651–15656. [CrossRef] [PubMed]
12. Berg, J.; Hung, Y.P.; Yellen, G. A genetically encoded fluorescent reporter of ATP: ADP ratio. *Nat. Methods* **2009**, *6*, 161–166. [CrossRef] [PubMed]
13. Miesenböck, G. Optogenetic control of cells and circuits. *Annu. Rev. Cell Dev. Biol.* **2011**, *27*, 731–758. [CrossRef]
14. Haubensak, W.; Kunwar, P.S.; Cai, H.; Ciocchi, S.; Wall, N.R.; Ponnusamy, R.; Biag, J.; Dong, H.W.; Deisseroth, K.; Callaway, E.M.; et al. Genetic dissection of an amygdala microcircuit that gates conditioned fear. *Nature* **2010**, *468*, 270–276. [CrossRef] [PubMed]
15. Rossi, M.A.; Calakos, N.; Yin, H.H. Spotlight on movement disorders: What optogenetics has to offer. *Mov. Disord.* **2015**, *30*, 624–631. [CrossRef] [PubMed]

16. Häusser, M. Optogenetics—The Might of Light. *N. Engl. J. Med.* **2021**, *385*, 1623–1626. [CrossRef]
17. Berry, M.H.; Holt, A.; Broichhagen, J.; Donthamsetti, P.; Flannery, J.G.; Isacoff, E.Y. Photopharmacology for vision restoration. *Curr. Opin. Pharmacol.* **2022**, *65*, 102259. [CrossRef]
18. Tye, K.M.; Deisseroth, K. Optogenetic investigation of neural circuits underlying brain disease in animal models. *Nat. Rev. Neurosci.* **2012**, *13*, 251–266. [CrossRef]
19. Häusser, M. Optogenetics: The age of light. *Nat. Methods* **2014**, *11*, 1012–1014. [CrossRef]
20. Linders, L.E.; Supiot, L.; Du, W.; D'Angelo, R.; Adan, R.A.; Riga, D.; Meye, F.J. Studying Synaptic Connectivity and Strength with Optogenetics and Patch-Clamp Electrophysiology. *Int. J. Mol. Sci.* **2022**, *23*, 11612. [CrossRef]
21. Fougère, M.; van der Zouwen, C.I.; Boutin, J.; Neszvecsko, K.; Sarret, P.; Ryczko, D. Optogenetic stimulation of glutamatergic neurons in the cuneiform nucleus controls locomotion in a mouse model of Parkinson's disease. *Proc. Natl. Acad. Sci. USA* **2021**, *118*, e2110934118. [CrossRef] [PubMed]
22. Parker, K.L.; Kim, Y.; Alberico, S.L.; Emmons, E.B.; Narayanan, N.S. Optogenetic approaches to evaluate striatal function in animal models of Parkinson disease. *Dialogues Clin. Neurosci.* **2016**, *18*, 99–107. [CrossRef] [PubMed]
23. Chaffiol, A.; Duebel, J. Mini-review: Cell type-specific optogenetic vision restoration approaches. *Adv. Exp. Med. Biol.* **2018**, *1074*, 69–73. [PubMed]
24. McClements, M.E.; Staurenghi, F.; MacLaren, R.E.; Cehajic-Kapetanovic, J. Optogenetic gene therapy for the degenerate retina: Recent advances. *Front. Neurosci.* **2020**, *14*, 570909. [CrossRef]
25. Sahel, J.A.; Boulanger-Scemama, E.; Pagot, C.; Arleo, A.; Galluppi, F.; Martel, J.N.; Esposti, S.D.; Delaux, A.; de Saint Aubert, J.B.; de Montleau, C.; et al. Partial recovery of visual function in a blind patient after optogenetic therapy. *Nat. Med.* **2021**, *27*, 1223–1229. [CrossRef]
26. Nussinovitch, U.; Gepstein, L. Optogenetics for in vivo cardiac pacing and resynchronization therapies. *Nat. Biotechnol.* **2015**, *33*, 750–754. [CrossRef]
27. Kastanenka, K.V.; Calvo-Rodriguez, M.; Hou, S.S.; Zhou, H.; Takeda, S.; Arbel-Ornath, M.; Lariviere, A.; Lee, Y.F.; Kim, A.; Hawkes, J.M.; et al. Frequency-dependent exacerbation of Alzheimer's disease neuropathophysiology. *Sci. Rep.* **2019**, *9*, 8964. [CrossRef]
28. Ko, H.; Yoon, S.P. Optogenetic neuromodulation with gamma oscillation as a new strategy for Alzheimer disease: A narrative review. *J. Yeungnam Med. Sci.* **2022**, *39*, 269–277. [CrossRef]
29. Ye, H.; Fussenegger, M. Optogenetic medicine: Synthetic therapeutic solutions precision-guided by light. *Cold Spring Harb. Perspect. Med.* **2019**, *9*, a034371. [CrossRef]
30. Patrono, E.; Svoboda, J.; Stuchlík, A. Schizophrenia, the gut microbiota, and new opportunities from optogenetic manipulations of the gut-brain axis. *Behav. Brain Funct.* **2021**, *17*, 7. [CrossRef]
31. Bolbat, A.; Schultz, C. Recent developments of genetically encoded optical sensors for cell biology. *Biology of the Cell* **2017**, *109*, 1–23. [CrossRef] [PubMed]
32. Belousov, V.V.; Fradkov, A.F.; Lukyanov, K.A.; Staroverov, D.B.; Shakhbazov, K.S.; Terskikh, A.V.; Lukyanov, S. Genetically encoded fluorescent indicator for intracellular hydrogen peroxide. *Nat. Methods* **2006**, *3*, 281–286. [CrossRef]
33. Ermakova, Y.G.; Bilan, D.S.; Matlashov, M.E.; Mishina, N.M.; Markvicheva, K.N.; Subach, O.M.; Subach, F.V.; Bogeski, I.; Hoth, M.; Enikolopov, G.; et al. Red fluorescent genetically encoded indicator for intracellular hydrogen peroxide. *Nat. Commun.* **2014**, *5*, 5222. [CrossRef] [PubMed]
34. Ponsioen, B.; Zhao, J.; Riedl, J.; Zwartkruis, F.; van der Krogt, G.; Zaccolo, M.; Moolenaar, W.H.; Bos, J.L.; Jalink, K. Detecting cAMP-induced Epac activation by fluorescence resonance energy transfer: Epac as a novel cAMP indicator. *EMBO Rep.* **2004**, *5*, 1176–1180. [CrossRef] [PubMed]
35. Deuschle, K.; Okumoto, S.; Fehr, M.; Looger, L.L.; Kozhukh, L.; Frommer, W.B. Construction and optimization of a family of genetically encoded metabolite sensors by semirational protein engineering. *Protein Sci.* **2005**, *14*, 2304–2314. [CrossRef] [PubMed]
36. Takanaga, H.; Chaudhuri, B.; Frommer, W.B. GLUT1 and GLUT9 as major contributors to glucose influx in HepG2 cells identified by a high sensitivity intramolecular FRET glucose sensor. *Biochim. Biophys. Acta BBA Biomembr.* **2008**, *1778*, 1091–1099. [CrossRef] [PubMed]
37. Okumoto, S.; Looger, L.L.; Micheva, K.D.; Reimer, R.J.; Smith, S.J.; Frommer, W.B. Detection of glutamate release from neurons by genetically encoded surface-displayed FRET nanosensors. *Proc. Natl. Acad. Sci. USA* **2005**, *102*, 8740–8745. [CrossRef]
38. Hires, S.A.; Zhu, Y.; Tsien, R.Y. Optical measurement of synaptic glutamate spillover and reuptake by linker optimized glutamate-sensitive fluorescent reporters. *Proc. Natl. Acad. Sci. USA* **2008**, *105*, 4411–4416. [CrossRef]
39. San Martin, A.; Ceballo, S.; Baeza-Lehnert, F.; Lerchundi, R.; Valdebenito, R.; Contreras-Baeza, Y.; Alegria, K.; Barros, L.F. Imaging mitochondrial flux in single cells with a FRET sensor for pyruvate. *PLoS ONE* **2014**, *9*, e85780. [CrossRef]
40. San Martín, A.; Ceballo, S.; Ruminot, I.; Lerchundi, R.; Frommer, W.B.; Barros, L.F. A genetically encoded FRET lactate sensor and its use to detect the Warburg effect in single cancer cells. *PLoS ONE* **2013**, *8*, e57712. [CrossRef]
41. Tian, L.; Hires, S.A.; Mao, T.; Huber, D.; Chiappe, M.E.; Chalasani, S.H.; Petreanu, L.; Akerboom, J.; McKinney, S.A.; Schreiter, E.R.; et al. Imaging neural activity in worms, flies and mice with improved GCaMP calcium indicators. *Nat. Methods* **2019**, *6*, 875–881. [CrossRef] [PubMed]
42. Markova, O.; Mukhtarov, M.; Real, E.; Jacob, Y.; Bregestovski, P. Genetically encoded chloride indicator with improved sensitivity. *J. Neurosci. Methods* **2008**, *170*, 67–76. [CrossRef] [PubMed]

43. Shen, Y.; Nasu, Y.; Shkolnikov, I.; Kim, A.; Campbell, R.E. Engineering genetically encoded fluorescent indicators for imaging of neuronal activity: Progress and prospects. *Neurosci. Res.* **2002**, *152*, 3–14. [CrossRef] [PubMed]
44. McKenzie, C.K.; Sanchez-Romero, I.; Janovjak, H. Flipping the Photoswitch: Ion Channels Under Light Control. *Adv. Exp. Med. Biol.* **2015**, *869*, 101–117.
45. Broichhagen, J.; Frank, J.A.; Trauner, D. A roadmap to success in photopharmacology. *Acc. Chem. Res.* **2015**, *48*, 1947–1960. [CrossRef]
46. Hull, K.; Morstein, J.; Trauner, D. In vivo photopharmacology. *Chem. Rev.* **2018**, *118*, 10710–10747. [CrossRef]
47. Bregestovski, P.D.; Maleeva, G.V. Photopharmacology: A brief review using the control of potassium channels as an example. *Neurosci. Behav. Physiol.* **2019**, *49*, 184–191. [CrossRef]
48. Castagna, R.; Kolarski, D.; Durand-de Cuttoli, R.; Maleeva, G. Orthogonal Control of Neuronal Circuits and Behavior Using Photopharmacology. *J. Mol. Neurosci.* **2022**, *72*, 1433–1442. [CrossRef]
49. Szymanski, W.; Ourailidou, M.E.; Velema, W.A.; Dekker, F.J.; Feringa, B.L. Light-controlled histone deacetylase (HDAC) inhibitors: Towards photopharmacological chemotherapy. *Chem. Eur. J.* **2015**, *21*, 16517–16524. [CrossRef]
50. Udasin, R.; Sil, A.; Zomot, E.; Achildiev Cohen, H.; Haj, J.; Engelmayer, N.; Lev, S.; Binshtok, A.M.; Shaked, Y.; Kienzler, M.A.; et al. Photopharmacological modulation of native CRAC channels using azoboronate photoswitches. *Proc. Natl. Acad. Sci. USA* **2022**, *119*, e2118160119. [CrossRef]
51. Bispo, M.; van Dijl, J.M.; Szymanski, W. Molecular Photoswitches in Antimicrobial Photopharmacology. *Mol. Photoswitches Chem. Prop. Appl.* **2022**, *2*, 843–871.
52. Gazerani, P. Shedding light on photo-switchable analgesics for pain. *Pain Manag.* **2017**, *7*, 71–74. [CrossRef] [PubMed]
53. Iseppon, F.; Arcangeletti, M. Optogenetics and photopharmacology in pain research and therapeutics. *STEMedicine* **2020**, *1*, e43. [CrossRef]
54. Caporale, N.; Kolstad, K.D.; Lee, T.; Tochitsky, I.; Dalkara, D.; Trauner, D.; Kramer, R.; Dan, Y.; Isacoff, E.Y.; Flannery, J.G. LiGluR restores visual responses in rodent models of inherited blindness. *Mol. Ther.* **2011**, *19*, 1212–1219. [CrossRef] [PubMed]
55. Prischich, D.; Gomila, A.M.J.; Milla-Navarro, S.; Sangüesa, G.; Diez-Alarcia, R.; Preda, B.; Matera, C.; Batlle, M.; Ramírez, L.; Giralt, E.; et al. Adrenergic Modulation With Photochromic Ligands. *Angew. Chem. Int. Ed. Engl.* **2021**, *60*, 3625–3631. [CrossRef]
56. Gerasimov, E.; Erofeev, A.; Borodinova, A.; Bolshakova, A.; Balaban, P.; Bezprozvanny, I.; Vlasova, O.L. Optogenetic Activation of Astrocytes-Effects on Neuronal Network Function. *Int. J. Mol. Sci.* **2021**, *22*, 9613. [CrossRef] [PubMed]
57. Abd El-Aziz, T.M.; Kaur, A.; Shapiro, M.S.; Stockand, J.D.; Archer, C.R. Optogenetic Control of PIP2 Interactions Shaping ENaC Activity. *Int. J. Mol. Sci.* **2022**, *23*, 3884. [CrossRef]
58. Lilja, A.; Didio, G.; Hong, J.; Heo, W.D.; Castrén, E.; Umemori, J. Optical Activation of TrkB (E281A) in Excitatory and Inhibitory Neurons of the Mouse Visual Cortex. *Int. J. Mol. Sci.* **2022**, *23*, 10249. [CrossRef]
59. Salmina, A.B.; Kapkaeva, M.R.; Vetchinova, A.S.; Illarioshkin, S.N. Novel Approaches Used to Examine and Control Neurogenesis in Parkinson's Disease. *Int. J. Mol. Sci.* **2021**, *22*, 9608. [CrossRef]
60. Ivashkina, O.I.; Gruzdeva, A.M.; Roshchina, M.A.; Toropova, K.A.; Anokhin, K.V. Imaging of C-fos Activity in Neurons of the Mouse Parietal Association Cortex during Acquisition and Retrieval of Associative Fear Memory. *Int. J. Mol. Sci.* **2021**, *22*, 8244. [CrossRef]
61. Zhilyakov, N.; Arkhipov, A.; Malomouzh, A.; Samigullin, D. Activation of Neuronal Nicotinic Receptors Inhibits Acetylcholine Release in the Neuromuscular Junction by Increasing Ca2+ Flux through Cav1 Channels. *Int. J. Mol. Sci.* **2021**, *22*, 9031. [CrossRef] [PubMed]
62. Ponomareva, D.; Petukhova, E.; Bregestovski, P. Simultaneous Monitoring of pH and Chloride (Cl-) in Brain Slices of Transgenic Mice. *Int. J. Mol. Sci.* **2021**, *22*, 13601. [CrossRef] [PubMed]
63. Sotskov, V.P.; Pospelov, N.A.; Plusnin, V.V.; Anokhin, K.V. Calcium Imaging Reveals Fast Tuning Dynamics of Hippocampal Place Cells and CA1 Population Activity during Free Exploration Task in Mice. *Int. J. Mol. Sci.* **2022**, *23*, 638. [CrossRef] [PubMed]
64. Matera, C.; Calvé, P.; Casadó-Anguera, V.; Sortino, R.; Gomila, A.M.J.; Moreno, E.; Gener, T.; Delgado-Sallent, C.; Nebot, P.; Costazza, D.; et al. Reversible Photocontrol of Dopaminergic Transmission in Wild-Type Animals. *Int. J. Mol. Sci.* **2022**, *23*, 10114. [CrossRef]
65. Nin-Hill, A.; Mueller, N.P.F.; Molteni, C.; Rovira, C.; Alfonso-Prieto, M. Photopharmacology of Ion Channels through the Light of the Computational Microscope. *Int. J. Mol. Sci.* **2021**, *22*, 12072. [CrossRef]

International Journal of
Molecular Sciences

Article

Optogenetic Activation of Astrocytes—Effects on Neuronal Network Function

Evgenii Gerasimov [1],*, Alexander Erofeev [1], Anastasia Borodinova [2], Anastasia Bolshakova [1], Pavel Balaban [2], Ilya Bezprozvanny [1,3] and Olga L. Vlasova [1],*

1   Laboratory of Molecular Neurodegeneration, Peter the Great St. Petersburg Polytechnic University, Khlopina St. 11, 194021 St. Petersburg, Russia; alexandr.erofeew@gmail.com (A.E.); bolshakova.av@spbstu.ru (A.B.); Ilya.Bezprozvanny@UTSouthwestern.edu (I.B.)
2   Cellular Neurobiology of Learning Lab, Institute of Higher Nervous Activity and Neurophysiology of the Russian Academy of Science, Butlerova St. 5A, 117485 Moscow, Russia; borodinova.msu@mail.ru (A.B.); pmbalaban@gmail.com (P.B.)
3   Department of Physiology, UT Southwestern Medical Center at Dallas, Dallas, TX 75390, USA
*   Correspondence: evgeniigerasimov1997@gmail.com (E.G.); olvlasova@yandex.ru (O.L.V.)

Abstract: Optogenetics approach is used widely in neurobiology as it allows control of cellular activity with high spatial and temporal resolution. In most studies, optogenetics is used to control neuronal activity. In the present study optogenetics was used to stimulate astrocytes with the aim to modulate neuronal activity. To achieve this goal, light stimulation was applied to astrocytes expressing a version of ChR2 (ionotropic opsin) or Opto-α1AR (metabotropic opsin). Optimal optogenetic stimulation parameters were determined using patch-clamp recordings of hippocampal pyramidal neurons' spontaneous activity in brain slices as a readout. It was determined that the greatest increase in the number of spontaneous synaptic currents was observed when astrocytes expressing ChR2(H134R) were activated by 5 s of continuous light. For the astrocytes expressing Opto-α1AR, the greatest response was observed in the pulse stimulation mode (T = 1 s, t = 100 ms). It was also observed that activation of the astrocytic Opto-a1AR but not ChR2 results in an increase of the fEPSP slope in hippocampal neurons. Based on these results, we concluded that Opto-a1AR expressed in hippocampal astrocytes provides an opportunity to modulate the long-term synaptic plasticity optogenetically, and may potentially be used to normalize the synaptic transmission and plasticity defects in a variety of neuropathological conditions, including models of Alzheimer's disease and other neurodegenerative disorders.

Keywords: optogenetics; astrocytes; hippocampal neurons; patch-clamp; channelrhodopsin-2; opto-α1-adrenoreceptor

Citation: Gerasimov, E.; Erofeev, A.; Borodinova, A.; Bolshakova, A.; Balaban, P.; Bezprozvanny, I.; Vlasova, O.L. Optogenetic Activation of Astrocytes—Effects on Neuronal Network Function. *Int. J. Mol. Sci.* **2021**, *22*, 9613. https://doi.org/10.3390/ijms22179613

Academic Editors: Piotr D. Bregestovski and Carlo Matera

Received: 28 July 2021
Accepted: 1 September 2021
Published: 4 September 2021

Publisher's Note: MDPI stays neutral with regard to jurisdictional claims in published maps and institutional affiliations.

1. Introduction

Astrocytes play an integral role in the maintenance and regulation of neural networks in the brain. They are able to influence neuronal activity by regulating the extracellular concentration of potassium ions, as well as neurotransmitters, due to the expression on their membrane of a large number of transporters of electrogenic transmitters such as glutamate [1,2], gamma-aminobutyric acid [3,4], and glycine [5,6]. By releasing gliotransmitters, astrocytes act on neuronal receptors, modulating neuronal excitability, synaptic transmission, and synaptic plasticity. Astrocytes do not generate action potentials in response to a stimulus, but respond with intracellular increase of [Ca^{2+}] [7]. When Ca^{2+} waves propagate in astrocyte cytoplasm, serine, cytokines, and lactate are released, which can modulate activity of neighboring neurons [8]. The ability of astrocytes to release glutamate allows regulation of the function of NMDA receptors, thereby controlling the excitation of neuronal network [9].

Astrocytes are also an irreplaceable part of the tripartite synapse, which involves coordinated activity of pre- and postsynaptic membrane and astrocytes [10,11]. Their activity is closely related to synaptic potency [12] and is controlled via several types of metabotropic receptors linked to calcium levels. Release of gliotransmitters from astrocytes occurs in both calcium-dependent and calcium-independent ways. The ability of astrocytes to release gliotransmitters in a millisecond time scale is critical for their role in integration of information in neuronal networks [13].

Astrocytes are closely related to the pathogenesis and pathological processes in the brain and, in particular, in the context of Alzheimer's disease (AD) [14–19]. AD [20] is characterized by a progressive memory loss and cognitive dysfunctions, accumulation of a significant number of Aβ-amyloid plaques [21], abnormal neuronal calcium homeostasis [22], and accumulation of neurofibrillary tangles [23]. Shift in excitation and inhibition balance in neuronal network is often considered one of the causes of AD pathology [24,25]. It has been proposed that regulation of neuronal network activity in AD by stimulation of astrocytes may lead to beneficial effects by stabilizing activity of the network [26].

Optogenetic techniques allow selective and precise regulation of cellular activity [27,28]. In the present study, the optogenetic approach was used to stimulate activity of astrocytes. For optogenetic activation of astrocytes, two different opsins were used: ChR2 [29] that acts as an ion channel and metabotropic opsin Opto-a1AR [30], stimulation of which leads to activation of IP3 receptor and elevation of cytosolic calcium concentration. In this study, we performed comparison of effects on neuronal function resulting from activation of astrocytes by these two optogenetic tools. Obtained results are useful for future experimental evaluation of astrocyte activation in the context of AD and other neurodegenerative disease models.

2. Results

2.1. Specificity of Expression of AAV2/5 GfaABC1D_ChR2(H134R)-mCherry and AAV2/5 GfaABC1D_Opto-a1AR-EYFP

Genetic constructs of ChR2(H134R) [29] and Opto-a1AR [30] were obtained and packaged into AAV2/5 adeno-associated viruses (AAV) under control of astrocyte-specific GfaABC1D promoter (see Materials and Methods). AAV-ChR2(H134R)-mCherry and AAV-Opto-a1AR-EYFP viruses were stereotaxically injected into hippocampal region of the mice (C57BL/6J strain) and immunohistochemical experiments were performed 3 weeks after injection to verify specificity of transgene expression. Obtained results confirmed co-localization of ChR2-mCherry and astrocytic marker glial fibrillary acid protein (GFAP) (Figure 1A). Similar results were obtained with Opto-a1AR-EYFP (Figure 1B). Obtained results confirmed astrocyte-specific expression of both constructs, in agreement with known specificity of GfaABC1D promoter [31]. Astrocyte-specific expression of Opto-a1AR-EYFP construct was further confirmed by high resolution confocal imaging of mouse hippocampal slices combined with nuclei labeling by DAPI staining (Figure 1C).

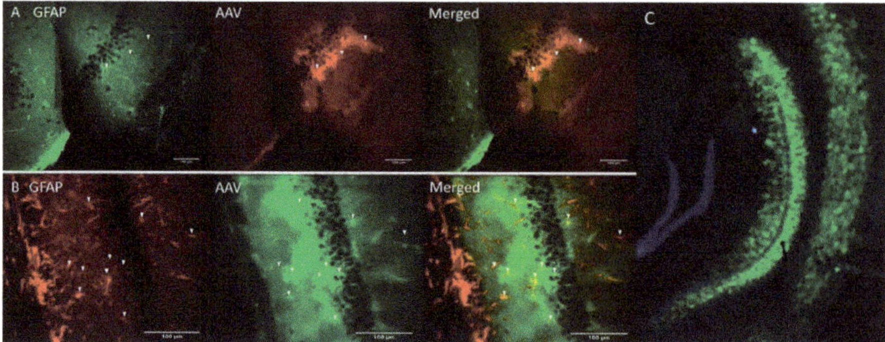

Figure 1. Astrocyte-specific expression of ChR2(H134R)-mCherry and Opto-a1AR-EYFP constructs. (**A**) Confocal images of

fixed slices of mouse brain tissue three weeks after unilateral administration of AAV2/5 GfaABC1D_ChR2(H134R)–mCherry (red). GFAP staining (green) was used to label astrocytes. Scale bar is 100 μm. (**B**) Confocal images of fixed slices of mouse brain tissue three weeks after unilateral administration of AAV2/5 GfaABC1D_Opto-a1AR_EYFP (green). GFAP staining (red) was used to label astrocytes. Scale bar is 100 μm. (**C**) Confocal image of fixed slice of mouse brain tissue three weeks after unilateral an administration of AAV2/5 GfaABC1D_Opto-a1AR_EYFP (green), nuclei are labeled with DAPI (blue), 4× magnification.

2.2. Activation of Astrocytes Expressing ChR2(H134R) or Opto-α1AR Leads to Enhancement of Pyramidal Neuron's Activity

To evaluate functional effects of astrocyte activation on neuronal function, spontaneous excitatory postsynaptic currents (sEPSC) were recorded in a whole-cell configuration from hippocampal slices 3 weeks after unilateral injection of ChR2(H134R)-mCherry and Opto-a1AR-EYFP expressing viral constructs. Intracellular recording was performed in the presence of 1 μM tetrodotoxin (TTX) in the pipette solution to prevent neurons from action potential generation in normal ACSF as extracellular media. Patch-clamp recordings were performed from neurons located at approximate depth 100 μm from the surface of a slice, where all processes of neurons and astrocytes were undamaged. To determine optimal parameters of optogenetic activation of astrocytes that expressed ChR2(H134R)-mCherry, different protocols of light administration ($\lambda = 473$ nm) were applied with variable intervals of stimulation (T) and duration of the light pulses (t). The conditions that were tested: (t = 20 ms, T = 20 ms); (t = 20 ms, T = 1 s); and (t = 100 ms, T = 1 s) and continuous light (t = 5 s). In each experiment, frequency of sEPSC following light stimulation was normalized to the frequency of sEPSC in the same neuron prior to light stimulation. Highest increase in the frequency of sEPSC currents was observed in a group with 5 s of continuous light stimulation (mean increase to 1.81 ± 0.15, $n = 4$) and in a t = 100 ms, T = 1 s group (the mean increased to 1.39 ± 0.14, $n = 5$) (Figure 2A). We further found that optogenetic stimulation of hippocampal astrocytes did not affect amplitudes of spontaneous currents in neurons, with no difference before and after the optogenetic stimulation in distribution of EPSCs amplitudes at all stimulation protocols ($n = 4$, $p > 0.05$, Mann–Whitney U test) (Figure 2B). To rule out potential effects of phototoxic damage and/or heating of a slice by pulses of light, control recordings were performed from hippocampal slices of non-injected hemisphere of the same mice. In these control experiments, we discovered that continuous light stimulation with t = 5 s led to non-significant changes in sEPSC frequency compared to baseline level value (mean 0.91 ± 0.08, $n = 6$, $p > 0.01$, Mann–Whitney U test) (Figure 2A). Increase of sEPSC's frequency suggests that in ChR2(H134R) expressing astrocytes the optogenetic activation may lead to activation of spontaneous glutamate release in pyramidal neuron synapses, as has been previously shown for visual cortex neurons [32].

To determine an effect of Gq-coupled opsin activation, the hippocampal CA1 astrocytes expressing the Opto-α1AR-EYFP were optogenetically stimulated, and sEPSC recorded using the same approach as described above for ChR2(H134R)-mCherry. To identify optimal parameters of stimulation, the following optogenetic protocols were applied: (t = 20 ms, T = 1 s); (t = 100 ms, T = 1 s) and continuous light (t = 5 s). The greatest increase in the frequency of currents was observed in the group with (t = 100 ms, T = 1 s) (the mean increased to 1.71 ± 0.36, $n = 4$) (Figure 3A). No difference in sEPSC frequency was observed in control experiments in slices from non-injected hemisphere under the same stimulation conditions (t = 100 ms, T = 1 s) (Figure 3A). Similar to experiments with ChR2(H134R)-mCherry, optogenetic stimulation of Opto-a1AR-EYFP had no significant effect on distribution of sEPSC amplitudes (Figure 3B).

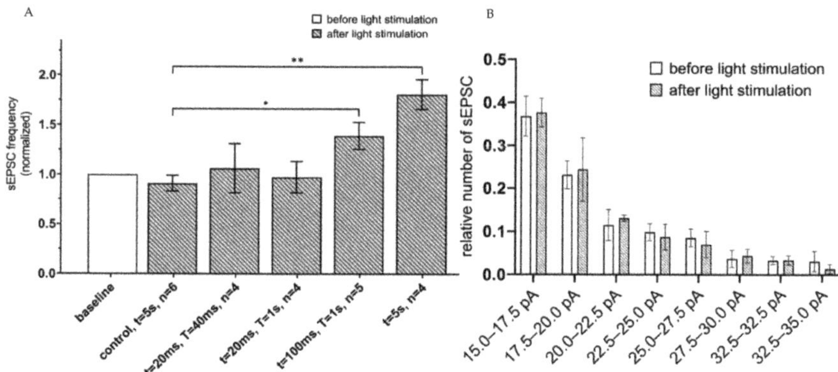

Figure 2. sEPSC changes in response to astrocyte activation using ChR2(H134R)–mCherry. (**A**) Values of the normalized frequencies of sEPSC of hippocampal neurons in the CA1 region after light activation (t = 20 ms, T = 20 ms; t = 20 ms, T = 1 s; t = 100 ms, T = 1 s; continuous light t = 5 s) of astrocytes expressing ChR2(H134R)—mCherry. The data are presented as the mean ± SEM, **: $p < 0.01$, *: $p < 0.05$. (**B**) distribution of sEPSC amplitudes before and after optogenetic (t = 5 s) activation of astrocytes expressing ChR2(H134R)—mCherry on the membrane. The data are presented as the mean ± SEM, $n = 4$.

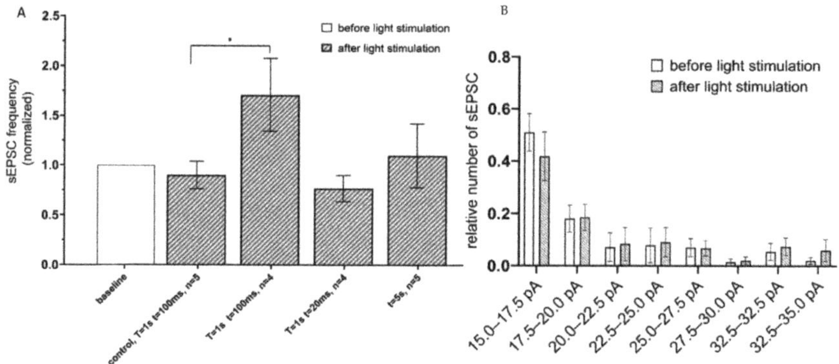

Figure 3. sEPSC changes in response to astrocyte activation using Opto-a1AR-EYFP. (**A**) Values of the normalized frequencies of sEPSC in the CA1 hippocampal neurons after light activation (t = 20 ms, T = 20 ms; t = 20 ms, T = 1 s; t = 100 ms, T = 1 s; continuously t = 5 s) of astrocytes expressing Opto-a1AR-EYFP. The data are presented as the mean ± SEM, *: $p < 0.05$. (**B**) Distribution of sEPSC amplitudes before and after optogenetic activation of astrocytes expressing Opto-a1AR-EYFP on the membrane. The data are presented as an average ± SEM.

The obtained results suggest that optogenetic stimulation of astrocytes expressing ChR2(H134R)-mCherry or Opto-α1AR-EYFP constructs elicited gliotransmitters release that increased frequency of spontaneous neuronal activity of hippocampal pyramidal neurons but had no effect on the amplitudes of sEPSC currents.

2.3. Activation of Astrocytes Expressing ChR2(H134R)-mCherry Had No Effect on the Field Excitatory Postsynaptic Potentials

In the next series of experiments, the effects of optogenetic activation of astrocytes on CA1 hippocampal field excitatory postsynaptic potentials (fEPSPs) were evaluated. To achieve fEPSP recordings, hippocampal Shaffer collaterals were stimulated via twisted bipolar electrodes before and after optogenetic activation of astrocytes expressing ChR2(H134R)-

mCherry by 5 s of continuous light stimulation. As in previous experiments, hippocampal slices from non-injected hemispheres were used as a control. On average, the value of normalized fEPSP slope in the last 5 min prior to light stimulation was equal to 92.4 ± 8.3 ($n = 3$) in experimental group and 99.1 ± 4.8 ($n = 4$) in control group (Figure 4A). We discovered that two light pulses of 5 s light duration had no significant effect on the average value of the normalized fEPSP slope (%) in both control and experimental groups (ChR2-group ($n = 3$) vs. control group ($n = 4$), $p > 0.05$, Mann–Whitney U test) (Figure 4B). We further noticed that fEPSPs recorded in the experimental (ChR2) group appears to be reduced after light stimulation when compared to the control group (Figure 4A), but the difference has not reached a level of statistical significance (Figure 4B).

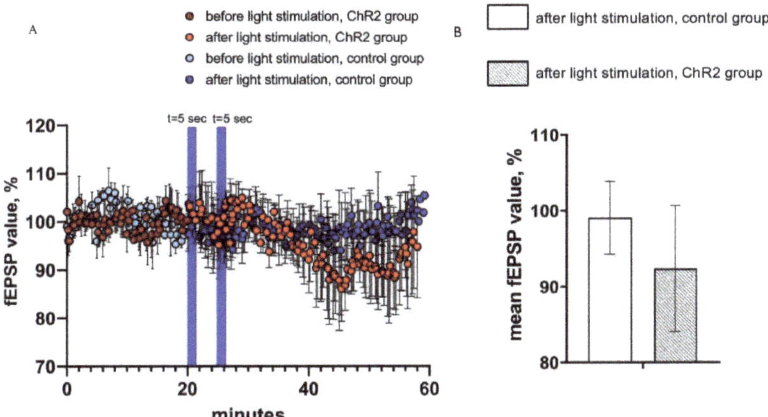

Figure 4. fEPSP changes in response to astrocyte activation using ChR2(H134R)–mCherry. (**A**) the normalized values of fEPSP slope (%) before and after light stimulation of astrocytes in the experimental (ChR2 expressing) and control groups. In the figure, the blue bars show the time of continuous light stimulation. (**B**) Average value of the normalized fEPSP value (%) after light stimulation (2 pulses t = 5 s) in the control and experimental groups, $n = 4$ for control measurements, $n = 3$ in the experimental group. The data are presented as an average $\pm$ SEM.

2.4. Activation of Opto-α1AR-EYFP-Expressing Astrocytes Leads to Increase of Field Excitatory Postsynaptic Potential

By using the same approach as for ChR2(H134R)-mCherry, we evaluated changes in hippocampal fEPSP in response to optogenetic activation of astrocytes transduced with Opto-α1AR-EYFP viral construct. T = 1 s and t = 100 ms were chosen as the light stimulation protocol in these experiments, as this protocol resulted in the highest increase in sEPSC (Figure 3A). As in previous experiments, hippocampal slices from non-injected hemispheres were used as a control. In these experiments, the average normalized slope (%) in last 5 min of recordings in experimental (Opto-α1AR) group was 130.0 ± 13.6 ($n = 3$), while in control group it was 102.3 ± 3.5 ($n = 3$). Obtained results demonstrated that light stimulation of astrocytes expressing Opto-a1AR-EYFP leads to significant potentiation of fEPSP (Figure 5A,B). From these results we concluded that Opto-α1AR-EYFP-mediated activation of astrocytes is able to potentiate the hippocampal synaptic transmission significantly, resulting in enhanced fEPSP.

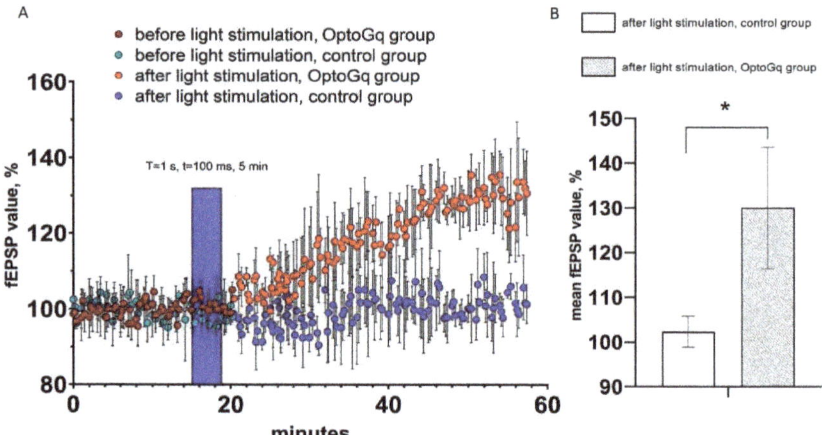

Figure 5. fEPSP changes in response to astrocyte activation using Opto-a1AR-EYFP. (**A**) Normalized value of slope (%) before and after light stimulation of astrocytes in the control and experimental (Opto-a1AR expressing) groups of mice. In the figure, the blue bar marks the timing of light stimulation. (**B**) Average value of the normalized slope value after light stimulation (T = 1s, t = 100ms) in the control and experimental groups mice, n = 3 for control measurements, n = 3 in the experimental group. The data are presented as an average ± SEM, *: $p < 0.05$.

3. Discussion

In this study, the parameters of optogenetic activation of astrocytes that lead to enhancement of neuronal activity were defined. It was determined that the mode of 5 s continuous light stimulation of ChR2(H134R), expressed in astrocytes, provides the highest increase in spontaneous neuronal activity. Impulse mode with T = 1 s, t = 100 ms parameters had maximal effect on spontaneous neuronal activity when Opto-a1AR was used to activate astrocytes. Furthermore, our results and the data from the literature [33] suggest that stimulation of Opto-a1AR expressed in hippocampal astrocytes has a potential for enhancing the long-term synaptic plasticity in mice. In contrast, no significant effect on fEPSP was observed in experiments with astrocytes expressing ChR2(H134R). Possible mechanism responsible for synaptic plasticity changes following activation of Opto-a1AR in astrocytes may be related to secretion of glutamate and D-serine, both of which can potentiate synaptic plasticity [34].

Ionotropic opsins are well studied in different types of cells in the nervous system. Effects of ChR2 activation in astrocytes have been previously described [35–38]. It was demonstrated that activation of ChR2 in astrocytes triggers a release of glutamate and increase in the frequency of spontaneous excitatory postsynaptic currents in pyramidal neurons [38], in agreement with our findings (Figure 2). Activation of ChR2 in astrocytes can also have positive influence on interneurons' excitability and negative influence on pyramidal neuron activity, what reduces their action potential frequency. It is possible that different and even opposite effects of astrocytic ChR2 activation may be related to various regimes of illumination [36,38] and to different neuronal activity patterns [39].

It was previously shown that chemogenetic activation of hM3Dq-expressing astrocytes by CNO in the hippocampal CA1 neurons leads to increase in mEPSC and synaptic potentiation [33]. Optogenetic activation of Gq-signaling in astrocytes in our experiments did not result in such strong and fast changes in synaptic plasticity. As an explanation, it can be proposed that due to quite low membrane resistance in astrocytes (due to the presence of gap-junctions [40]) more time is needed for more gentle activation by light in comparison with DREADD (Designer Receptors Exclusively Activated by Designer Drugs). In another study, activation of G-protein coupled receptors in astrocytes was performed by expression of melanopsin. In this case, the optogenetic low-frequency stimulation led to a robust EPSC potentiation that persisted after 30 min of recording [37]. Moreover, authors showed that expression of melanopsin in hippocampal astrocytes caused an elevation of IP3-dependent Ca^{2+} signal in their fine processes. It was also reported that light stimulation of astrocytes transfected with Opto-a1AR resulted in a raise of the Ca^{2+} concentration [30]. Increase in astrocytic Ca^{2+} concentration and release of glutamate is most likely an explanation for increased sEPSC frequency following activation of Opto-a1AR in our experiments.

The potentiation of field potentials during optogenetic activation of astrocytes expressing Opto-a1AR is observed in connection with the release of not only glutamate, but also gliotransmitters—for example, D-serine, which is necessary for the formation of long-term changes in plasticity. The release of D-serine occurs through Ca^{2+}—and SNARE-dependent exocytosis along with that occurring through alternative non-exocytotic pathways [41]. According to the literature data [42], activation of the metabotropic opsin Opto-a1AR leads to a significant increase in the intracellular concentration of Ca^{2+}, which can increase the excretion of serine from astrocytes into the intracellular space. This increase is associated with the activation of intracellular calcium depots. This gliotransmitter may not be released by astrocytes when they activate ChR2, since there is not such a strong increase in the concentration of intra-astrocyte calcium, which is mediated both by its flow through the plasma membrane and by the involvement of intracellular calcium stores. It might be the reason for the lack of a field potential potentiation effect after the activation of ChR2. Obtained results suggest a possibility for regulating the neuronal networks functioning by using light stimulation of Opto-a1AR opsin expressed in astrocytes with parameters defined in this study. Potentially, this approach can be used for correcting the neuronal network dysfunction and improving the synaptic plasticity in a variety of neuropathological conditions, including models of Alzheimer's disease and other neurodegenerative disorders. Importantly, stimulation of astroglia can also convert it into reactive glia with cytotoxic activity, which can provoke the death of neurons and exacerbate neuroinflammation. In vivo experiments with optogenetic stimulation of astrocytes in mouse models of AD and other disorders are needed to evaluate validity of stimulation parameters defined in our study with brain slices and to refine the proposed experimental approach in order to avoid potential negative effects of astrocyte stimulation on brain function.

4. Materials and Methods

4.1. Animals

The breeding colony of C57BL/6J mice obtained from the Jackson Laboratory was established and maintained in a vivarium with 4–5 mice per cage and a 12 h light/dark cycle in the animal facility, and were used for the fEPSP experiments. This line was taken because the 5xFAD mice with Alzheimer's disease genetic model were made using this genetic line, and future experiments are planned to be conducted on them. Albino outbred mice (Rappolovo farm, Leningradsky District, Russia) were used for patch-clamp experiments on acute hippocampal slices. All procedures were approved by principles of the European convention (Strasburg, 1986) and the Declaration of International Medical Association regarding the humane treatment of animals (Helsinki, 1996) and approved by the Bioethics Committee of the Peter the Great St. Petersburg Polytechnic University at St. Petersburg, Russia (Ethical permit number 2-n-b from 25 January 2021).

4.2. Plasmids and Production of Viral Constructs

For selective expression of channel rhodopsin in astrocytes we used plasmid AAV pZac2.1 GfaABC1D_ChR2(H134R)-mCherry (Addgene, #112496) that contains short version of astrocyte-specific GFAP promoter GfaABC1D [43]. Opto-a1AR encoding plasmid was generated on the basis of pZac2.1 as follows: Opto-a1AR-EYFP fragment was amplified by PCR using pcDNA3.1/opto-a1AR-EYFP plasmid (Addgene #20947) as a template, and then cloned into pZac2.1 plasmid using NheI/XbaI restriction enzymes to replace the ChR2(H134R)-mCherry fragment. Resulting plasmid AAV GfaABC1D_Opto-a1AR-EYFP was verified by sequencing. The payloads were packaged using commercially available plasmid with AAV5 serotype (Addgene, #104964) to generate recombinant AAV2/5 viruses.

Production of viral particles was carried out according to the standard protocols used for AAV preparation. Briefly, HEK293T cells were plated on polylysine-coated Petri dishes and grown in a DMEM medium supplemented with 10% of FBS until a density of 70–80% was reached. Next, HEK293T cell were subjected to the triple-plasmid transfection using PEI reagent. At the day 3 post-transfection, media and cells were collected and processed separately. Cells were harvested and subjected to freeze/thaw cycles in liquid nitrogen. Supernatants were treated with polyethylene glycol (PEG, Sigma-Aldrich, St. Louis, MO, USA) 8000, the PEG-precipitated AAVs were collected by centrifugation. Viral particles, extracted from cells and supernatant, were combined and treated with benzonase nuclease to destroy any unpacked DNA. Then AAV particles were purified by iodixanol gradient ultracentrifugation. The required fraction, enriched with viral particles, was collected, filtered, and transferred to the Amicon Ultra-15 centrifugal filter unit for buffer exchange and concentration of virus suspension to the final volume of 110–130 µL. Virus titer was then determined by quantitative PCR using primer pair targeting AAV2 ITR sequence in the construct (Forward: GGAACCCCTAGTGATGGAGTT; Reverse: CGGCCTCAGTGAGCGA). To remove any extra-viral DNA before qPCR measurements, virus aliquots were treated with DNase I. Resulting virus titer was equal to 9.1×10^{14} vg/mL for AAV2/5 GfaABC1D_ChR2(H134R)_mCherry, and 2.93×10^{14} vg/mL for AAV2/5 GfaABC1D_opto-a1AR_EYFP.

4.3. Viral Constructs Delivery via Stereotaxic Surgery

For viral constructs (AAV2/5 pZac2.1 GfaABC1D_ChR2(H134R)_mCherry and AAV2/5 GfaABC1D_Opto-a1AR_EYFP) delivery to the hippocampus, mice aged ~2 months and weighing 24–26 g were used. Injections of viral constructs were performed using a stereotaxic device (68001, RWD Life Science, Guangdong, China), a syringe with a thin needle (84,853, 7758-02, Hamilton, Reno, NV, USA), as well as a heated mat and a temperature controller (69,002, RWD Life Science, Guangdong, China). Surgery was carried out under anesthesia of the animals by anesthetizing 1.5–2.5% with a gas mixture of isoflurane. After a control check of the depth of anesthesia in the animal, the viral constructs were administered according to the standard protocol [44] at the following coordinates: AP−2.1, DV−1.8, ML+2.4, with a volume of 1.5 µL at a rate of 0.1 µL/min.

4.4. Immunohistochemistry

To test the specificity of injected viruses, three weeks after injection the immunohistochemical staining of mouse brain tissue sections was performed according to the standard protocol [45]. For this purpose, mice were anesthetized by intraperitoneal injection of urethane solution (250 mg/mL in 0.9% NaCl, Sigma-Aldrich, St. Louis, MO, USA). Then transcardial perfusion was performed with PBS followed by standard 1.5% paraformaldehyde solution (30–50 mL, PFA, Sigma-Aldrich, USA). The brain was removed and placed in a 4% PFA solution for post-fixation for 1 week at +4 °C. Fixed brain slices with a thickness of 20–50 microns were obtained using a microtome (5100 MZ, Campden Instruments, United Kingdom) in a PBS solution and stored in a 24-well plate filled with 0.5% PFA.

Permeabilization of fixed tissue was performed using 0.1% Triton X-100 in PBS, then slices were placed in a blocking buffer (5% BSA in PBS) for 6 h at room temperature. After 6 h, the slices were placed in a solution with primary antibodies (2.5% BSA in PBS-1 mL, 0.2% Tween20-20 µL, antibodies-1 µL, dilution 1/1000) for 8 h on a shaker at +4 °C. Primary antibodies—Anti-GFAP (644701, Biolegend, San Diego, CA, USA)—were used for astrocyte staining. After 8 h primary antibodies were washed and slices were incubated with secondary antibodies (Goat anti-Mouse IgG (H + L) Cross-Adsorbed Secondary Antibody, Alexa Fluor 488, Invitrogen A-11001, dilution 1/2000) for immunostaining of astrocytes expressing ChR2-mCherry, and for experimental group of astrocytes expressing Opto-a1AR-EYFP, the following secondary antibodies were used (Goat anti-Mouse IgG

(H + L) Cross-Adsorbed Ready probes Secondary Antibody, Alexa Fluor 594, Invitrogen R37121, dilution 1/1000). Within 8 h after the addition of secondary antibodies, the sections were incubated on a shaker at a low speed at +4 °C.

4.5. Slice Electrophysiology

Transcardial perfusion was performed with saturated carbogen (95% O_2 / 5% CO_2) modified 0–2 °C solution of ACSF (92 mM NMDG, 2.5 mM KCl, 1.25 mM NaH_2PO_4, 30 mM $NaHCO_3$, 20 mM HEPES, 25 mM D-glucose, 2 mM Thiourea, 5 mM Na-ascorbate, 3 mM Na-pyruvate, 0.5 mM $CaCl_2$, 10 mM $MgSO_4$) and decapitation was performed. Horizontal slices of the brain with a thickness of 350 microns for the patch-clamp and 400 microns for recording excitatory field potentials were made using a microtome (Leica VT1200S (Leica Biosystems Division of Leica Microsystems Inc., Buffalo Grove, IL, USA)) from 3-month-old mice and maintained in a 0–2 °C solution of NMDG-ACSF saturated with carbogen. The hippocampus was isolated from each slice and incubated in modified ACSF solution (92 mM NaCl (Sigma-Aldrich, St. Louis, MO, USA), 2.5 mM KCl (Sigma-Aldrich, St. Louis, MO, USA), 1.25 NaH_2PO_4 (Sigma-Aldrich, St. Louis, MO, USA), 30 mM $NaHCO_3$ (Sigma-Aldrich, St. Louis, MO, USA), 20 mM HEPES (Sigma-Aldrich, St. Louis, MO, USA), 25 mM D-glucose (Sigma-Aldrich, St. Louis, MO, USA), 2 mM thiourea (Sigma-Aldrich, St. Louis, MO, USA), 5 mM Na-ascorbate (Sigma-Aldrich, St. Louis, MO, USA), 3 mM Na-pyruvate (Sigma-Aldrich, St. Louis, MO, USA), 2 mM $CaCl_2$ (Sigma-Aldrich, St. Louis, MO, USA), 2 mM $MgSO_4$ (Sigma-Aldrich, St. Louis, MO, USA)), saturated with carbogen with a controlled temperature (32–35 °C) for 15 min, after which the slices were incubated at room temperature of 23–25 °C. After 60 min of incubation, the sEPSC were recorded using the patch-clamp techniques (acute slices from outbreed albino-mice) or the field excitatory postsynaptic potentials were recorded (acute slices from C57BL/6J background mice) in standard ACSF (119 m NaCl, 2.5 mKCl, 1.25 $mNaH_2PO_4$, 24 $mNaHCO_3$, 5 mHEPES, 12.5 mD-glucose, 2 $mCaCl_2$, 2 $mMgSO_4$). Patch electrodes were fabricated from borosilicate glass (2–3 MΩ), and filled with internal solution (120 mK-gluconate, 20 mKCl, 10 mHEPES, 0.2 mEGTA, 2 $mMgCl_2$, 0.3 mNa_2GTP, 2 mMgATP).

In all voltage-clamp experiments, the neurons were held at −70 mV and sEPSC were recorded in presence of 1 μM TTX in internal solution. Only pyramidal neurons were included into statistics. They were identified by their shape and unique pattern of action potential generated using a step-protocol in current clamp mode.

For extracellular field EPSP recordings, the 400 μm horizontal hippocampal slices were used, and the Schaffer collaterals were stimulated by a twisted bipolar electrode. fEPSPs were recorded in the CA1 stratum radiatum using a glass pipette containing ACSF (250–450 KΩ). fEPSPs were low-pass filtered at 400 Hz.

For patch-clamp experiments with ChR_2 opsin 7 AAV-injected mice were taken into experiments and for Opto-a1AR group 6 mice were injected and studied. Only one neuron per slice was patched, and if there were problems with recording after optogenetic stimulation slice was changed and never used again. For all fEPSPs experiments one slice for experimental and one slice for control group for each mouse were taken, so number of mice is equal to amount of slices (four mice in ChR2-expressed group and three in Opto-a1AR group). Also, one slice for recording was used and only one optogenetic stimulation was performed per slice.

Optogenetic light stimulation was performed by means of the blue LED (LED4D067, 470 nm, Thorlabs Inc., Newton, NJ, USA) with maximum intensity of 35 mW mm^{-2} with a maximum photo flux of 250 mW.

Author Contributions: Conceptualization, E.G., O.L.V., and I.B.; Methodology, O.L.V. and P.B.; Validation O.L.V.; Formal analysis, E.G. and A.E.; Investigation, E.G. and A.E.; Resources, A.B. (Anastasia Borodinova) and A.B. (Anastasia Bolshakova); Writing—original draft preparation, E.G. and O.L.V.; Writing—review and editing, O.L.V., I.B., and P.B.; Visualization, E.G.; Supervision, O.L.V. and I.B.; Project administration, O.L.V., P.B., and I.B.; Funding acquisition, O.L.V. and I.B. All authors have read and agreed to the published version of the manuscript.

Funding: This research was funded by the Russian Science Foundation grant no. 20-65-46004 (O.L.V.) and by the Ministry of Science and Higher Education of the Russian Federation as part of World-class Research Center program: Advanced Digital Technologies (contract No. 075-15-2020-934 to I.B). The financial support was divided in the following way: experiments depicted on Figures 1 and 2 were supported by the Ministry of Science and Higher Education of the Russian Federation No. 075-15-2020-934, experiments depicted on Figures 3 and 4 were supported by the Russian Science Foundation 20-65-46004.

Institutional Review Board Statement: The study was conducted according to the guidelines of the Declaration of Helsinki, and approved by the Bioethics Committee of the Peter the Great St. Petersburg Polytechnic University at St. Petersburg, Russia and followed the principles of European convention (Strasbourg, 1986).

Informed Consent Statement: Not applicable.

Data Availability Statement: The presented data that is shown in this study is available by a request from the corresponding author.

Acknowledgments: pcDNA3.1/opto-a1AR-EYFP was a gift from Karl Deisseroth (Addgene plasmid # 20947); pZac2.1 GfaABC1D ChR2(H134R) mCherry SV40 was a gift from Baljit Khakh (Addgene plasmid # 112496); pAAV2/5 was a gift from Melina Fan (Addgene plasmid no. 104964). We would like to thank S.A. Pushkareva for technical support and T.P. Norekian for his help with the confocal images of the Opto-a1AR provided in the current paper.

Conflicts of Interest: The authors declare no conflict of interest.

References

1. Huang, Y.H.; Bergles, D.E. Glutamate transporters bring competition to the synapse. *Curr. Opin. Neurobiol.* **2004**, *14*, 346–352. [CrossRef]
2. Mahmoud, S.; Gharagozloo, M.; Simard, C.; Gris, D. Astrocytes Maintain Glutamate Homeostasis in the CNS by Controlling the Balance between Glutamate Uptake and Release. *Cells* **2019**, *8*, 184. [CrossRef]
3. Schousboe, A. Pharmacological and Functional Characterization of Astrocytic GABa Transport: A Short Review. *Neurochem. Res.* **2000**, *25*, 1241–1244. [CrossRef] [PubMed]
4. Ishibashi, M.; Egawa, K.; Fukuda, A. Diverse actions of astrocytes in GABAergic signaling. *Int. J. Mol. Sci.* **2019**, *20*, 2964. [CrossRef]
5. Shibasaki, K.; Hosoi, N.; Kaneko, R.; Tominaga, M.; Yamada, K. Glycine release from astrocytes via functional reversal of GlyT1. *J. Neurochem.* **2017**, *140*, 395–403. [CrossRef]
6. Sofroniew, M.V.; Vinters, H.V. Astrocytes: Biology and pathology. *Acta Neuropathol.* **2010**, *119*, 7–35. [CrossRef] [PubMed]
7. Kuga, N.; Sasaki, T.; Takahara, Y.; Matsuki, N.; Ikegaya, Y. Large-scale calcium waves traveling through astrocytic networks in vivo. *J. Neurosci.* **2011**, *31*, 2607–2614. [CrossRef]
8. Salmina, A.B.; Gorina, Y.V.; Erofeev, A.I.; Balaban, P.M.; Bezprozvanny, I.B.; Vlasova, O.L. Optogenetic and chemogenetic modulation of astroglial secretory phenotype. *Rev. Neurosci.* **2021**, *32*, 459–479. [CrossRef] [PubMed]
9. Halassa, M.M.; Haydon, P.G. Integrated brain circuits: Astrocytic networks modulate neuronal activity and behavior. *Annu. Rev. Physiol.* **2009**, *72*, 335–355. [CrossRef]
10. Perea, G.; Navarrete, M.; Araque, A. Tripartite synapses: Astrocytes process and control synaptic information. *Trends Neurosci.* **2009**, *32*, 421–431. [CrossRef] [PubMed]
11. Araque, A.; Parpura, V.; Sanzgiri, R.P.; Haydon, P.G. Tripartite synapses: Glia, the unacknowledged partner. *Trends Neurosci.* **1999**, *22*, 208–215. [CrossRef]
12. Perea, G.; Araque, A. Glial calcium signaling and neuron-glia communication. *Cell Calcium* **2005**, *38*, 375–382. [CrossRef]
13. Santello, M.; Calì, C.; Bezzi, P. Gliotransmission and the tripartite synapse. *Adv. Exp. Med. Biol.* **2012**, *970*, 307–331. [CrossRef] [PubMed]
14. Rappold, P.M.; Tieu, K. Astrocytes and Therapeutics for Parkinson's Disease. *Neurotherapeutics* **2010**, *7*, 413–423. [CrossRef] [PubMed]
15. Palpagama, T.H.; Waldvogel, H.J.; Faull, R.L.M.; Kwakowsky, A. The Role of Microglia and Astrocytes in Huntington's Disease. *Front. Mol. Neurosci.* **2019**, *12*, 258. [CrossRef]
16. Frost, G.R.; Li, Y.M. The role of astrocytes in amyloid production and Alzheimer's disease. *Open Biol.* **2017**, *7*, 170228. [CrossRef]
17. Perez-Nievas, B.G.; Serrano-Pozo, A. Deciphering the astrocyte reaction in Alzheimer's disease. *Front. Aging Neurosci.* **2018**, *10*, 114. [CrossRef] [PubMed]
18. González-Reyes, R.E.; Nava-Mesa, M.O.; Vargas-Sánchez, K.; Ariza-Salamanca, D.; Mora-Muñoz, L. Involvement of astrocytes in Alzheimer's disease from a neuroinflammatory and oxidative stress perspective. *Front. Mol. Neurosci.* **2017**, *10*, 427. [CrossRef]

19. Prà, I.D.; Armato, U.; Chiarini, A. Astrocytes' Role in Alzheimer's Disease Neurodegeneration. *Astrocyte Physiol. Pathol.* **2018**. [CrossRef]
20. Raghavan, A.; Shah, Z.A. Neurodegenerative disease. *Diet Exerc. Chronic Dis. Biol. Basis Prev.* **2014**, 339–390. [CrossRef]
21. Murphy, M.P.; Levine, H. Alzheimer's disease and the amyloid-β peptide. *J. Alzheimer's Dis.* **2010**, *19*, 311–323. [CrossRef] [PubMed]
22. Popugaeva, E.; Pchitskaya, E.; Bezprozvanny, I. Dysregulation of Intracellular Calcium Signaling in Alzheimer's Disease. *Antioxid. Redox Signal.* **2018**, *29*, 1176–1188. [CrossRef] [PubMed]
23. Binder, L.I.; Guillozet-Bongaarts, A.L.; Garcia-Sierra, F.; Berry, R.W. Tau, tangles, and Alzheimer's disease. *Biochim. Biophys. Acta Mol. Basis Dis.* **2005**, *1739*, 216–223. [CrossRef]
24. Eichler, S.A.; Meier, J.C. E-I balance and human diseases—From molecules to networking. *Front. Mol. Neurosci.* **2008**, *1*, 1–5. [CrossRef] [PubMed]
25. Vico Varela, E.; Etter, G.; Williams, S. Excitatory-inhibitory imbalance in Alzheimer's disease and therapeutic significance. *Neurobiol. Dis.* **2019**, *127*, 605–615. [CrossRef]
26. Toniolo, S.; Sen, A.; Husain, M. Modulation of brain hyperexcitability: Potential new therapeutic approaches in alzheimer's disease. *Int. J. Mol. Sci.* **2020**, *21*, 9318. [CrossRef] [PubMed]
27. Boyden, E.S.; Zhang, F.; Bamberg, E.; Nagel, G.; Deisseroth, K. Millisecond-timescale, genetically targeted optical control of neural activity. *Nat. Neurosci.* **2005**, *8*, 1263–1268. [CrossRef]
28. Nagel, G.; Szellas, T.; Huhn, W.; Kateriya, S.; Adeishvili, N.; Berthold, P.; Ollig, D.; Hegemann, P.; Bamberg, E. Channelrhodopsin-2, a directly light-gated cation-selective membrane channel. *Proc. Natl. Acad. Sci. USA* **2003**, *100*, 13940–13945. [CrossRef]
29. Octeau, J.C.; Gangwani, M.R.; Allam, S.L.; Tran, D.; Huang, S.; Hoang-Trong, T.M.; Golshani, P.; Rumbell, T.H.; Kozloski, J.R.; Khakh, B.S. Transient, Consequential Increases in Extracellular Potassium Ions Accompany Channelrhodopsin2 Excitation. *Cell Rep.* **2019**, *27*, 2249–2261.e7. [CrossRef]
30. Airan, R.D.; Thompson, K.R.; Fenno, L.E.; Bernstein, H.; Deisseroth, K. Temporally precise in vivo control of intracellular signalling. *Nature* **2009**, *458*, 1025–1029. [CrossRef]
31. Griffin, J.M.; Fackelmeier, B.; Fong, D.M.; Mouravlev, A.; Young, D.; O'Carroll, S.J. Astrocyte-selective AAV gene therapy through the endogenous GFAP promoter results in robust transduction in the rat spinal cord following injury. *Gene Ther.* **2019**, *26*, 198–210. [CrossRef]
32. Perea, G.; Yang, A.; Boyden, E.S.; Sur, M. Optogenetic astrocyte activation modulates response selectivity of visual cortex neurons in vivo. *Nat. Commun.* **2014**, *5*. [CrossRef]
33. Adamsky, A.; Kol, A.; Kreisel, T.; Doron, A.; Ozeri-Engelhard, N.; Melcer, T.; Refaeli, R.; Horn, H.; Regev, L.; Groysman, M.; et al. Astrocytic Activation Generates De Novo Neuronal Potentiation and Memory Enhancement. *Cell* **2018**, *174*, 59–71.e14. [CrossRef]
34. Henneberger, C.; Papouin, T.; Oliet, S.H.R.; Rusakov, D.A. Long-term potentiation depends on release of d-serine from astrocytes. *Nature* **2010**, *463*, 232–236. [CrossRef]
35. Chen, J.; Tan, Z.; Zeng, L.; Zhang, X.; He, Y.; Gao, W.; Wu, X.; Li, Y.; Bu, B.; Wang, W.; et al. Heterosynaptic long-term depression mediated by ATP released from astrocytes. *Glia* **2013**, *61*, 178–191. [CrossRef]
36. Tan, Z.; Liu, Y.; Xi, W.; Lou, H.; Zhu, L.; Guo, Z.; Mei, L.; Duan, S. Glia-derived ATP inversely regulates excitability of pyramidal and CCK-positive neurons. *Nat. Commun.* **2017**, *8*, 13772. [CrossRef]
37. Mederos, S.; Hernández-Vivanco, A.; Ramírez-Franco, J.; Martín-Fernández, M.; Navarrete, M.; Yang, A.; Boyden, E.S.; Perea, G. Melanopsin for precise optogenetic activation of astrocyte-neuron networks. *Glia* **2019**, *67*, 915–934. [CrossRef]
38. Courtney, C.D.; Sobieski, C.; Ramakrishnan, C.; Ingram, R.J.; Wojnowski, N.M.; DeFazio, R.A.; Deisseroth, K.; Christian-Hinman, C.A. Optogenetic activation of Gq signaling in astrocytes yields stimulation-specific effects on basal hippocampal synaptic excitation and inhibition. *bioRxiv* **2021**, *105825*, 1–32. [CrossRef]
39. Covelo, A.; Araque, A. Neuronal activity determines distinct gliotransmitter release from a single astrocyte. *Elife* **2018**, *7*, e32237. [CrossRef]
40. McNeill, J.; Rudyk, C.; Hildebrand, M.E.; Salmaso, N. Ion Channels and Electrophysiological Properties of Astrocytes: Implications for Emergent Stimulation Technologies. *Front. Cell. Neurosci.* **2021**, *15*, 1–22. [CrossRef]
41. Martineau, M.; Parpura, V.; Mothet, J.P. Cell-type specific mechanisms of D-serine uptake and release in the brain. *Front. Synaptic Neurosci.* **2014**, *6*, 12. [CrossRef] [PubMed]
42. Figueiredo, M.; Lane, S.; Stout, R.F.; Liu, B.; Parpura, V.; Teschemacher, A.G.; Kasparov, S. Comparative analysis of optogenetic actuators in cultured astrocytes. *Cell Calcium* **2014**, *56*, 208–214. [CrossRef]
43. Borodinova, A.A.; Balaban, P.M.; Bezprozvanny, I.B.; Salmina, A.B.; Vlasova, O.L. Genetic Constructs for the Control of Astrocytes' Activity. *Cells* **2021**, *10*, 1600. [CrossRef]
44. Osten, P.; Cetin, A.; Komai, S.; Eliava, M.; Seeburg, P.H. Stereotaxic gene delivery in the rodent brain. *Nat. Protoc.* **2007**, *1*, 3166–3173. [CrossRef]
45. Sun, S.; Zhang, H.; Liu, J.; Popugaeva, E.; Xu, N.-J.; Feske, S.; White, C.L.; Bezprozvanny, I. Reduced Synaptic STIM2 Expression and Impaired Store-Operated Calcium Entry Cause Destabilization of Mature Spines in Mutant Presenilin Mice. *Neuron* **2014**, *82*, 79–93. [CrossRef]

International Journal of
Molecular Sciences

Article

Optogenetic Control of PIP2 Interactions Shaping ENaC Activity

Tarek Mohamed Abd El-Aziz [1,2], Amanpreet Kaur [3], Mark S. Shapiro [1], James D. Stockand [1] and Crystal R. Archer [1,*]

1 Department of Cellular and Integrative Physiology, University of Texas Health Science Center at San Antonio, San Antonio, TX 78228, USA; mohamedt1@uthscsa.edu (T.M.A.E.-A.); shapirom@uthscsa.edu (M.S.S.); stockand@uthscsa.edu (J.D.S.)
2 Faculty of Science, Zoology Department, Minia University, El-Minia 61519, Egypt
3 Department of Chemistry, University of Washington, Seattle, WA 98195, USA; amanpreet.kaur1223@gmail.com
* Correspondence: archerc@uthscsa.edu

Abstract: The activity of the epithelial Na$^+$ Channel (ENaC) is strongly dependent on the membrane phospholipid phosphatidylinositol 4,5-bisphosphate (PIP2). PIP2 binds two distinct cationic clusters within the N termini of β- and γ-ENaC subunits (βN1 and γN2). The affinities of these sites were previously determined using short synthetic peptides, yet their role in sensitizing ENaC to changes in PIP2 levels in the cellular system is not well established. We addressed this question by comparing the effects of PIP2 depletion and recovery on ENaC channel activity and intracellular Na$^+$ levels [Na$^+$]$_i$. We tested effects on ENaC activity with mutations to the PIP2 binding sites using the optogenetic system CIBN/CRY2-OCRL to selectively deplete PIP2. We monitored changes of [Na$^+$]$_i$ by measuring the fluorescent Na$^+$ indicator, CoroNa Green AM, and changes in channel activity by performing patch clamp electrophysiology. Whole cell patch clamp measurements showed a complete lack of response to PIP2 depletion and recovery in ENaC with mutations to βN1 or γN2 or both sites, compared to wild type ENaC. Whereas mutant βN1 also had no change in CoroNa Green fluorescence in response to PIP2 depletion, γN2 did have reduced [Na$^+$]$_i$, which was explained by having shorter CoroNa Green uptake and half-life. These results suggest that CoroNa Green measurements should be interpreted with caution. Importantly, the electrophysiology results show that the βN1 and γN2 sites on ENaC are each necessary to permit maximal ENaC activity in the presence of PIP2.

Keywords: ENaC; phosphoinositides; PIP2; optogenetic; CRY2; sodium channel; CoroNa Green

Citation: Abd El-Aziz, T.M.; Kaur, A.; Shapiro, M.S.; Stockand, J.D.; Archer, C.R. Optogenetic Control of PIP2 Interactions Shaping ENaC Activity. *Int. J. Mol. Sci.* **2022**, 23, 3884. https://doi.org/10.3390/ijms23073884

Academic Editors: Piotr D. Bregestovski and Carlo Matera

Received: 14 March 2022
Accepted: 29 March 2022
Published: 31 March 2022

Publisher's Note: MDPI stays neutral with regard to jurisdictional claims in published maps and institutional affiliations.

1. Introduction

The epithelial Na$^+$ channel, ENaC, is a trimeric channel that closely resembles the chalice-like structure of the closely related acid sensing ion channel (ASIC) [1,2]. ENaC is comprised of 3 independent proteins subunits, called α, β, and γ, which are encoded by three distinct genes [3–5]. ENaC conducts Na$^+$ across tight epithelia such as those lining the lungs and kidney tubules [3]. ENaC is necessary for liquid clearance in the lungs and consequently, the knockout of α-ENaC in mice is lethal. In contrast, the overexpression of β-ENaC follows a pattern of cystic fibrosis [6]. In the kidney, ENaC is the final arbiter of Na$^+$ reabsorption. Pathological disturbance of ENaC results in blood pressure disorders such as Liddle's syndrome [7,8]. The body's dependence on Na$^+$ homeostasis underscores the importance of the proper function of the mechanisms that regulate ENaC. Many of these mechanisms regulate ENaC by acting on its intracellular domains. Each ENaC subunit resembles a hairpin structure with a large extracellular globular domain anchored by two transmembrane domains [1]. Each transmembrane domain connects to relatively short intracellular N and C terminal tails comprising 55–85 amino acids; thus, ENaC has 6 intracellular domains. Although the bulk of ENaC has been elucidated using cryo-electron microscopy, the structure of the intracellular domains remains obscure [1]. Structure prediction software and experimental observations indicate that these intracellular domains are likely largely disordered yet they may adopt helical structure when interacting with other regulatory cofactors, such as the membrane phospholipid, phosphatidylinositol 4,5-bisphosphate (PIP2) [9]. PIP2 is a low abundance phospholipid in the plasma membrane that is necessary for ENaC to be maximally open [10–12]. Earlier studies showed that mutation of polybasic clusters in ENaC subunits reduces its ability to respond to changes in PIP2 [11]. More recently, we showed that PIP2 binds two of these distinct clusters located on the intracellular termini of β and γ subunits of ENaC, referred to here as the βN1 and γN2 sites [13]. PIP2 consists of a phosphorylated, cytoplasmic inositol ring (called the PIP2 headgroup) which is anchored to the plasma membrane via phosphodiester linkage to two fatty acyl chains embedded within the inner leaflet of the membrane [14]. The PIP2 headgroup bears two phosphoryl groups covalently bound to its carbons at positions 4 and 5 (C4 and C5) [14–16]. Those phosphoryl groups can be reversibly removed or added by site-specific lipid phosphatases and kinases [17]. This study takes advantage of the ability of the phosphatase OCRL to deplete PIP2 by removing the C5 phosphoryl group [18]. The anionic PIP2 headgroup forms electrostatic interactions with cationic amino acid residues of proteins [14] and is hypothesized to bind the βN1 and γN2 sites on ENaC. βN1 is located on the extreme, intracellular N terminus of β-ENaC (Figure 1a, left). γN2 is located on the intracellular N terminus of γ-ENaC, near the inner plasma membrane (Figure 1a, right). In our previous work, we reported that the equilibrium constant of PIP2 for βN1 is K_d ~5 μM, and for γN2 is K_d ~13 μM [13]. The γC site of γ-ENaC (Figure 1a, right) is a potential PIP2 interaction site because it demonstrated an interesting profile in binding experiments but had very weak affinity for PIP2 [13].

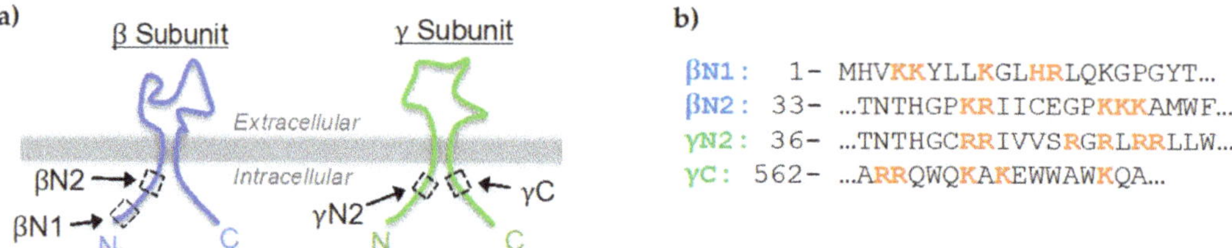

Figure 1. Cationic sites in ENaC tested for PIP2 binding. (**a**) Schematic of the ENaC subunits with the sites tested for PIP2 binding, βN1, γN2, γC, and βN2, indicated by dashed boxes. Adapted from Archer et al., 2020 [13]. (**b**) Human ENaC sequences corresponding to βN1, γN2, γC, and βN2. The cationic residues neutralized to alanine are indicated in boldface orange font.

In this study, we investigated how mutations made to βN1, γN2 and γC affect the activity of the human ENaC transfected in mammalian cells. We mutagenized the channel by substituting alanine for each cationic residue within these domains, indicated in bold orange letters in Figure 1b. We used the optogenetic dimerization pair CIBN-CAAX and mCherry-CRY2-OCRL (mCh-CRY2-OCRL), which has been previously used to examine the effects of PIP2 depletion on ion channel activity [13,18]. CIBN-CAAX is localized to the plasma membrane via the CAAX moiety. mCh-CRY2-OCRL is rapidly recruited from the cytoplasm to the plasma membrane by dimerization with CIBN, in response to blue light illumination (BLI) [18]. The 5-phosphatase OCRL is a protein that selectively removes the carbon 5 phosphoryl group of PIP2 [17,19]. Membrane-localized mCh-CRY2-OCRL rapidly depletes PIP2. mCh-CRY2-OCRL reversibly translocates to the cytoplasm in the absence of blue light within ~5–7 min, after which PIP2 levels are immediately replenished [20–22]. We previously used the CIBN/mCh-CRY2-OCRL system in HEK 293 cells to show that BLI-induced PIP2 depletion caused a reduction of intracellular Na^+ levels ($[Na^+]_i$) in cells expressing wild type ENaC (wtENaC) [13]. Here, we used the CIBN/mCh-CRY2-OCRL system to show that the βN1 and γN2 sites play a strong role in sensitizing ENaC to changes in PIP2 levels in the cellular system. Channel activity was measured by two different methods. First, we used the Na^+ selective fluorescent indicator, CoroNa Green AM, to track changes in $[Na^+]_i$ of cells expressing mutant ENaC, in response to BLI. We compared those results to the direct measurements of mutant ENaC activity in response to BLI using whole cell patch clamp electrophysiology. The purpose of this study was to determine whether each independent PIP2-ENaC interaction site is sufficient to abolish the response of ENaC to changes in PIP2 levels, and to evaluate CoroNa Green as a tool for studying Na^+ ion channel activity.

2. Results

2.1. Mutations of the PIP2-ENaC Binding Sites Do Not Affect Membrane Expression of ENaC

We measured the membrane expression of ENaC using Total Internal Reflection Fluorescence (TIRF) microscopy. In TIRF, the excitation light is nearly totally and internally reflected at the interface between two media transparent to light with different refractive indices, in this case glass (~1.52) and aqueous solution (1.33). However, an "evanescent wave" does propagate into the second medium normal to the plane of the interface, but decays exponentially over a short distance with a length constant of ~300 nm, thus allowing only fluorescent proteins located within that distance from cells adherant to a glass coverslip to be excited [23,24]. We used TIRF microscopy to determine if the mutations used in this study (Figure 1b) affected the membrane expression of α-, β- and γ-ENaC subunits. HEK 293 cells were transiently transfected with cDNA encoding ECFP-tagged α-ENaC (CFP-αENaC), along with EYFP-tagged β-ENaC (βENaC-YFP) and mCherry-tagged γ-ENaC (γENaC-mCherry) constructs containing wild-type sequences or mutations to the indicated cationic clusters. The data shown in Supplemental Figure S1a,b show that all three subunits were expressed at the plasma membrane in the same cell. Cells expressing the γN2 mutants had reduced membrane expression of γ-ENaC, whereas α- and β-ENaC expression was similar regardless of whether it was part of the wt or mutant channel. Moreover, the membrane expression and activity of ENaC was unaffected by the expression of mCh-CRY2-OCRL under dark (normal PIP2) conditions (Supplemental Figure S1c,d).

2.2. PIP2 Depletion Reduces $[Na^+]_i$ and Current Density in Cells with wtENaC

To quantify changes of $[Na^+]_i$, we incubated cells with the fluorescent cell-permeable Na^+ indicator, CoroNa Green AM, where high intensity of emission corresponds to an increase of $[Na^+]_i$ and decreased intensity corresponds to decreased $[Na^+]_i$. CoroNa Green is excited between 440–514 nm with peak emission at 490 nm [25]. CIBN-CRY2 dimerization occurs between 405 and 500 nm [20]. Therefore, we used brief laser excitation at 514 nm to excite CoroNa Green with minimal impact on PIP2 levels [22]. TIRF microscopy was used to capture these intracellular changes in local $[Na^+]_i$ which are expected to occur just inside the membrane near ENaC while changes in global $[Na^+]_i$ would not be easily detected. HEK cells expressing CFP-αENaC (excited with a 445 nm laser), untagged β-ENaC, untagged γ-ENaC, untagged CIBN-CAAX and mCh-CRY2-OCRL (excited with a 561 nm laser) were used for the ENaC-PIP2-response experiments. The cells were then pulsed with blue light (BLI, 445 nm laser) to stimulate mCh-CRY2-OCRL translocation to the membrane and subsequent PIP2 depletion. The presence of CFP-αENaC was confirmed in each cell during the BLI step. The representative micrographs and summary graph of CoroNa Green intensity in Figure 2a,b show that there was no change in the mean $[Na^+]_i$ in control cells with no ENaC, in response to BLI. In contrast, BLI-induced depletion of PIP2 caused a significant reduction of $[Na^+]_i$ (~34% decrease) in cells expressing wtENaC, indicating that ENaC is strongly dependent on PIP2 (Figure 2c,d). These results are consistent with our previous findings that the ENaC-specific channel blocker amiloride blocked the reduction of $[Na^+]_i$ caused by BLI-induced depletion of PIP2 in cells expressing wtENaC [13].

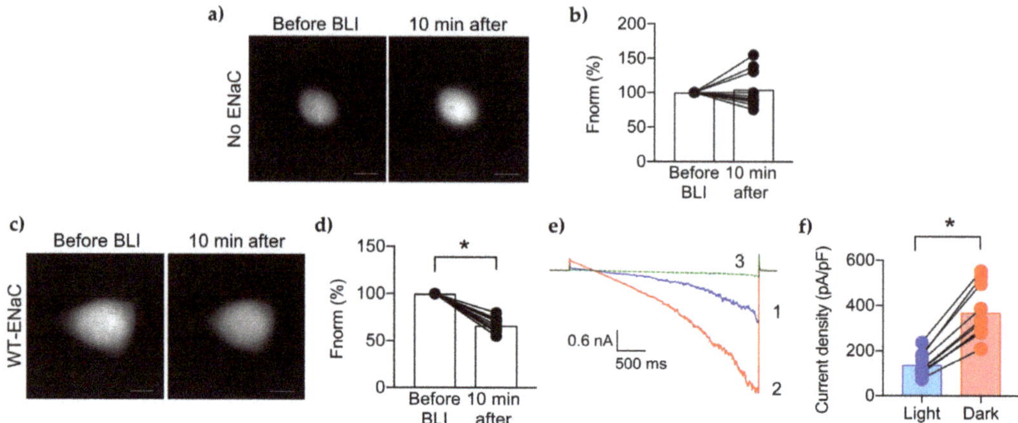

Figure 2. PIP2 depletion decreases [Na$^+$]$_i$ and current density of wtENaC. Representative micrographs show the CoroNa Green uptake in HEK 293 cells without transfection (**a**) or transfected with wtENaC (**c**). Summary graphs of changes in the mean [Na$^+$]$_i$ of cells with no ENaC (n = 11 cells) (**b**) or wtENaC (n = 11) (**d**), as measured by CoroNa intensity excited at 514 nm, 10 min after BLI, normalized to CoroNa levels before BLI. (**e**) Representative current trace showing wtENaC activity under BLI (1, blue trace), followed by Dark (2, red trace), and finally with application of 10 µM amiloride (3, green trace). (**f**) Summary graph of the mean current density of wtENaC at −100 mV under low PIP2 levels (Light, blue bar and blue circles) vs. maximum PIP2 levels after 10 min (Dark, red bar and blue circles) (n = 10). For each graph, the line drawn between 2 circles represents an independent cell under different PIP2 levels. * Indicates *p* < 0.0001 determined by the paired *t* test.

For patch clamp experiments, we used Chinese hamster ovary (CHO) cells because they have no detectable background ion channel activity. The cells were grown on glass chips coated with poly-l-lysine (PLL). Since PLL has been used previously as a PIP2 scavenger in excised patches, we compared the activity of wtENaC on glass chips coated with PLL to glass chips coated with gelatin. Supplemental Figure S2 shows that the PIP2-dependent activity of ENaC in cells grown on PLL was similar to those grown on gelatin. We started patch clamp recordings in the presence of transmitted light and BLI-induced PIP2 depletion in order to find the cells expressing both mCherry and ECFP and to visually guide the patch pipette to the cell surface. BLI was performed using an epifluorescence microscope equipped with a mercury lamp and standard ECFP and mCherry filter cubes. Macroscopic ENaC currents were recorded during voltage ramp from 0 to −100 mV. The current density at −100 mV after 10 minutes was reported. Under this protocol, wtENaC exhibited low current density under BLI (~137 ± 50 pA/pF) after 10 min, shown in Figure 2e, blue trace; Figure 2f, blue bar. After 3 min of recording under BLI, we turned off the lights to both the microscope and the room, to allow translocation of mCh-CRY2-OCRL to the cytoplasm followed by recovery of PIP2. The activity of wtENaC immediately began to increase, reaching a peak current density of 371 ± 117 pA/pF at 10 min (Figure 2e, red trace; Figure 2f, red bar). Figure 2f summarizes the voltage ramp recordings taken at 10 min under depleted PIP2 (blue bar) and recovered PIP2 level (red bar). The time course of this PIP2-dependent ENaC recovery is described in further detail in Figure 3c, and later in Section 2.5. Perfusion with 10 µM of the selective ENaC blocker, amiloride, abolished wtENaC activity demonstrating that the PIP2-dependent currents are solely from ENaC (Figure 2e, green trace). The peak current density of wtENaC under recovered PIP2 was similar to wtENaC under normal conditions without mCh-CRY2-OCRL (Supplemental Figure S3). This result shows that the CIBN/mCh-CRY2-OCRL system is effective for measuring ENaC response to PIP2 levels. These results further indicate that the presence of cytoplasmic CRY2-OCRL does not affect normal PIP2 levels and PIP2 is replenished to its normal levels in the dark.

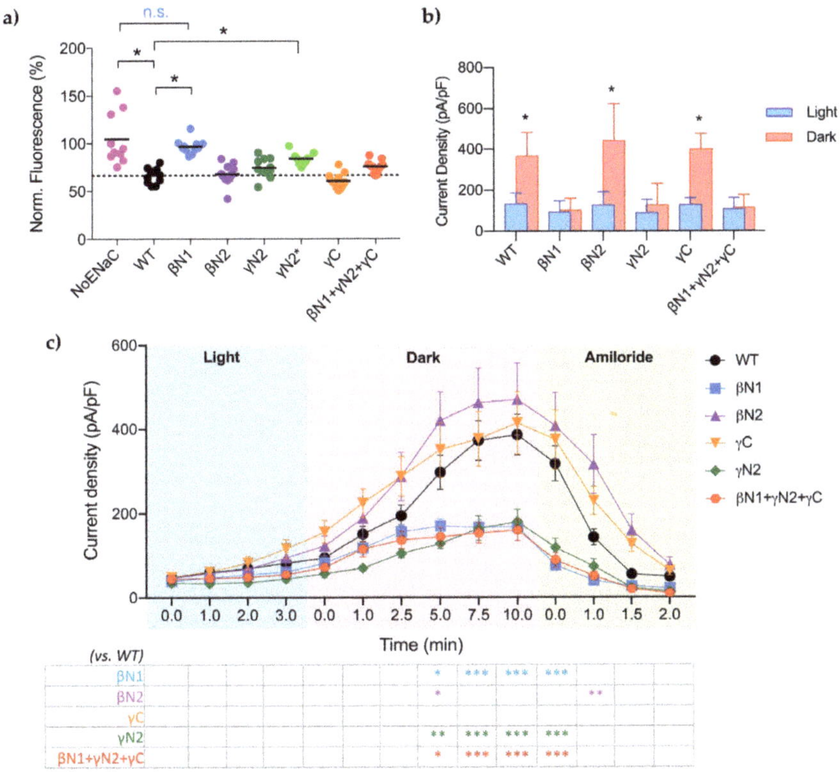

Figure 3. Summary of mutant-ENaC responses to PIP2 depletion and recovery. (**a**) Summary graph comparing the CoroNa Green intensity of cells expressing each mutant 10 min after BLI. *, $p \leq 0.0001$, n.s., no significant difference. The dashed line indicates the mean CoroNa Green intensity of wtENaC 10 min after BLI. (**b**) Summary graph comparing the mean current densities $\pm$ SD of each ENaC under BLI-induced PIP2 depletion (Light/blue) and PIP2 recovered (Dark/red). *, $p < 0.0001$, comparing light to dark. (**c**) Summary timeline of the current density of each mutant ENaC in response to recovery of PIP2 levels after CRY2-OCRL relocation to the cytoplasm, expressed as the mean of 5–7 cells $\pm$ SEM. Two-way ANOVA (column factor, mutant; row factor, time) indicated significant differences between 5 min dark and 1 min after amiloride, where ***, $p < 0.0001$, **, $p < 0.005$, and *, $p < 0.05$, compared to wtENaC, noted in the table below the time plot.

2.3. Cellular-Mutagenesis Validates the βN1 Site as a PIP2 Binding Site

We previously reported that a peptide corresponding to the βN1 site of ENaC has moderate affinity for PIP2 (K_d ~5 μM) [13]. To determine if βN1 is critical for ENaC activity in a PIP2-dependent manner, we tested the mutant βN1-ENaC in the optogenetic cellular assay. After BLI, CoroNa Green emission remained unchanged 10 min after BLI-induced PIP2 depletion (Figure 4a,b). For whole cell patch clamp electrophysiology, the mean current density of mutant βN1-ENaC was similar to wtENaC under BLI-induced PIP2 depletion (95 $\pm$ 53, Figure 4d, vs. 137 $\pm$ 50 pA/pF, Figure 2f blue bars). Switching to dark conditions allowed for full PIP2 recovery, yet the current density of mutant βN1-ENaC only slightly increased to 106 $\pm$ 57 pA/pF up to 10 min in the dark (Figure 4c,d, red; Figure 3a,c, blue circles). Mutant βN1-ENaC exhibited the same basal activity with normal PIP2 levels in the absence of CIBN/mCh-CRY2-OCRL (Supplemental Figure S3). In both of these experiments, mutant βN1-ENaC did not respond to the controlled changes in PIP2 levels, suggesting that the βN1 site is necessary for the PIP2-ENaC interactions.

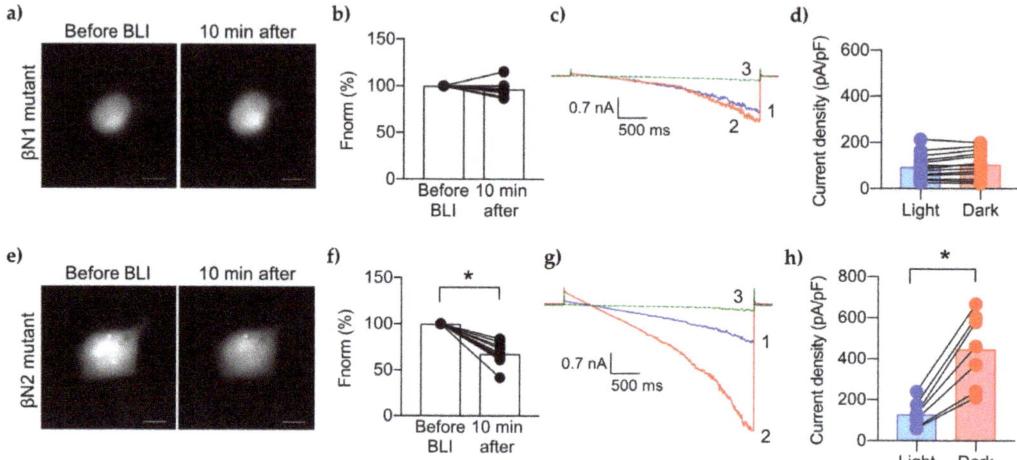

Figure 4. Effects of PIP2 depletion on mutant β-ENaC. Representative micrographs showing the CoroNa Green uptake in HEK 293 cells transfected with mutant βN1-ENaC (**a**) or mutant βN2-ENaC (**e**). The graphs in (**b**,**f**) summarize changes in the mean $[Na^+]_i$ as determined by CoroNa Green fluorescence levels (n = 11 per group). Representative current traces of the βN1 mutant (**c**), or the βN2 mutant (**g**), under BLI (1, blue trace), then dark (2, red trace), then 10 μM amiloride (3, green trace). Summary graphs of the mean current density for the βN1 mutant (n = 14) (**d**), or the βN2 mutant (n = 7) (**h**), at −100 mV under low PIP2 levels ("Light", blue bar and blue circles) vs. maximum PIP2 levels ("Dark", red bar and red circles). For each graph, the line drawn between 2 circles represents an independent cell under different PIP2 levels. *, $p < 0.0001$ determined by paired t test.

Mutations were also made to the βN2 site of ENaC. Although this site also contains several cationic residues, the peptide corresponding to this domain did not directly bind PIP2 in our earlier study, and was used here as a control [13]. Figure 4e,f shows that the CoroNa Green intensity decreased by ~32%, similar to the wtENaC response. The current traces also followed the same pattern as wtENaC, with very low activity during BLI-induced PIP2 depletion followed by a large increase in channel activity after PIP2 recovery in the dark (Figure 4g). The summary graph in Figure 4h shows the mean current density under PIP2 depletion was 129 ± 63 pA/pF and increased to 447 ± 180 pF/pA under PIP2 recovery, which is similar to wtENaC activity. We also observed high ENaC activity in cells without CIBN/mCh-CRY2-OCRL (Supplement Figure S3). These results are consistent with the binding data showing βN2 site is not a PIP2 binding site.

2.4. Cellular-Mutagenesis Assays Confirm That γN2, but Not γC, Binds PIP2

The γN2 peptide also has a moderate biochemical affinity for PIP2 (K_d ~13 μM) [13]. Therefore, we expected the mutant γN2-ENaC would produce results similar to the mutant βN1-ENaC. Surprisingly, BLI-induced PIP2 depletion caused a ~26% decrease of CoroNa Green intensity in cells expressing mutant γN2-ENaC (Figure 5a,b). The mean CoroNa Green intensity of mutant γN2-ENaC after 10 min was similar to wt ENaC (Figure 3a). However, the mutant γN2-ENaC had low current density under BLI-induced PIP2 depletion, which was similar to mutant βN1-ENaC (Figure 5c, blue trace and Figure 5d, blue bar). The current density did not significantly increase after full PIP2 recovery in the dark (92 ± 63 pA/pF vs. 131 ± 102 pA/pF, respectively) (Figure 5c, red trace and

Figure 5d, red bar). The activity of mutant γN2-ENaC also remained low in cells without CIBN/mCh-CRY2-OCRL (Supplemental Figure S3).

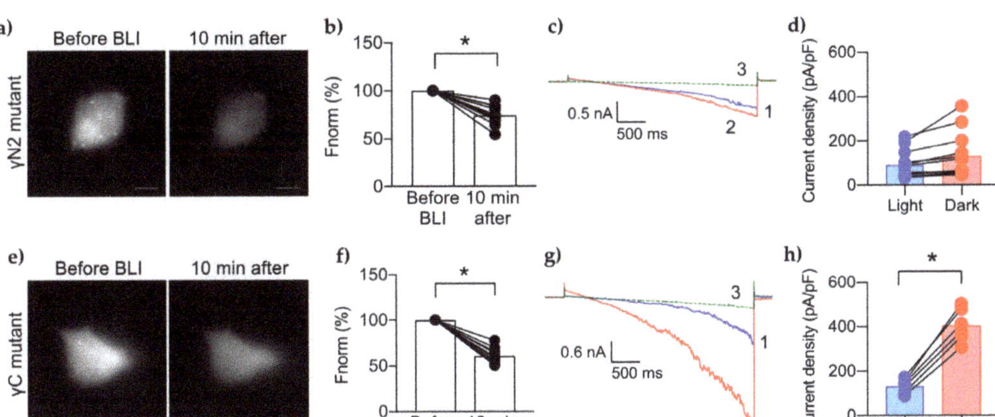

Figure 5. Effects of PIP2 depletion on mutant γ-ENaC. Representative micrographs showing the CoroNa Green fluorescence levels in HEK 293 cells transfected with ENaC containing alanine substitutions to the γN2 site (**a**) or the γC site (**e**). Summary graphs (**b,f**) of changes in the mean $[Na^+]_i$ as determined by CoroNa Green fluorescence levels. Representative current traces of the γN2 mutant (**c**) or the γC mutant (**g**) under BLI (1, blue trace), then dark (2, red trace), then 10 μM amiloride (3, green trace). Summary graphs of the mean current density for the γN2 mutant (**d**) or the γC mutant (**h**) at −100 mV under low PIP2 levels ("Light", blue bar and blue circles) vs. maximum PIP2 levels ("Dark", red bar and red circles). *, $p < 0.0001$ determined by paired *t* test.

To better understand why cells expressing the mutant γN2-ENaC had a strong decrease of CoroNa Green but no change in current density in response to changes in PIP2, we examined the CoroNa Green emission under normal PIP2 levels. As shown in Supplemental Figure S4, the CoroNa Green emission of mutant γN2 cells was low CoroNa Green at the beginning of each experiment compared to wtENaC (1024 ± 1044 vs. 9197 ± 1267 AU) (Figure S4a). Nether had a significant difference in CoroNa Green emission after 10 min. Moreover, CoroNa Green emission of mutant γN2-ENaC cells were ~83% lower than cells expressing wtENaC or mutant βN1-ENaC (Figure S4b). These data suggest that CoroNa Green had less uptake in cells expressing mutant γN2-ENaC and that the Na^+ indicator may have leaked from those cells a faster rate than the experimental time. This reduced CoroNa Green emission could explain why cells expressing γN2-ENaC do not correspond with patch clamp data. Since the starting values of the Na^+ indicator were so low compared to the other cells tested, we hesitate to strongly rely on the CoroNa data for this mutant.

We also examined the CoroNa Green response in a partial mutant, γN2*, which contained the substitutions R48A/R50A/R52A/R53A of human γ-ENaC, in contrast to the γN2 mutant that included the substitutions R42A and R43A. We found that cells expressing the γN2* mutant had a smaller decrease of CoroNa Green intensity in response to BLI (16 ± 7%) after BLI-induced PIP2 depletion, which was also significantly less change of intensity than wtENaC (Supplemental Figure S5 and Figure 3a). Cells expressing mutant γN2* had normal starting levels of CoroNa Green (Supplemental Figure S4b), but the mutant γN2* subunit also had reduced membrane expression (Supplemental Figure S1). These results suggests that PIP2 interacts with the "*. . . SRGRLRRL. . .*" sequence of the γN2 site, which is closer to the transmembrane domain TM1 of γ-ENaC. Together, these results indicate the γN2 site is critical for PIP2 regulation of ENaC.

The γC domain of ENaC was previously reported to be a PIP2-binding candidate. Its weak affinity (K_d ~800 μM) suggested that it might play a moderate to small role in the ENaC response to changes in PIP2 levels. However, we found that PIP2 depletion caused a significant decrease of CoroNa Green intensity ~40 ± 8% in cells expressing mutant γC-ENaC (Figure 5e,f). This decrease was similar to wtENaC. In addition, the current density of mutant γC-ENaC fully recovered to 406 ± 75 pA/pF within 10 min after PIP2 recovery in the dark (Figures 3c and 5g,h). These results are consistent with an earlier study from our lab showing that the current density from mutant mouse γC-ENaC is similar to wtENaC [11]. These results strongly suggest that the γC domain is not involved in PIP2 binding or regulation of ENaC.

2.5. Mutation of All 3 PIP2 Binding Candidates Does Not Further Increase PIP2 Sensitivity

We mutagenized the three candidate PIP2 sites on ENaC to create the triple mutant, βN1-γN2γC-ENaC. Since cells expressing mutant βN1-ENaC had no change of CoroNa Green intensity in response to BLI-induced PIP2 depletion, we expected to see similar results with this triple mutant. Instead, the CoroNa Green intensity of cells expressing the triple mutant were reduced by 25% ± 7% (Figure 6a,b). Cells expressing the triple mutant had low starting levels of CoroNa in contrast to the other cells, but similar to mutant γN2-ENaC (Supplemental Figure S4b). In patch clamp experiments, the triple mutant exhibited low mean current density (111 ± 52 pA/pF) under BLI-induced PIP2 depletion, with no change in response to PIP2 recovery in the dark (118 ± 60 pA/pF) (Figure 6c,d). The activity of this triple mutant was also low in cells without CIBN/mCh-CRY2-OCRL (Supplemental Figure S3). This low basal activity of both wtENaC and mutant βN1-ENaC suggests that ENaC has a small, constitutive activity that is not dependent on PIP2. These results show that mutating βN1 and γN2 sites together results in a lack of ENaC response to PIP2 depletion that is not different compared to mutating the βN1 and γN2 sites on their own.

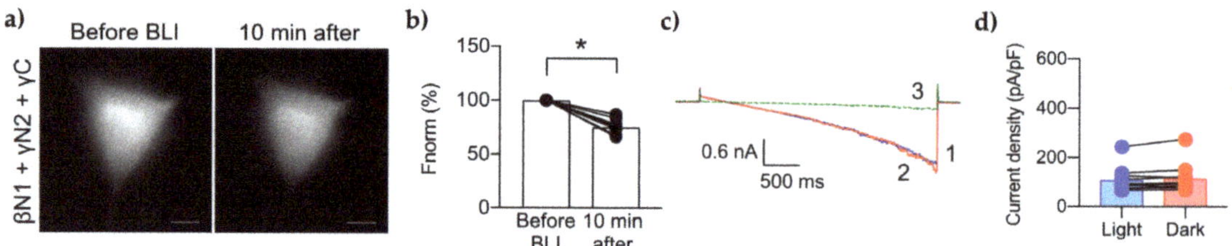

Figure 6. Effects of PIP2 depletion on the triple mutant. Representative micrographs showing the CoroNa Green fluorescence levels in HEK 293 cells transfected with triple mutant ENaC containing alanine substitutions to the βN1, γN2, and γC sites (**a**). Summary graph (**b**) of changes in the mean [Na$^+$]$_i$ as determined by CoroNa Green fluorescence levels. Representative current traces of the triple mutant (**c**) under BLI (1, blue trace), then dark (2, red trace), then 10 μM amiloride (3, green trace). Summary graph of the mean current density for the triple mutant (**d**) at −100 mV under low PIP2 levels ("Light", blue bar and blue circles) vs. maximum PIP2 levels ("Dark", red bar and red circles). *, $p < 0.0001$ determined by paired *t* test.

The graphs in Figure 3 summarize the effects of PIP2 depletion and recovery on the mutants tested in this study. Figure 3a shows that the Corona Green emission of cells expressing mutant βN1 were not different from the "No ENaC" control levels 10 min after BLI. Both βN1 and γN2* mutants had significantly less change of CoroNa Green than wtENaC. The mean change of CoroNa Green of cells expressing mutant γN2 and triple mutant were not found to be significantly different than wtENaC. The patch clamp results in Figure 3b revealed a clear distinction in the response of each mutant to changes in PIP2 levels. Each wt and mutant tested had very low residual activity ~100pA/pF under BLI-induced PIP2 depletion (blue bars), whereas wtENaC, mutant βN2-ENaC and mutant γC-ENaC displayed robust activity in response to PIP2 recovery in the dark (red bars). The maximal current density of mutant βN1- and γN2-ENaC and the triple mutant all remained low, with ~37% ENaC activity compared to wtENaC at normal PIP2 levels, after 10 min PIP2 recovery (Figure 3b,c—blue, green and red lines). This low activity under both normal (Supplemental Figure S3) and recovered PIP2 levels indicates that βN1- and γN2 are both necessary for PIP2 regulation of ENaC.

In contrast, the PIP2-responsive ENaC constructs (wtENaC, and βN2 and γC mutants) had maximal activity after 7–10 min of recovery time in the dark (Figure 3c-black, purple and orange lines). The difference between the mutants at each time point, compared to wtENaC, are noted in the table below the chart. This timing is consistent with another study which reported a $t_{1/2}$ = ~6.8 min for the mCh-CRY2-OCRL to dissociate from CIBN and redistribute back to the cytoplasm, followed by the resynthesis of PIP2 [18]. We also observed that all ENaC constructs in this study exhibited a minimal level of current density of ~100 pA/pF that was not dependent on PIP2 (Figure 3b and Supplemental Figure S3). These results indicate that the maximal current density of ENaC facilitated by PIP2 is dependent on the βN1- and γN2 sites; and ENaC produces a low basal level of activity that is not dependent on PIP2. This observation is in agreement with Pochynyuk et al, 2007 [13], which also reported that ENaC has basal activity that is independent of PIP2, and that PIP2 may serve a permissive role for increasing the open probability of ENaC.

3. Discussion

The results of this study validate our previous study showing that peptides corresponding to the βN1 and γN2 sites of ENaC bind to PIP2 [14]. The present study expands on this by investigating the effects of mutant βN1 and γN2 on the ENaC response to changes in PIP2 levels. Here, we show that mutating either βN1 or γN2 equally inhibit the ability of ENaC to reach maximal activity compared to wtENaC. The results suggest that PIP2 may interact with both βN1 and γN2 at the same time to permit ENaC maximal activity. Mutating either site alone prevents ENaC from responding to changes in PIP2 levels. We also observed a low residual ENaC activity of ~100 pA/pF in all mutants and wtENaC, even when PIP2 was depleted. This is consistent with earlier observations in our lab showing that PIP2 serves a permissive role for maximizing the P_o of ENaC [13]. These data indicate that PIP2 depletion would reduce ENaC activity rather than completely abolish its activity.

Based on our results, CoroNa Green may be a useful method for testing Na^+ ion channel activity or examining changes in $[Na^+]_i$, but should be paired with patch clamp or other more rigorous methods. Careful attention should be made for designing control experiments to understand the efficiency of loading and retention of the Na^+ indicator prior to starting experiments, and caution be made for interpreting the results. Indeed, mutation of the γN2 site of ENaC corresponded to reduced CoroNa Green emission, possibly due to less uptake into the cells or faster leakage from the cells, although it is unclear why (Supplemental Figure S4).

Earlier studies suggest there may be interplay between PIP2 and ubiquitin in controlling ENaC. The βN1 site of ENaC carries several lysines that were previously reported to be targets of ubiquitination-induced protein turnover [26–28]. However, we observed no change of membrane ENaC levels with mutations to those lysines, compared to the membrane levels of wtENaC (Supplemental Figure S1). These data are consistent with a recent finding that ubiquitination of the β-ENaC subunit was minimal [29]. It was also previously reported that deletion of the βN1 site of mouse ENaC also had no change in membrane expression [11]. If mutating βN1 prevented the ubiquitination of ENaC, then we would expect increased expression at the membrane, however there was no change (Supplemental Figure S1). The results from the present study suggest that the βN1 site of ENaC may only be a PIP2 binding site, although it does not rule out the possibility of ubiquitin having other impact on ENaC activity involving the βN1, to include competition or interplay with PIP2.

ENaC has moderate biochemical affinity for PIP2 of K_d ~5 μM for βN1 and K_d ~13 μM for the γN2 site. Our data show that these moderate affinity levels are responsible for the high sensitivity of ENaC to PIP2 levels. A very high affinity would probably result in less response to PIP2. For example, Kv7.3 channels which have high apparent affinity and is less sensitive to depletion of PIP2 compared with Kv7.2 channels having lower apparent affinity and thus very sensitive to depletion of PIP2 [30,31]. It is still unclear whether both PIP2 sites form a single "binding pocket" by simultaneously binding PIP2 or if each site binds PIP2 independently of each other. Having insight to this question will further impact our understanding of how other signaling cofactors or genetic mutations modulate the PIP2-ENaC interactions. PIP2 regulates the activity of many K^+ and Ca^{2+} ion channels, and many of these channels exhibit a defined PIP2-binding pocket comprised of distant cationic residues within their intracellular domains [16,32,33]. This makes a strong case for ENaC to also have a PIP2 binding pocket comprised of its βN1 and γN2 sites. We anticipate better clarification on this question as more structural studies emerge.

The γN2 site overlaps a nuclear localization sequence (NLS), that includes R42 and R43 within the human ENaC sequence " ... HGCRR ... ", shown in Figure 1b [34,35]. Our lab has reported that the N terminus of γ-ENaC becomes cleaved, then translocates to the nucleus or nucleolus [34]. While the cleavage site and function of free γN is unclear, an increased level of free γN leads to reduced macroscopic current density of ENaC. Mutating this NLS site appeared to play some role in CoroNa Green uptake and half life in the cells, given that cells expressing the γN2 mutant including R42A and R43A had reduced CoroNa Green uptake but cells expressing the γN2* mutant (without those mutations) had normal CoroNa Green uptake. These data could serve as the rationale for furthering the study of cleaved γN in regulating ENaC activity and how that might impact ENaC response to PIP2 depletion.

Acid sensing ion channels (ASICs) are within the same superfamily as ENaC [3,36]. A recent study of chicken ASIC1 reveals another interesting role for how PIP2 might facilitate Na^+ permeation through ENaC [37]. As with ASIC1, each α-, β- and γ-ENaC subunit bears an "HG motif" in the N terminal "pre-TM1" domain adjacent to the N-terminal transmembrane domain (TM1) [38]. Genetic mutation of the HG motif (G37S) on β-ENaC results in pseudohypoaldosteronism I (PHA1B) by reducing the P_o of ENaC [39], underscoring the importance of the HG motif. In the homo-trimeric ASIC1, the preTM1 domains of each subunit loop back toward the TM1 domains in the plasma membrane and line the pore of the channel protein [37]. The histidines of the HG motifs lean towards the pore, constricting the lower permeation pathway in the desensitized and resting configurations of the channel. Interestingly, the PIP2 binding domain of γ-ENaC (γN2) is located between the HG motif and TM1 that closely align to this region of the ASIC sequence. If ENaC adopts the same pre-TM1 pore loop structure to allow its HG motifs to control the size of the lower permeation pathway, one could predict a direct role for PIP2 controlling ion permeation of ENaC by tugging this pre-TM1 loop to create an "open pore" configuration. We look forward to seeing future structural studies of ENaC shed more light on the interactions between PIP2 and ENaC.

4. Materials and Methods

4.1. Plasmids, Synthetic Peptides and Reagents

Plasmids containing mCherry-CRY2-OCRL and CIBN-CAAX were gifts from Pietro De Camilli (Addgene plasmid #66836 and #79574). Human ENaC (hENaC) constructs of untagged β and γ subunits in the pMT3 vector capable of expressing channel subunits in mammalian systems have been described previously [40,41]. The cDNA sequence for ⟨-hENaC was cloned in frame into the pECFP-C1 (Takara Bio USA, Inc.; San Jose, CA, USA) plasmid. The resulting ECFP-tagged-⟨ENaC fusion protein (CFP-⟨ENaC) was used to track ENaC expression in patch clamp and CoroNa green experiments. For membrane expression experiments, the cDNA sequence for β-hENaC was cloned into the pEYFP-N1 vector (Takara Bio USA, Inc.; San Jose, CA, USA) to create "βENaC-YFP" and and γ-hENaC was cloned into a mCherry vector to create "γ-ENaC-mCherry". EYFP and mCherry were fused to the C terminus of ENaC. Unless otherwise indicated, all three subunits were simultaneously transfected to cells with the indicated mutation to form a hetero-trimeric ENaC channel.

Site directed mutagenesis was performed by TOP Gene Technologies (St-Laurent, QC, Canada) or in the lab using the Quickchange Lightning Site-directed Mutagenesis kit (Agilent, Santa Clara, CA, USA) and standard desalted primers from Thermo Fisher (Waltham, NA, USA). The following mutants were created by substituting alanine for the indicated cationic residues: βN1 (K4A, K5A, K9A, H12A, R13A); βN2 (K39A, R40A, K47A, K48A, K49A); γN2 (R42A, R43A, R48A, R50A, R52A, R53A); γN2* (R48A, R50A, R52A, R53A); γC (R563A, R564A, K568A, K570A, K576A); and the triple mutant βN2γN2γC containing substitutions corresponding to the βN2, γN2, and γC mutants. All sequences were confirmed by standard sequencing (Psomagen, Rockville, MD, USA).

HEK 293 (CRL-1573) and CHO-K1 (CCL-61) cells were from ATCC (American Type Culture Collection, ATCC, Manassas, VA, USA). The cell permeant Na^+ indicator, CoroNa Green AM (ThermoFisher Scientific, Waltham, MA, USA), was used to monitor intracellular $[Na^+]_i$ levels.

4.2. Quantificaton of Intracellular Na^+

The intracellular Na^+ measurements were recorded following a protocol previously described in detail [14]. Briefly, HEK 293 cells were grown on uncoated glass bottom dishes (MatTek, No. 1.5 coverglass) and transfected with 0.25 μg of each plasmid using Fugene HD (Promega, Madison, WI, USA). Plasmids containing untagged β and γ-ENaC (wt or mutant) were transfected along with CFP-αENaC, untagged CIBN and mCh-CRY2-OCRL plasmids. Cells were kept in the dark and in the presence of 10 μM amiloride for ~48 h. Prior to imaging, cells were incubated in the dark 30 min in serum-free DMEM containing 10 μM amiloride (Sigma, Burlington, MA, USA), 5 μM CoroNa Green AM (Molecular Probes, Eugene, OR, USA) and 0.04% Pluronic F-127 (Biotium, Fremont, CA, USA), then replaced with PBS. A 60×/1.45 TIRF oil objective with 1.5× amplification on an inverted Nikon Eclipse TE2000-U fluorescence microscope was used for the CoroNa Green experiments. Imaging and blue light illumination (BLI) were performed using OBIS FP fiber pigtailed lasers (Coherent, Inc., Santa Clara, CA, USA) at 2% power. Membrane levels of mCherry-CRY2-OCRL and ENaC were quantified under TIRF settings, excitation at 561 nm and 445 nm, respectively. CoroNa Green was also examined under TIRF microscopy by excitation at 514 nm. To induce CRY2 dimerization with CIBN, we used BLI pulses (excitation at 445 nm) at 300 ms for 30 sec at a frequency of 1 Hz. Images were captured before BLI, 0 min after, 5 min after and 10 min after BLI, with an Andor iXon Ultra camera. Images were evaluated using Metamorph software (Molecular Devices, San Jose, CA, USA) and ImageJ [42]. The change in fluorescence was plotted and normalized to the starting fluorescence of each experiment. Statistical significance for each experiment was determined using GraphPad Prism 7. To compare changes in membrane expression, ordinary one-way ANOVA was performed followed by Dunnett's multiple comparison test compared to wtENaC membrane expression on at least n = 6 cells per test. For comparison of CoroNa Green intensity changes within the same cells, significance was determined by a paired, two-tailed t test, and each analysis was performed on n = 11 individual cells. *, $p < 0.0001$; **, $p < 0.005$; #, $p < 0.05$; n.s., no significant difference.

4.3. Membrane Expression of ENaC

The relative membrane expression levels of wt and mutant ENaC were determined using TIRF microscopy. Plasmids containing CFP-α-ENaC, βENaC-YFP (wt or mutant) and γ-ENAC-mCherry (wt or mutant) were each transfected (0.25 μg each) to HEK 293 cells using Fugene HD (Promega, Madison, WI, USA). Cells were grown on uncoated glass bottom dishes (MatTek, No. 1.5 coverglass) for ~48 h in the presence of 10 μM amiloride in DMEM + 10% FBS. Prior to imaging, the media was replaced with PBS. TIRF imaging was carried out as described above. ECFP was excited at 445 nm. EYFP was excited at 514 nm, and mCherry was excited at 561 nm. Data were collected and analyze as described above. Mean intensities were measured in ImageJ and analyzed and plotted in GraphPad Prism 7. One-way ANOVA was performed followed by Dunnett's multiple comparison test compared to wtENaC membrane expression on at least n = 8–14 cells for each of 2 experiments. **, $p < 0.005$; *, $p < 0.05$.

4.4. Electrophysiology

CHO-K1 cells were grown on glass chips coated with poly-l-lysine (Sigma, St. Louis, MO, USA). Cells were transfected with 0.125 μg each α, β and γ ENaC plasmid (wt or mutant) and 0.25 μg each CIBN and mCh-CRY2-OCRL plasmid, using Lipofectamine 3000 (Invitrogen; Thermofisher, Waltham, MA, USA) as described in the manufacturer's protocol. In brief, 60% confluent cells in a 35 mm dish were transiently transfected with the same ENaC constructs as described above, in the presence or absence of the CIBN/CRY2 constructs. Cells were kept in the dark with 10 μM amiloride. Cells were patched within 24–48 h after transfection. Whole-cell currents of ENaC were recorded under voltage clamp using the extracellular solution, (in mM) 150 NaCl, 1 CaCl$_2$, 2 MgCl$_2$, and 10 HEPES (pH 7.4). The pipette solution contained (in mM) 120 CsCl, 5 NaCl, 2 MgCl$_2$, 5 EGTA, 10 HEPES, 2 ATP, and 0.1 GTP (pH 7.4). The current densities were recorded on an Axopatch 200B (Molecular Devices, San Jose, CA, USA) patch-clamp amplifier interfaced via a Digidata 1550B (Molecular Devices) to a PC running the pClamp 11 suite of software (Molecular Devices). All currents were filtered at 1 kHz. Cells were clamped to a 40 mV holding potential with voltage ramps (500 ms) from 60 mV down to −100 mV used to elicit current. Whole-cell capacitance, on average 8–10 pF, was compensated. Series resistance, on average 3–6 megaohms, were also compensated.

Cells were visualized with a 10× or 40× objective under BLI conditions with transmitted light. Cells expressing CFP and mCherry were found using a mercury lamp with 445 nm and 561 nm filter cubes. The starting ENaC recordings were made under 445 nm BLI-induced PIP2 depletion conditions. After approximately 3 min under BLI, the microscope and ambient room lights were switched off and the recordings were made for an additional 10 min, followed by perfusion of 10 μM amiloride. Two-way ANOVA was used to determine statistical significance of n = 6–15 cells, with the row factor defined as the mutant and the column factor defined as light vs. dark, followed by Sidak Holm's multiple comparisons test. For the time-course experiment, two-way ANOVA was used, where row factor = time and column factor = mutant, followed by Dunnett's multiple comparisons test.

Supplementary Materials: The following supporting information can be downloaded at: https://www.mdpi.com/article/10.3390/ijms23073884/s1.

Author Contributions: Conceptualization, C.R.A. and J.D.S.; methodology, C.R.A. and T.M.A.E.-A.; validation, C.R.A., A.K. and T.M.A.E.-A.; formal analysis, C.R.A. and T.M.A.E.-A.; investigation, C.R.A.; resources, J.D.S., C.R.A. and M.S.S.; data curation, C.R.A., A.K. and T.M.A.E.-A.; writing—original draft preparation, C.R.A.; writing—review and editing, C.R.A. and J.D.S.; visualization, C.R.A.; supervision, C.R.A.; project administration, C.R.A. and J.D.S.; funding acquisition, C.R.A. and J.D.S. All authors have read and agreed to the published version of the manuscript.

Funding: This research was funded by The American Heart Association, grant numbers 20POST35210746 and 857310 (to C.R.A.); and National Institutes of Health NIDDK R01, grant number DK113816 (to J.D.S.). This content is solely the responsibility of the authors and does not necessarily represent the official views of the American Heart Association or National Institutes of Health.

Data Availability Statement: The data supporting the findings of this study are contained within the contents of this article. The datasets generated during this study will be freely provided by the corresponding author upon request.

Acknowledgments: We would like to thank Fabio Borges Vigil for assisting with the design of figures.

Conflicts of Interest: The authors declare no conflict of interest.

References

1. Noreng, S.; Bharadwaj, A.; Posert, R.; Yoshioka, C.; Baconguis, I. Structure of the human epithelial sodium channel by cryo-electron microscopy. *eLife* **2018**, *7*, e39340. [CrossRef] [PubMed]
2. Noreng, S.; Posert, R.; Bharadwaj, A.; Houser, A.; Baconguis, I. Molecular principles of assembly, activation, and inhibition in epithelial sodium channel. *eLife* **2020**, *9*, e59038. [CrossRef] [PubMed]
3. Hanukoglu, I.; Hanukoglu, A. Epithelial sodium channel (ENaC) family: Phylogeny, structure-function, tissue distribution, and associated inherited diseases. *Gene* **2016**, *579*, 95–132. [CrossRef]
4. Canessa, C.M.; Schild, L.; Buell, G.; Thorens, B.; Gautschi, I.; Horisberger, J.-D.; Rossier, B.C. Amiloride-sensitive epithelial Na$^+$ channel is made of three homologous subunits. *Nature* **1994**, *367*, 463–467. [CrossRef] [PubMed]
5. Firsov, D.; Gautschi, I.; Mérillat, A.; Rossier, B.C.; Schild, L. The heterotetrameric architecture of the epithelial sodium channel (ENaC). *EMBO J.* **1998**, *17*, 344–352. [CrossRef] [PubMed]
6. Bhalla, V.; Hallows, K.R. Mechanisms of ENaC regulation and clinical implications. *J. Am. Soc. Nephrol.* **2008**, *19*, 1845–1854. [CrossRef] [PubMed]
7. Enslow, B.T.; Stockand, J.D.; Berman, J.M. Liddle's syndrome mechanisms, diagnosis and management. *Integr. Blood Press. Control* **2019**, *12*, 13–22. [CrossRef] [PubMed]
8. Tetti, M.; Monticone, S.; Burrello, J.; Matarazzo, P.; Veglio, F.; Pasini, B.; Jeunemaitre, X.; Mulatero, P. Liddle Syndrome: Review of the Literature and Description of a New Case. *Int. J. Mol. Sci.* **2018**, *19*, 812. [CrossRef] [PubMed]
9. Kota, P.; Buchner, G.; Chakraborty, H.; Dang, Y.L.; He, H.; Garcia, G.; Kubelka, J.; Gentzsch, M.; Stutts, M.J.; Dokholyan, N.V. The N-terminal domain allosterically regulates cleavage and activation of the epithelial sodium channel. *J. Biol. Chem.* **2014**, *289*, 23029–23042. [CrossRef] [PubMed]
10. Pochynyuk, O.; Bugaj, V.; Vandewalle, A.; Stockand, J.D. Purinergic control of apical plasma membrane PI(4,5)P2 levels sets ENaC activity in principal cells. *Am. J. Physiol. Renal. Physiol.* **2008**, *294*, F38–F46. [CrossRef] [PubMed]
11. Pochynyuk, O.; Tong, Q.; Medina, J.; Vandewalle, A.; Staruschenko, A.; Bugaj, V.; Stockand, J.D. Molecular determinants of PI(4,5)P2 and PI(3,4,5)P3 regulation of the epithelial Na+ channel. *J. Gen. Physiol.* **2007**, *130*, 399–413. [CrossRef] [PubMed]
12. Yue, G.; Malik, B.; Yue, G.; Eaton, D.C. Phosphatidylinositol 4,5-bisphosphate (PIP2) stimulates epithelial sodium channel activity in A6 cells. *J. Biol. Chem.* **2002**, *277*, 11965–11969. [CrossRef]
13. Archer, C.R.; Enslow, B.; Carver, C.M.; Stockand, J.D. Phosphatidylinositol 4,5-bisphosphate directly interacts with the β and γ subunits of the sodium channel ENaC. *J. Biol. Chem.* **2020**, *295*, 7958–7969. [CrossRef] [PubMed]
14. McLaughlin, S.; Wang, J.; Gambhir, A.; Murray, D. PIP(2) and proteins: Interactions, organization, and information flow. *Annu. Rev. Biophys. Biomol. Struct.* **2002**, *31*, 151–175. [CrossRef] [PubMed]
15. Dickson, E.J.; Hille, B. Understanding phosphoinositides: Rare, dynamic, and essential membrane phospholipids. *Biochem. J.* **2019**, *476*, 1–23. [CrossRef] [PubMed]
16. Suh, B.C.; Hille, B. PIP2 is a necessary cofactor for ion channel function: How and why? *Annu. Rev. Biophys.* **2008**, *37*, 175–195. [CrossRef] [PubMed]
17. Mandal, K. Review of PIP2 in Cellular Signaling, Functions and Diseases. *Int. J. Mol. Sci.* **2020**, *21*, 8342. [CrossRef]
18. Idevall-Hagren, O.; Dickson, E.J.; Hille, B.; Toomre, D.K.; De Camilli, P. Optogenetic control of phosphoinositide metabolism. *Proc. Natl. Acad. Sci. USA* **2012**, *109*, E2316–E2323. [CrossRef] [PubMed]
19. Trésaugues, L.; Silvander, C.; Flodin, S.; Welin, M.; Nyman, T.; Graslund, S.; Hammarström, M.; Berglund, H.; Nordlund, P. Structural basis for phosphoinositide substrate recognition, catalysis, and membrane interactions in human inositol polyphosphate 5-phosphatases. *Structure* **2014**, *22*, 744–755. [CrossRef] [PubMed]
20. Banerjee, R.; Schleicher, E.; Meier, S.; Viana, R.M.; Pokorny, R.; Ahmad, M.; Bittl, R.; Batschauer, A. The signaling state of Arabidopsis cryptochrome 2 contains flavin semiquinone. *J. Biol. Chem.* **2007**, *282*, 14916–14922. [CrossRef] [PubMed]
21. Kennedy, M.J.; Hughes, R.; Peteya, L.A.; Schwartz, J.W.; Ehlers, M.D.; Tucker, C.L. Rapid blue-light-mediated induction of protein interactions in living cells. *Nat. Methods* **2010**, *7*, 973–975. [CrossRef] [PubMed]
22. Tucker, C.L.; Vrana, J.D.; Kennedy, M.J. Tools for controlling protein interactions using light. *Curr. Protoc. Cell Biol.* **2014**, *64*, 17. [CrossRef] [PubMed]
23. Fish, K.N. Total internal reflection fluorescence (TIRF) microscopy. *Curr. Protoc. Cytom.* **2009**, *50*, 12.18.1–12.18.13. [CrossRef] [PubMed]
24. Zaika, O.; Hernandez, C.C.; Bal, M.; Tolstykh, G.P.; Shapiro, M.S. Determinants within the turret and pore-loop domains of KCNQ3 K+ channels governing functional activity. *Biophys. J.* **2008**, *95*, 5121–5137. [CrossRef] [PubMed]
25. Meier, S.D.; Kovalchuk, Y.; Rose, C.R. Properties of the new fluorescent Na+ indicator CoroNa Green: Comparison with SBFI and confocal Na+ imaging. *J. Neurosci. Methods* **2006**, *155*, 251–259. [CrossRef]
26. Eaton, D.C.; Malik, B.; Bao, H.-F.; Yu, L.; Jain, L. Regulation of epithelial sodium channel trafficking by ubiquitination. *Proc. Am. Thorac. Soc.* **2010**, *7*, 54–64. [CrossRef] [PubMed]
27. Zhou, R.; Patel, S.V.; Snyder, P.M. Nedd4-2 catalyzes ubiquitination and degradation of cell surface ENaC. *J. Biol. Chem.* **2007**, *282*, 20207–20212. [CrossRef] [PubMed]

28. Wiemuth, D.; Ke, Y.; Rohlfs, M.; Mc Donald, F.J. Epithelial sodium channel (ENaC) is multi-ubiquitinated at the cell surface. *Biochem. J.* **2007**, *405*, 147–155. [CrossRef] [PubMed]

29. Frindt, G.; Ke, Y.; Rohlfs, M.; Mc Donald, F.J. Ubiquitination of renal ENaC subunits in vivo. *Am. J. Physiol. Renal. Physiol.* **2020**, *318*, F1113–F1121. [CrossRef] [PubMed]

30. Li, Y.; Gamper, N.; Hilgemann, D.W.; Shapiro, M.S. Regulation of Kv7 (KCNQ) K$^+$ channel open probability by phosphatidylinositol 4,5-bisphosphate. *J. Neurosci.* **2005**, *25*, 9825–9835. [CrossRef] [PubMed]

31. Choveau, F.S.; De la Rosa, V.; Bierbower, S.M.; Hernandez, C.C.; Shapiro, M.S. Phosphatidylinositol 4,5-bisphosphate (PIP(2)) regulates KCNQ3 K(+) channels by interacting with four cytoplasmic channel domains. *J. Biol. Chem.* **2018**, *293*, 19411–19428. [CrossRef] [PubMed]

32. Hansen, S.B. Lipid agonism: The PIP2 paradigm of ligand-gated ion channels. *Biochim. Biophys. Acta* **2015**, *1851*, 620–628. [CrossRef] [PubMed]

33. Hansen, S.B.; Tao, X.; MacKinnon, R. Structural basis of PIP2 activation of the classical inward rectifier K$^+$ channel Kir2.2. *Nature* **2011**, *477*, 495–498. [CrossRef] [PubMed]

34. Mironova, E.; Stockand, J.D. Activation of a latent nuclear localization signal in the NH2 terminus of γ-ENaC initiates feedback regulation of channel activity. *Am. J. Physiol. Renal. Physiol.* **2010**, *298*, F1188–F1196. [CrossRef] [PubMed]

35. Chelsky, D.; Ralph, R.; Jonak, G. Sequence requirements for synthetic peptide-mediated translocation to the nucleus. *Mol. Cell Biol.* **1989**, *9*, 2487–2492. [PubMed]

36. Boscardin, E.; Alijevic, O.; Hummler, E.; Frateschi, S.; Kellenberger, S. The function and regulation of acid-sensing ion channels (ASICs) and the epithelial Na(+) channel (ENaC): IUPHAR Review 19. *Br. J. Pharmacol.* **2016**, *173*, 2671–2701. [CrossRef]

37. Yoder, N.; Gouaux, E. The His-Gly motif of acid-sensing ion channels resides in a reentrant 'loop' implicated in gating and ion selectivity. *eLife* **2020**, *9*, e56527. [CrossRef]

38. Gründer, S.; Jaeger, N.F.; Gautschi, I.; Schild, L.; Rossier, B.C. Identification of a highly conserved sequence at the N-terminus of the epithelial Na$^+$ channel alpha subunit involved in gating. *Pflug. Arch.* **1999**, *438*, 709–715. [CrossRef]

39. Gründer, S.; Firsov, D.; Chang, S.S.; Jaeger, N.F.; Gautschi, I.; Schild, L.; Lifton, R.P.; Rossier, B.C. A mutation causing pseudohypoaldosteronism type 1 identifies a conserved glycine that is involved in the gating of the epithelial sodium channel. *EMBO J.* **1997**, *16*, 899–907. [CrossRef]

40. McDonald, F.J.; Price, M.P.; Snyder, P.M.; Welsh, M.J. Cloning and expression of the beta- and gamma-subunits of the human epithelial sodium channel. *Am. J. Physiol.* **1995**, *268 Pt 1*, C1157–C1163. [CrossRef]

41. Tong, Q.; Menon, A.G.; Stockand, J.D. Functional polymorphisms in the alpha-subunit of the human epithelial Na$^+$ channel increase activity. *Am. J. Physiol. Renal. Physiol.* **2006**, *290*, F821–F827. [CrossRef] [PubMed]

42. Schneider, C.A.; Rasband, W.S.; Eliceiri, K.W. NIH Image to ImageJ: 25 years of image analysis. *Nat. Methods* **2012**, *9*, 671–675. [CrossRef] [PubMed]

International Journal of
Molecular Sciences

Article

Optical Activation of TrkB (E281A) in Excitatory and Inhibitory Neurons of the Mouse Visual Cortex

Antonia Lilja [1,2], Giuliano Didio [2], Jongryul Hong [3], Won Do Heo [3,4,5], Eero Castrén [2,*]
and Juzoh Umemori [2,6,*]

[1] Faculty of Psychology and Neuroscience, Maastricht University, 6229 ER Maastricht, The Netherlands
[2] Neuroscience Center, Helsinki Institute of Life Science (HiLIFE), University of Helsinki,
00100 Helsinki, Finland
[3] Department of Biological Science, Korea Advanced Institute of Science and Technology (KAIST),
Daejeon 305-701, Korea
[4] Center for Cognition and Sociality, Institute for Basic Science (IBS), Daejeon 305-701, Korea
[5] KAIST Institute for the BioCentury, KAIST, Daejeon 305-701, Korea
[6] Gene and Cell Technology, A.I. Virtanen Institute, University of Eastern Finland, 70211 Kuopio, Finland
* Correspondence: eero.castren@helsinki.fi (E.C.); juzoh.umemori@uef.fi (J.U.)

Abstract: The activation of tropomyosin receptor kinase B (TrkB), the receptor of brain-derived neurotrophic factor (BDNF), plays a key role in induced juvenile-like plasticity (iPlasticity), which allows restructuring of neural networks in adulthood. Optically activatable TrkB (optoTrkB) can temporarily and spatially evoke iPlasticity, and recently, optoTrkB (E281A) was developed as a variant that is highly sensitive to light stimulation while having lower basal activity compared to the original optoTrkB. In this study, we validate optoTrkB (E281A) activated in alpha calcium/calmodulin-dependent protein kinase type II positive (CKII$^+$) pyramidal neurons or parvalbumin-positive (PV$^+$) interneurons in the mouse visual cortex by immunohistochemistry. OptoTrkB (E281A) was activated in PV$^+$ interneurons and CKII$^+$ pyramidal neurons with blue light (488 nm) through the intact skull and fur, and through a transparent skull, respectively. LED light stimulation significantly increased the intensity of phosphorylated ERK and CREB even through intact skull and fur. These findings indicate that the highly sensitive optoTrkB (E281A) can be used in iPlasticity studies of both inhibitory and excitatory neurons, with flexible stimulation protocols in behavioural studies.

Keywords: neural plasticity; tropomyosin kinase receptor B; parvalbumin-positive interneurons; calcium/calmodulin positive pyramidal neurons; V1/visual cortex; multiple regression with interaction and simple slopes analysis

Citation: Lilja, A.; Didio, G.; Hong, J.; Heo, W.D.; Castrén, E.; Umemori, J. Optical Activation of TrkB (E281A) in Excitatory and Inhibitory Neurons of the Mouse Visual Cortex. *Int. J. Mol. Sci.* **2022**, *23*, 10249. https://doi.org/10.3390/ijms231810249

Academic Editors: Carlo Matera and Piotr Bregestovski

Received: 31 July 2022
Accepted: 2 September 2022
Published: 6 September 2022

Publisher's Note: MDPI stays neutral with regard to jurisdictional claims in published maps and institutional affiliations.

1. Introduction

Neural plasticity refers to the processes through which environmental factors influence neuronal structures and functions [1]. Two key players in neural plasticity are brain-derived neurotrophic factor (BDNF) and its high-affinity receptor, tropomyosin receptor kinase B (TrkB), which supports the development of neurons and neuroplasticity at a cellular, synaptic, and network level. When BDNF binds to TrkB, two TrkB receptors form a dimer that autophosphorylates [2]. This sets in motion several signalling cascades, including induction of long term potentiation (LTP), which is crucial for memory and learning [2].

After a critical period of heightened neural plasticity in childhood [3,4], neural networks become consolidated and neurotrophin signalling patterns change [5–7]. However, various interventions, such as antidepressant treatment, have been shown to induce a state of increased neuroplasticity in the adult brain that is similar to that of the juvenile critical period [1,8]. This artificially induced juvenile-like neuroplasticity in the adult brain has been coined iPlasticity [1,9].

iPlasticity has frequently been studied in the visual cortex. Eye-specific ocular dominance develops in the visual cortex during the juvenile critical period [4]. In adult age, this ocular dominance can be changed by restoring visual cortex plasticity. This can be done using antidepressants, such as fluoxetine. Chronic administration of fluoxetine has been shown to successfully induce amblyopia, shift ocular dominance, and increase levels of BDNF in the visual cortex [10]. Further insight into the reinstatement of visual cortex plasticity has been provided by studies investigating parvalbumin-positive (PV^+) interneurons. Our group recently showed that fluoxetine failed to induce visual cortex plasticity when PV^+ interneurons expressed a reduced amount of TrkB [11].

Optically activatable TrkB (optoTrkB) has been developed to study the effects of TrkB activation with high spatiotemporal specificity [12]. In optoTrkB, a photolyase homology region (PHR) of cryptochrome 2, a blue light receptor, is fused with a TrkB receptor, and blue light stimulation successfully activates TrkB signalling pathways [12]. OptoTrkB is preferable to other approaches to manipulate TrkB signalling, since optoTrkB can be activated in specific cells and brain areas at specific times. This allows us to establish causal relationships between TrkB signalling, neural network changes, and behavioural effects with temporal and spatial precision [13]. Recently, a new type of optoTrkB (E281A) was developed and successfully activated by blue light stimulation through the intact skull and fur of mice [13]. OptoTrkB (E281A) also showed a low level of basal activity, allowing for more experimental control as spontaneous optoTrkB signalling was reduced. OptoTrkB (E281A) was successfully activated and the phosphorylation of extracellular signal-regulated kinase (pERK) was increased in $CKII^+$ neurons of the dentate gyrus in the hippocampus. However, it is unknown if optoTrkB (E281A) can be activated in other neuronal subpopulations than $CKII^+$ pyramidal neurons and in other brain regions than in the hippocampus.

OptoTrkB is introduced in a double-inverted open reading frame structure (DIO), whereby the optoTrkB sequence is inverted and flanked by two incompatible lox-P sites. This structure allows to turn on TrkB expression via Cre recombinase-mediated recombination [14,15]. Thus, optoTrkB can be expressed in a specific neuronal subpopulation by injecting it into a certain region of genetically modified mice expressing Cre recombinase in specific cells, such as CKII- and PV- positive neurons. Adeno-associated virus (AAV) is one of the most widely used viral vehicles of genetic information and has lower immunogenicity than other viruses, making it less likely to be eliminated by the immune system before gene delivery [16].

Thus, in this study, we infected AAV-DIO-optoTrkB (E281A) into the visual cortex of mice lines that express Cre recombinase specifically in $CKII^+$ pyramidal neurons and PV^+ interneurons, in order to express optoTrkB in the specific neuronal subpopulations (Figure 1). OptoTrkB (E281A) activation was then triggered with blue light stimulation through the transparent skull or atop the intact fur head. Then, we validated the activation by immunohistochemistry of two TrkB downstream signals, pERK and cAMP-calcium response element binding protein (CREB), a transcription factor that plays a key role in LTP [2].

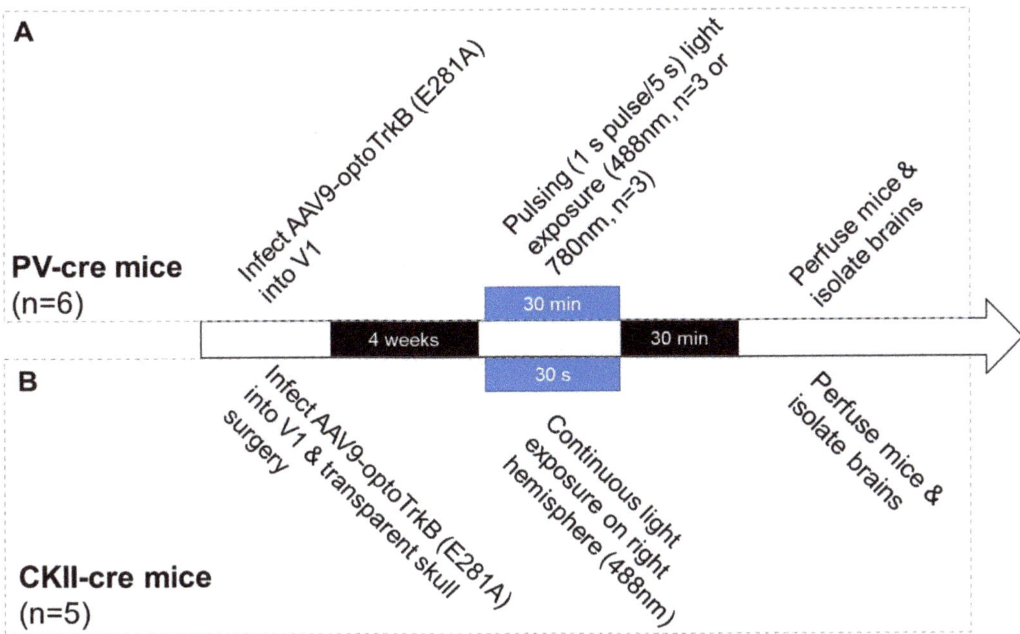

Figure 1. Timeline of optoTrkB (E281) stimulation and perfusion. (**A**) Stimulation protocol for PV-Cre mice. (**B**) Stimulation protocol for CKII-Cre mice.

2. Results

2.1. Activation of OptoTrkB (E281A) in PV^+ Interneurons

Expression of optoTrkB (E281A) can be inferred by the intensity of human influenza hemagglutinin (HA) tagged at the C-terminal of optoTrkB. Immunohistochemistry by co-staining with anti-PV and anti-HA antibodies in PV-Cre mice infected with a pAAV-hSyn1-DIO-OptocytTrkB(E281A)-HA construct showed that HA-positive neurons overlapped with PV^+ interneurons, indicating successful infection of the visual cortex in PV-Cre mice (Figure 2A–F). Closer inspection of individual neurons showed that HA staining surrounded PV-stained cells (Figure 2D–F), indicating that optoTrkB (E281A) was successfully infected in PV^+ interneurons of PV-Cre mice.

Image analysis revealed that cells in the infected area of PV-Cre mice expressed HA more strongly than in the uninfected area (Figure 2A–C). However, anti-HA partially stains background particles, and there are also some non-infected PV^+ interneurons in the infected area. To ensure that only neurons infected with optoTrkB (E281A) tagged with HA were included for further analysis, a cutoff point for minimum HA intensity was set by comparing the distribution of HA intensity values in the cells from the infected area (N = 88) to those of an uninfected adjacent area (N = 60). The point of overlap of these distribution curves was selected as the cutoff point for minimum HA intensity (Figure S1). Neurons from the infected area that had lower HA intensity than this cutoff were excluded from further analysis (Figure S2). After applying this cutoff value, 44 stimulated neurons and 41 unstimulated neurons remained for the final analysis.

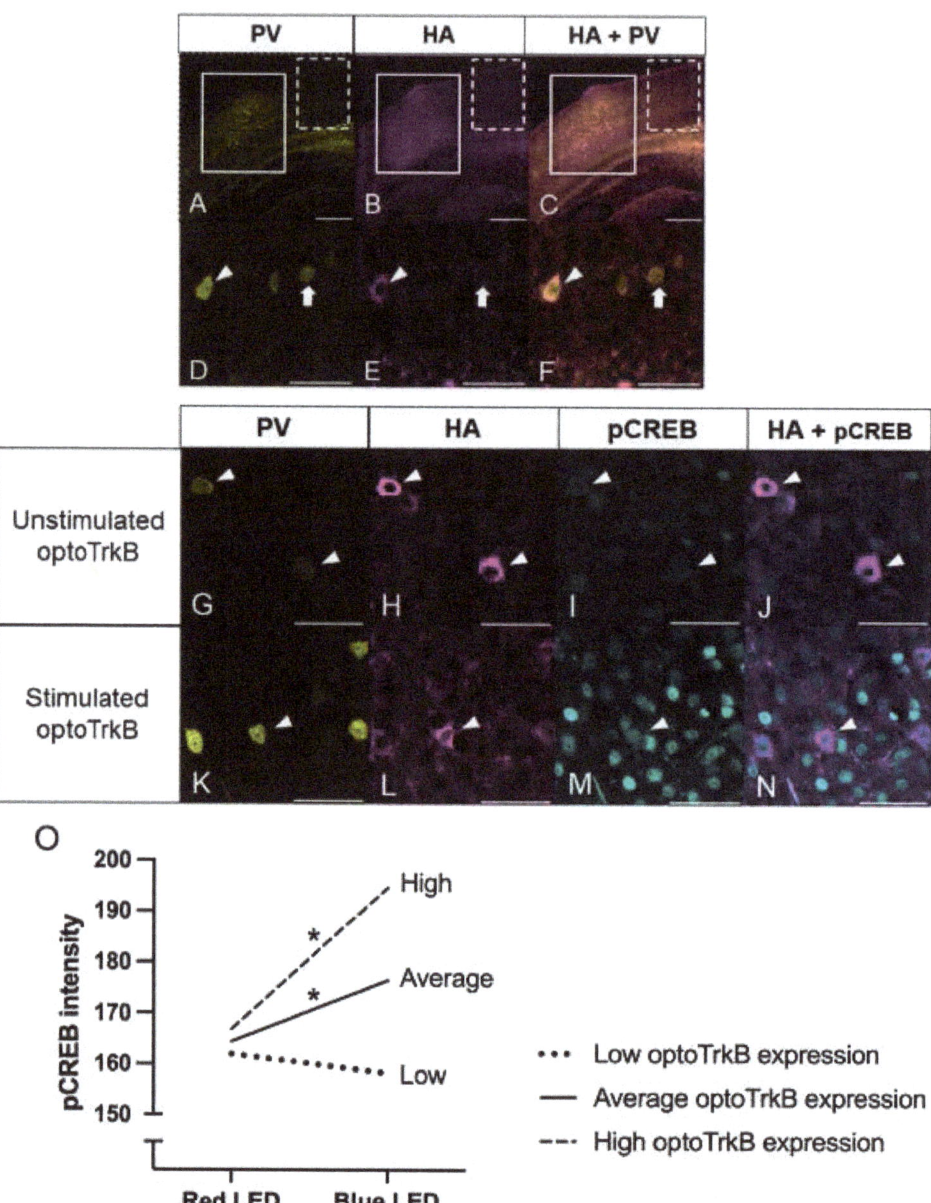

Figure 2. (**A–F**) OptoTrkB (E281A) was successfully expressed and activated in PV⁺ interneurons in the visual cortex of PV-Cre mice. Representative images of V1 stained with anti-PV antibody (**A,D**), HA antibody (**B,E**), and merged (**C,F**). HA-expressing neurons also express PV, indicating that optoTrkB (E281A) is specifically expressed in PV⁺ interneurons. Solid outline, dashed outline, arrowheads,

and arrows indicate the infected area, non-infected area, PV$^+$ interneurons infected with optoTrkB (E281A), and PV$^+$ interneurons not infected with optoTrkB (E281A), respectively. (**G–N**) Light-stimulated PV$^+$ interneurons show higher pCREB intensity than unstimulated PV$^+$ interneurons. Representative images of V1 staining with antibodies of anti-PV (**G,K**), HA (**H,L**), pCREB (**I,M**), and merged HA and pCREB (**J,N**). Neurons in unstimulated (**G–J**), and stimulated brain (**K–N**). Scale bar (**A–C**): 500 um; scale bar (**D–N**): 50 um. Arrowheads point to PV$^+$ interneurons infected with optoTrkB. (**O**) Regression slopes for the effect of LED light stimulation on pCREB for high (dash line), average (solid line), and low (dot line) levels of optoTrkB (E281A). OptoTrkB (E281A) stimulation significantly increased pCREB intensity for neurons expressing high and average levels of optoTrkB (E281A), but not for those with low expression levels of optoTrkB (E281A). The regression equations for cells expressing optoTrkB (E281A) at high, average, and low level are as follows: *pCREB high HA* = 166.839 + 27.722 * *stimulation* + 0.040 * *HAintensity* + 0.257 * *stimulation* * *HAintensity* + *error* (p ($\beta_{stimulation}$) = 0.001); *pCREB average HA* = 164.402 + 11.962 * *stimulation* + 0.040 * *HAintensity* + 0.257 * *stimulation* * *HAintensity* + *error* (p ($\beta_{stimulation}$) = 0.007); *pCREB low HA* = 161.956 − 3.798 * *stimulation* + 0.040 * *HAintensity* + 0.257 * *stimulation* * *HAintensity* + *error* (p ($\beta_{stimulation}$) = 0.284). * $p < 0.017$ (Bonferroni-corrected threshold).

The intensity of pCREB was analysed in the cells expressing optoTrkB (E281) (Figure 2G–N). A multiple regression assessing the effect of optoTrkB (E281A) stimulation and HA intensity on pCREB intensity revealed a significant difference in pCREB between the groups ($F(3, 81) = 5.414$, $p = 0.002$). Inspection of the regression coefficients revealed a significant interaction effect between stimulation and HA intensity ($\beta = 0.257$, $p = 0.004$), suggesting that the effect of optoTrkB (E281A) stimulation on downstream signalling depends on the extent to which the neuron expressed optoTrkB (E281A) receptors. Simple slope analysis was then conducted by analysing the effect of optoTrkB (E281A) stimulation at average, low (one standard deviation below the average), and high (one standard deviation above the average) optoTrkB (E281A) intensity (Figure 2O). The analysis showed that at average HA intensity, optoTrkB (E281A) stimulation significantly increased pCREB intensity ($\beta = 11.962$, $p = 0.007$). Similarly, at one standard deviation above average HA intensity, optoTrkB (E281A) stimulation significantly increased pCREB intensity ($\beta = 27.722$, $p = 0.001$). However, at one standard deviation below average HA intensity, there was no effect of optoTrkB (E281A) stimulation on pCREB intensity ($\beta = 3.798$, $p = 0.506$). These results indicate that pCREB increased after optoTrkB stimulation only in PV$^+$ interneurons expressing average to high levels of optoTrkB (E281A).

2.2. Activation of OptoTrkB (E281A) in CKII$^+$ Pyramidal Neurons

CKII-Cre mice underwent transparent skull surgery prior to light stimulation. Immunohistochemical analysis of the brains of CKII-Cre mice infected with a pAAV-hSyn1-DIO-OptocytTrkB(E281A)-HA construct verified that they exhibited HA in the visual cortex (Figure 3A–H). Moreover, co-staining of HA and CKII antibodies in individual neurons affirmed that optoTrkB (E281A) was expressed specifically in CKII$^+$ neurons (Figure 3C–H).

As with PV$^+$ interneurons, a cutoff or minimum HA intensity value was set in order to ensure that only infected neurons were analysed. The distribution of HA intensity in cells from the infected area (N = 107) was compared to those of an uninfected adjacent area (N = 80) (Figure S3). The cutoff point was established in the same way as for PV$^+$ interneurons, leaving 50 stimulated neurons and 53 unstimulated neurons for the final analysis (Figure S4).

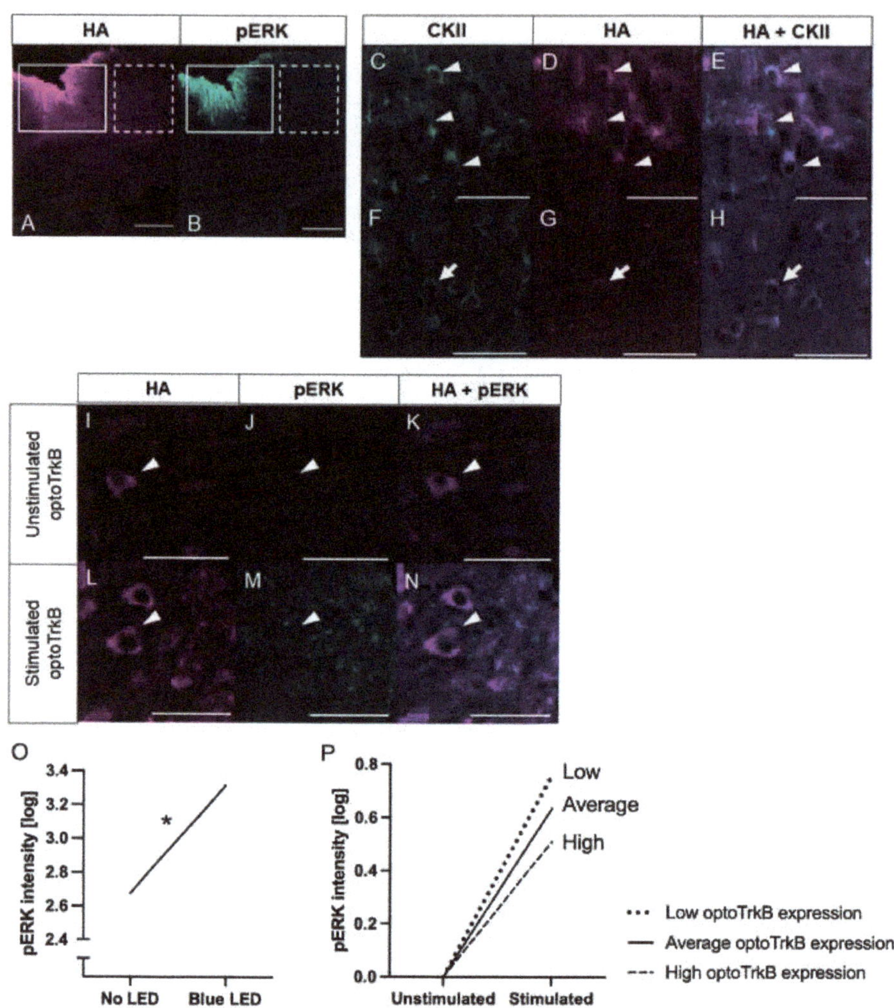

Figure 3. (**A–H**) OptoTrkB (E281A) was successfully expressed in CKII[+] neurons in the visual cortex of CKII-Cre mice. Representative images of V1 stained with anti-CKII antibody (**C,F**), HA antibody (**A,D,G**), pERK antibody (**B**), and merged CKII and HA (**E,H**). HA expressing neurons also express CKII, indicating that optoTrkB (E281A) was specifically expressed in CKII[+] neurons. Solid outline, dashed outline, arrowheads, and arrows indicate the infected area, non-infected area, CKII[+] neurons infected with optoTrkB (E281A), CKII[+] neurons not infected with optoTrkB (E281A), respectively. (**I–N**) LED Light stimulated CKII[+] neurons show higher pERK intensity than unstimulated CKII[+] neurons. Representative images of V1 neurons stained with antibodies of anti-HA (**I,L**), anti-pERK (**J,M**), and merged (**K,N**). Neurons in stimulated (**L–N**) and unstimulated (**I–K**) hemispheres. Arrowheads point to CKII[+]-neurons infected with optoTrkB (E281A). Scale bar (**A,B**) = 500 um; scale bar (**C–N**) = 50 um. (**O**) Regression slope for the effect of LED Light Stimulation on pERK. OptoTrkB (E281A) stimulation significantly increased pERK intensity for CKII[+] neurons. The regression equation is as follows: ($pCREB = 2.676 + 0.640 * LEDstimulation + 0.007 * HAintensity + error$). * $p < 0.05$. (**P**) LED light stimulation increases optoTrkB (E281A) downstream signalling in CKII[+] neurons to an equal extent in neurons expressing low ($\beta1 = 0.761$, dotted line), average ($\beta1 = 0.635$, solid line) and high ($\beta1 = 0.509$, dashed line) optoTrkB (E281A) levels.

The intensity of pERK was further analysed in neurons expressing optoTrkB (Figure 3I–N). The raw pERK intensity scores underwent natural logarithmic data transformation to correct for a violation of homoscedasticity. The transformed variable was used as the dependent variable in the data analysis. A multiple regression model for this data indicated an overall significant difference in pERK between the groups ($F(3, 99) = 34.002$, $p < 0.001$), but no significant interaction effect ($\beta = 0.001$, $p = 0.494$) (Figure 3P). After the non-significant interaction term was excluded from the model, a multiple regression assessing the effect of optoTrkB (E281A) stimulation and HA intensity on pERK intensity revealed a significant difference in pERK between the groups ($F(2, 100) = 51.039$, $p < 0.001$). pERK was significantly higher in the stimulated neurons ($\beta = 0.640$, $p = 0.001$) (Figure 3O). These results indicate that optoTrkB (E281A) stimulation increased pERK intensity to a similar extent for neurons expressing low, average, and high levels of optoTrkB (E281A).

3. Discussion

This study assesses the activation of a new, highly sensitive, optically activatable optoTrkB (E281A) receptor in PV^+ interneurons and $CKII^+$ pyramidal neurons of the visual cortex. We demonstrate that the highly sensitive optoTrkB (E281A) can be introduced in a specific neuron type in a specific area of interest, and activated using weak, short-term light stimulation in flexible contexts. The activation took place even when optoTrkB (E281A) was stimulated through intact fur and skull.

The analyses of pCREB and pERK intensity show that LED light stimulation results in an increase in downstream signalling of optoTrkB (E281A), both in inhibitory PV^+ interneurons, as well as in the excitatory $CKII^+$ neurons, as shown by a previous study [13]. Interestingly, the higher the expression of optoTrkB (E281A) in PV^+ interneurons, the more downstream signalling was promoted by light stimulation, while LED light stimulation failed to activate neurons with low levels of optoTrkB (E281A). In contrast, optoTrkB (E281A) activation is the same in $CKII^+$ pyramidal neurons expressing low and high levels of optoTrkB (E281A). In other words, LED light stimulation promotes downstream signalling independent of the level of optoTrkB (E281A) in $CKII^+$ neurons. There are some possible explanations for this effect. Most likely, transparent skull surgery in the case of $CKII^+$ neurons allowed the light to sufficiently stimulate all neurons, including ones expressing low levels of optoTrkB (E281A). Injecting a higher concentration of the AAV9-DIO-optoTrkB (E281A) may allow light stimulation through the intact skull and fur to activate all optoTrkB (E281A) expressing neurons.

Previous research suggests great potential of manipulating TrkB activity for treatments of maladaptive neuropsychiatric symptoms. Recently, our group showed that optoTrkB stimulation in PV^+ interneurons of the visual cortex shifted ocular dominance by increasing network plasticity [11]. OptoTrkB stimulation decreased PV^+ interneuron excitability and promoted LTP and oscillatory synchrony in the visual cortex due to the disinhibition of pyramidal neurons [11]. In fact, Donato and colleagues showed that pharmacogenetic inhibition of hippocampal PV^+ interneurons changed the configuration of PV^+ interneurons, and improved learning [6]. Our group also showed that optoTrkB stimulation of $CKII^+$ pyramidal neurons in the ventral hippocampus increased neural plasticity in the form of increased expression of FosB and potentiation of LTP [17]. In addition, the activation of optoTrkB combined with extinction training reduced fear response in fear-conditioned mice [17]. These findings support the theory that TrkB activation plays a key role in facilitating network plasticity [1,8,9], although the reasons why TrkB activation in both inhibitory and excitatory neurons increase LTP remains unclear. Consequences of promoted neural plasticity by TrkB activation may vary across neuronal subtypes and brain areas, especially in terms of their electrophysiological properties. Further studies are needed to elucidate the different downstream pathways of TrkB activation depending on neuronal subtypes and brain areas.

An interesting avenue for future research is to investigate the effects of increased neural plasticity in specific neural nuclei, where various types of neurons are interacting with each other [18]. For instance, the nucleus accumbens (NAc), a subcortical nucleus in the basal ganglia, has been shown to play a key role in reward, motivation, and drug use [19,20]. A recent review found that individual NAc synapses follow different mechanisms of plasticity [21], and optoTrkB (E281A) can help to illuminate such unique features of neural plasticity as well as aid understanding of the role that plasticity of single neuronal subtypes plays in brain nuclei.

It is known that inhibitory interneurons in the prefrontal cortex (PFC) are involved in several different neuronal plasticity-related behaviours, such as reversal learning [22], extinction of drug addiction [23], and experienced pain [24,25]. This wide array of different behavioural consequences of interneuron plasticity raises questions about what determines the behavioural outcome of interneuron plasticity in the PFC.

The highly sensitive optoTrkB (E281A) allows us to conduct wireless optogenetics and can be applied for high flexibility in different behavioural experimental paradigms, such as social interaction [26,27] and Intellicage [28], where multiple animals have to be simultaneously tested. Thus, optoTrkB (E281A) can be used to explore the behavioural consequences of increased neural plasticity in different neural circuits.

One limitation to note is that this study did not compare transparent and intact skull stimulation of both CKII-Cre and PV-Cre mice. As the sensitivity of optoTrkB (E281A) could differ in CKII$^+$ neurons and PV$^+$ interneurons, the neuron type used in each stimulation protocol may act as a confounding variable in this study. To eliminate this possibility, future studies should use both PV-Cre and CKII-Cre mice to compare LED light stimulation protocols through a transparent and intact skull.

These limitations notwithstanding, the optoTrkB (E281A) that was evaluated in this study appears sensitive enough to be activated through intact skull and fur. This high sensitivity allows for a lot of flexibility in light stimulation protocols, as no optic cannula or transparent skull is needed to ensure sufficient light intensity entering the brain. Moreover, the highly sensitive optoTrkB (E281A) may be stimulated by using a LED light source on the cage lid, which reduces human handling of the mice and thus the stress that animals experience via direct handling [29]. These features make optoTrkB (E281A) a highly flexible tool to further illuminate the role of TrkB in neuroplasticity.

4. Materials and Methods

4.1. OptoTrkB DNA

This study used pAAV-hSyn1-DIO-OptocytTrkB(E281A)-HA as previously described [13]. The hSyn1 (Human synapsin 1) promoter allows the virus to target only neurons, not other cell types [30]. DIO codes for double-inverted open reading frame structure. Two incompatible lox-P sites are present on each side of the optoTrkB (E281A) sequence, allowing the transcription of optoTrkB only in Cre-expressing neurons via the Cre-recombinase-mediated inversion of the open reading frame [14,15]. This allows selective expression of optoTrkB (E281A) in defined neuronal subpopulations. Furthermore, cyt refers to the DNA coding for only the cytoplasmic region of the TrkB receptor, which ensures that optoTrkB (E281A) is only stimulated by light and not BDNF-binding. The HA (human influenza hemagglutinin) tag is attached to the C-terminal of the PHR region of optoTrkB (E281A), allowing us to detect optoTrkB (E281A) by immunological methods. The plasmid was packed into AAV serotype 9 (AAV9) by the AAV core unit at the University of Helsinki. AAV9 was selected because it has been shown to allow for stable gene expression in cortical neurons [31,32].

4.2. Infection of AAV-OptoTrkB into Mice

AAV9-optoTrkB (E281A) was injected into five mice expressing Cre in CKII$^+$ pyramidal neurons, and six mice expressing Cre in PV$^+$ interneurons. The visual cortex (V1) was selected as the brain area of interest in this study because it has been extensively studied in iPlasticity studies to date [10,33,34]. For virus injection, the mice were anaesthetised with Isoflurane and fixed on a stereotaxic frame. The virus was injected bilaterally into the visual cortex, 2.8 mm caudally and 2.5 mm laterally from Bregma, at a depth of 0.8 mm. In total, 500 nL of virus solution was injected [13]. After injection, at least four weeks were allowed for expressing optoTrkB (E281A).

Immunohistochemical analysis later revealed that one CKII-Cre mouse had very few cells expressing optoTrkB (E281A) (due to poor sample preparation), and this mouse was excluded from data analysis, reducing the final sample size for CKII-Cre mice from five to four.

4.3. LED Light Stimulation

For stimulation of PV-Cre mice injected bilaterally with AAV9-optoTrkB (E281A), the blue light stimulation was applied atop the head with an intact skull and fur. A 488 nm blue LED light pulse was given for one second every five seconds, for a total duration of 30 min. Half of the mice underwent blue LED light stimulation ($n = 3$), while the rest of the mice were exposed to a red LED light as a control group ($n = 3$).

CKII-Cre mice injected bilaterally with AAV9-optoTrkB (E281A) underwent transparent skull surgery in order to ensure sufficient intensity of the blue LED light to pass through the skull (for detailed protocol, see [35]). The transparent skull was covered with black nail polish until light stimulation took place. For optoTrkB-stimulation, a continuous 30-s 488 nm blue LED light was exposed directly onto the transparent skull, atop the right hemisphere. The left hemisphere was left unstimulated and thus functioned as a control condition for each mouse ($n = 5$). The mice were perfused 30 min after light stimulation.

4.4. Perfusion

Thirty minutes after the end of light stimulation, the mice were perfused in order to obtain the brains for immunohistochemistry. The mice were anaesthetised using pentobarbital (Mebunat, Orion Pharma, Espoo, Finland) mixed with Lidocaine (Orion Pharma), and the mice were fixed via transcardiac perfusion with 4% paraformaldehyde in pH 7.4 phosphate-buffered saline (PBS) with 3–4 mL per minute for 15–20 min. The brains were then dissected and stored in 4% paraformaldehyde overnight, after which the brains were transferred to 0.04% sodium azide in PBS until cutting. The brains were then embedded in 3% agar gel and cut into 40 μm slices using a vibratome (Leica Biosystems, Deer Park, IL, USA).

4.5. Immunohistochemistry

For pCREB immunohistochemistry, antigen retrieval was used to ensure sufficient binding of the primary antibodies to their antigens of interest. Secondly, for pERK immunohistochemistry, Tyramide Signal Amplification (TSA) (Perkin Elmer, Shelton, CT, USA) was used to enhance the signal emitted by pERK as previously described [13]. Briefly, for signal amplification via antigen retrieval, slices from the visual cortex were incubated at 70 °C in 0.01 M citric acid and further treated with 0.1 M Glycine in PBS. The slices were then blocked with 3% bovine serum albumin, 10% normal donkey serum and 10% normal goat serum, to block nonspecific binding of secondary antibodies made in these animals, as well as with Mouse-anti-Mouse buffer (Vector Laboratories, Newark, CA, USA), to block nonspecific binding of antibodies to endogenous mouse Immunoglobulin G. The slices of PV-Cre mice were then successively stained with anti-HA (1:500, Cell Signalling Technology, Danvers, MA, USA, 6E2 #2367), anti-PV (1:1000, Synaptic Systems, Göttingen, Germany, #19500), and anti-pCREB (1:1000, Cell Signalling Technology, #9198) diluted in PBS with 0.3% TritonX for 24–72 h at 4 °C. For CKII-Cre mice, the slices were similarly stained with anti-HA (1:800, Cell Signalling Technology, CST3724), anti-CKII (1:250, Abcam, Cambridge, United Kingdom, #22609), and anti-pERK (1:10,000, Cell Signalling Technology, #9101). All antibodies have been validated by previous studies (see Table S1). After washing in TNT buffer (0.1 M Tris-HCl pH 7.5, 0.15 M NaCl, 0.05% Tween-20) the samples were reacted with secondary antibodies conjugated with fluorophores, or with horseradish peroxidase combined with TSA. A catalogue of the antibodies used is listed in Supplemental Table S1.

After washing again with TNT, the samples were transferred to 0.1 M phosphate buffer and mounted onto glass slides with anti-fade mounting medium (Agilent Dako Fluorescence, Santa Clara, CA, USA).

4.6. Data Analysis

Image analysis using confocal images was conducted by following previously reported methods [6,36–38]. Detailed descriptions of the data analysis can be found in the Supplementary material (Supplementary Note S1). The slices were imaged using the Andor Dragonfly 505 spinning disc confocal microscope. The imaging settings (e.g., pinhole size, laser power, gain, and scan speed) were established with samples and negative controls (not stained with the primary antibody) in order to avoid overexposure and oversaturation, which would limit the dynamic range of the detectors [36,37]. Imaging settings were kept exactly the same for all samples to allow for quantification of emitted fluorescence [36,37]. The images were analysed in ImageJ. Between 10 and 15 infected neurons were selected per image, and intensity values of HA, PV, and pERK or pCREB were recorded. Briefly, the intensity value was quantitatively recorded from each channel by outlining the neuron at its widest point in the z-stack, and the mean brightness value across all pixels within the outlined area was used as the intensity value. To ensure that only infected neurons were included in the final data analysis, a cutoff value of HA intensity was established by measuring HA intensity values from ten uninfected neurons adjacent to the infected area. A more detailed description of the image analysis can be found in the Supplementary material (Note S1).

After the selection of infected neurons was finalised, the change in pERK or pCREB intensity as a result of optoTrkB (E281A) stimulation was analysed with multiple regression and simple slope analysis [39,40]. Analyses were done separately for PV-Cre and CKII-Cre mice. HA intensity as well as an interaction term of HA intensity and stimulation condition was first included in the regression model, and where the interaction term was insignificant, it was removed from the model. When the interaction term was significant, follow-up analysis was conducted using simple slopes analysis with Bonferroni correction. A detailed description of the statistical analysis can be found in the Supplementary Materials (Note S1).

5. Conclusions

In this study, we evaluated the utility of a new, highly sensitive, optically activatable TrkB receptor. OptoTrkB (E281A) was successfully transduced in PV^+ interneurons and $CKII^+$ pyramidal neurons specifically. Light stimulation through transparent skulls or even through a high opacity barrier (intact skull and fur) promoted phosphorylation of ERK and CREB, downstream signals of TrkB, in the neurons expressing optoTrkB (E281A) at a certain level. Our results indicate that this highly sensitive optoTrkB (E281A) receptor can be activated using wireless optogenetic methods, and thus can be used for a broad range of behavioural studies, such as social interaction tests and Intellicage. With this tool, we can study the role of TrkB signalling in different neural subtypes in the short and long term, which will help with developing more effective treatments for neuropsychiatric disorders, such as depression, post-traumatic stress disorder, and drug addiction.

Supplementary Materials: The supporting information can be downloaded at: https://www.mdpi.com/article/10.3390/ijms231810249/s1. References [6,11,36,37,39–46] are cited in the supplementary materials.

Author Contributions: Experiments and data analysis, A.L., G.D. and J.U.; writing—original draft preparation, A.L.; writing—review and editing, J.U. and E.C.; supervision, J.H., W.D.H., E.C. and J.U. All authors have read and agreed to the published version of the manuscript.

Funding: Open access funding provided by University of Helsinki. This research was funded by the ERC grant—iPLASTICITY (#322742), the Sigrid Jusélius foundation, Jane & Aatos Erkko Foundation, Academy of Finland grants (#307416, #327192), EU Joint Programme-Neurodegenerative Disease Research (JPND) CircProt project co-funded by EU and Academy of Finland (#643417), the bilateral exchange program between Academy of Finland and JSPS (Japan Society for the Promotion of Science), the Erasmus+ grant, and the University of Helsinki Research Foundation.

Institutional Review Board Statement: The animal study protocols were conducted in accordance with guidelines from the Council of Europe. They were approved by the State Provincial Office of Southern and Eastern Finland under the license ID ESAVI/38503/2019.

Informed Consent Statement: Not applicable.

Data Availability Statement: Not applicable.

Acknowledgments: Open access funding provided by University of Helsinki. We thank our lab technicians Sulo Kolehmainen and Seija Lågas for technical and practical help. We also thank the animal caretakers in the Laboratory Animal Center of the University of Helsinki (UH) for their help and support with the animals.

Conflicts of Interest: E.C. has received speaker fees from Jansen Cilag. The other authors declare no conflict of interest.

References

1. Castrén, E.; Antila, H. Neuronal Plasticity and Neurotrophic Factors in Drug Responses. *Mol. Psychiatry* **2017**, *22*, 1085–1095. [CrossRef] [PubMed]
2. Minichiello, L. TrkB Signalling Pathways in LTP and Learning. *Nat. Rev. Neurosci.* **2009**, *10*, 850–860. [CrossRef] [PubMed]
3. Knudsen, E.I. Sensitive Periods in the Development of the Brain and Behavior. *J. Cogn. Neurosci.* **2004**, *16*, 1412–1425. [CrossRef] [PubMed]
4. Wiesel, T.N.; Hubel, D.H. Shape and Arrangement of Columns in Cat's Striate Cortex. *J. Physiol.* **1963**, *165*, 559–568. [CrossRef]
5. Di Lieto, A.; Rantamäki, T.; Vesa, L.; Yanpallewar, S.; Antila, H.; Lindholm, J.; Rios, M.; Tessarollo, L.; Castrén, E. The Responsiveness of TrkB to BDNF and Antidepressant Drugs Is Differentially Regulated during Mouse Development. *PLoS ONE* **2012**, *7*, e32869. [CrossRef]
6. Donato, F.; Rompani, S.B.; Caroni, P. Parvalbumin-Expressing Basket-Cell Network Plasticity Induced by Experience Regulates Adult Learning. *Nature* **2013**, *504*, 272–276. [CrossRef]
7. Webster, M.J.; Herman, M.M.; Kleinman, J.E.; Shannon Weickert, C. BDNF and TrkB MRNA Expression in the Hippocampus and Temporal Cortex during the Human Lifespan. *Gene Expr. Patterns* **2006**, *6*, 941–951. [CrossRef]
8. Castrén, E.; Hen, R. Neuronal Plasticity and Antidepressant Actions. *Trends Neurosci.* **2013**, *36*, 259–267. [CrossRef] [PubMed]
9. Umemori, J.; Winkel, F.; Didio, G.; Llach Pou, M.; Castrén, E. IPlasticity: Induced Juvenile-like Plasticity in the Adult Brain as a Mechanism of Antidepressants: Antidepressant-Induced Plasticity. *Psychiatry Clin. Neurosci.* **2018**, *72*, 633–653. [CrossRef]
10. Vetencourt, J.F.M.; Sale, A.; Viegi, A.; Baroncelli, L.; De Pasquale, R.; O'Leary, O.F.; Castrén, E.; Maffei, L. The Antidepressant Fluoxetine Restores Plasticity in the Adult Visual Cortex. *Science* **2008**, *320*, 385–388. [CrossRef]
11. Winkel, F.; Ryazantseva, M.; Voigt, M.B.; Didio, G.; Lilja, A.; Llach Pou, M.; Steinzeig, A.; Harkki, J.; Englund, J.; Khirug, S.; et al. Pharmacological and Optical Activation of TrkB in Parvalbumin Interneurons Regulate Intrinsic States to Orchestrate Cortical Plasticity. *Mol. Psychiatry* **2021**, *26*, 7247–7256. [CrossRef] [PubMed]
12. Chang, K.-Y.; Woo, D.; Jung, H.; Lee, S.; Kim, S.; Won, J.; Kyung, T.; Park, H.; Kim, N.; Yang, H.W.; et al. Light-Inducible Receptor Tyrosine Kinases That Regulate Neurotrophin Signalling. *Nat. Commun.* **2014**, *5*, 4057. [CrossRef] [PubMed]
13. Hong, J.; Heo, W.D. Optogenetic Modulation of TrkB Signaling in the Mouse Brain. *J. Mol. Biol.* **2020**, *432*, 815–827. [CrossRef]
14. Fenno, L.E.; Mattis, J.; Ramakrishnan, C.; Hyun, M.; Lee, S.Y.; He, M.; Tucciarone, J.; Selimbeyoglu, A.; Berndt, A.; Grosenick, L.; et al. Targeting Cells with Single Vectors Using Multiple-Feature Boolean Logic. *Nat. Methods* **2014**, *11*, 763–772. [CrossRef]
15. Nagy, A. Cre Recombinase: The Universal Reagent for Genome Tailoring. *Genesis* **2000**, *26*, 99–109. [CrossRef]
16. Naso, M.F.; Tomkowicz, B.; Perry, W.L.; Strohl, W.R. Adeno-Associated Virus (AAV) as a Vector for Gene Therapy. *BioDrugs* **2017**, *31*, 317–334. [CrossRef] [PubMed]
17. Umemori, J.; Didio, G.; Winkel, F.; Pou, M.L.; Harkki, J.; Russo, G.L.; Verie, M.; Antila, H.; Buj, C.; Taira, T.; et al. Optical Activation of TrkB Neurotrophin Receptor in Mouse Ventral Hippocampus Promotes Plasticity and Facilitates Fear Extinction. *bioRxiv* **2021**. [CrossRef]
18. Purves, D.; Augustine, G.J.; Fitzpatrick, D.; Katz, L.C.; LaMantia, A.-S.; McNamara, J.O.; Williams, S.M. The Generation of Neuronal Diversity. In *Neuroscience*, 2nd ed.; Sinauer Associates: Sunderland, WA, USA, 2001.
19. Pascoli, V.; Turiault, M.; Lüscher, C. Reversal of Cocaine-Evoked Synaptic Potentiation Resets Drug-Induced Adaptive Behaviour. *Nature* **2012**, *481*, 71–75. [CrossRef]

20. Pascoli, V.; Terrier, J.; Espallergues, J.; Valjent, E.; O'Connor, E.C.; Lüscher, C. Contrasting Forms of Cocaine-Evoked Plasticity Control Components of Relapse. *Nature* **2014**, *509*, 459–464. [CrossRef]

21. Turner, B.D.; Kashima, D.T.; Manz, K.M.; Grueter, C.A.; Grueter, B.A. Synaptic Plasticity in the Nucleus Accumbens: Lessons Learned from Experience. *ACS Chem. Neurosci.* **2018**, *9*, 2114–2126. [CrossRef]

22. Xu, Y.; Wang, M.-L.; Tao, H.; Geng, C.; Guo, F.; Hu, B.; Wang, R.; Hou, X.-Y. ErbB4 in Parvalbumin-Positive Interneurons Mediates Proactive Interference in Olfactory Associative Reversal Learning. *Neuropsychopharmacology* **2022**, *47*, 1292–1303. [CrossRef] [PubMed]

23. Lüscher, C.; Malenka, R.C. Drug-Evoked Synaptic Plasticity in Addiction: From Molecular Changes to Circuit Remodeling. *Neuron* **2011**, *69*, 650–663. [CrossRef]

24. Ong, W.-Y.; Stohler, C.S.; Herr, D.R. Role of the Prefrontal Cortex in Pain Processing. *Mol. Neurobiol.* **2019**, *56*, 1137–1166. [CrossRef] [PubMed]

25. Sang, K.; Bao, C.; Xin, Y.; Hu, S.; Gao, X.; Wang, Y.; Bodner, M.; Zhou, Y.-D.; Dong, X.-W. Plastic Change of Prefrontal Cortex Mediates Anxiety-like Behaviors Associated with Chronic Pain in Neuropathic Rats. *Mol. Pain* **2018**, *14*, 174480691878393. [CrossRef] [PubMed]

26. Toth, I.; Neumann, I.D. Animal Models of Social Avoidance and Social Fear. *Cell Tissue Res.* **2013**, *354*, 107–118. [CrossRef]

27. File, S.E.; Hyde, J.R.G. Can social interaction be used to measure anxiety? *Br. J. Pharmacol.* **1978**, *62*, 19–24. [CrossRef]

28. Kiryk, A.; Janusz, A.; Zglinicki, B.; Turkes, E.; Knapska, E.; Konopka, W.; Lipp, H.-P.; Kaczmarek, L. IntelliCage as a Tool for Measuring Mouse Behavior—20 Years Perspective. *Behav. Brain Res.* **2020**, *388*, 112620. [CrossRef]

29. Hurst, J.L.; West, R.S. Taming Anxiety in Laboratory Mice. *Nat. Methods* **2010**, *7*, 825–826. [CrossRef]

30. Kügler, S.; Kilic, E.; Bähr, M. Human Synapsin 1 Gene Promoter Confers Highly Neuron-Specific Long-Term Transgene Expression from an Adenoviral Vector in the Adult Rat Brain Depending on the Transduced Area. *Gene Ther.* **2003**, *10*, 337–347. [CrossRef]

31. Hammond, S.L.; Leek, A.N.; Richman, E.H.; Tjalkens, R.B. Cellular Selectivity of AAV Serotypes for Gene Delivery in Neurons and Astrocytes by Neonatal Intracerebroventricular Injection. *PLoS ONE* **2017**, *12*, e0188830. [CrossRef]

32. Royo, N.C.; Vandenberghe, L.H.; Ma, J.-Y.; Hauspurg, A.; Yu, L.; Maronski, M.; Johnston, J.; Dichter, M.A.; Wilson, J.M.; Watson, D.J. Specific AAV Serotypes Stably Transduce Primary Hippocampal and Cortical Cultures with High Efficiency and Low Toxicity. *Brain Res.* **2008**, *1190*, 15–22. [CrossRef]

33. Hanover, J.L.; Huang, Z.J.; Tonegawa, S.; Stryker, M.P. Brain-Derived Neurotrophic Factor Overexpression Induces Precocious Critical Period in Mouse Visual Cortex. *J. Neurosci.* **1999**, *19*, RC40. [CrossRef] [PubMed]

34. Pizzorusso, T.; Medini, P.; Berardi, N.; Chierzi, S.; Fawcett, J.W.; Maffei, L. Reactivation of Ocular Dominance Plasticity in the Adult Visual Cortex. *Science* **2002**, *298*, 1248–1251. [CrossRef] [PubMed]

35. Steinzeig, A.; Molotkov, D.; Castrén, E. Chronic Imaging through "Transparent Skull" in Mice. *PLoS ONE* **2017**, *12*, e0181788. [CrossRef] [PubMed]

36. North, A.J. Seeing Is Believing? A Beginners' Guide to Practical Pitfalls in Image Acquisition. *J. Cell Biol.* **2006**, *172*, 9–18. [CrossRef]

37. Shihan, M.H.; Novo, S.G.; Le Marchand, S.J.; Wang, Y.; Duncan, M.K. A Simple Method for Quantitating Confocal Fluorescent Images. *Biochem. Biophys. Rep.* **2021**, *25*, 100916. [CrossRef]

38. Donato, F.; Chowdhury, A.; Lahr, M.; Caroni, P. Early- and Late-Born Parvalbumin Basket Cell Subpopulations Exhibiting Distinct Regulation and Roles in Learning. *Neuron* **2015**, *85*, 770–786. [CrossRef]

39. Preacher, K.J. A Primer on Interaction Effects in Multiple Linear Regression. 2003. Available online: https://quantpsy.org/ (accessed on 1 July 2022).

40. Howell, D.C. Chapter 15: Multiple Regression. In *Statistical Methods for Psychology*; Thomson Wadsworth: Belmont, CA, USA, 2010; ISBN 978-0-495-59784-1.

41. Song, A.; Zhu, L.; Gorantla, G.; Berdysz, O.; Amici, S.A.; Guerau-de-Arellano, M.; Madalena, K.M.; Lerch, J.K.; Liu, X.; Quan, N. Salient Type 1 Interleukin 1 Receptor Expression in Peripheral Non-Immune Cells. *Sci. Rep.* **2018**, *8*, 723. [CrossRef]

42. Field, J.; Nikawa, J.; Broek, D.; MacDonald, B.; Rodgers, L.; Wilson, I.A.; Lerner, R.A.; Wigler, M. Purification of a RAS-Responsive Adenylyl Cyclase Complex from Saccharomyces Cerevisiae by Use of an Epitope Addition Method. *Mol. Cell Biol.* **1988**, *8*, 2159–2165. [CrossRef] [PubMed]

43. Erondu, N.; Kennedy, M. Regional Distribution of Type II Ca2+/Calmodulin-Dependent Protein Kinase in Rat Brain. *J. Neurosci.* **1985**, *5*, 3270–3277. [CrossRef]

44. Rhee, K.D.; Ruiz, A.; Duncan, J.L.; Hauswirth, W.W.; LaVail, M.M.; Bok, D.; Yang, X.-J. Molecular and Cellular Alterations Induced by Sustained Expression of Ciliary Neurotrophic Factor in a Mouse Model of Retinitis Pigmentosa. *Investig. Ophthalmol. Vis. Sci.* **2007**, *48*, 1389. [CrossRef] [PubMed]

45. Tan, Y.; Rouse, J.; Zhang, A.; Cariati, S.; Cohen, P.; Comb, M.J. FGF and Stress Regulate CREB and ATF-1 via a Pathway Involving P38 MAP Kinase and MAPKAP Kinase-2. *EMBO J.* **1996**, *15*, 4629–4642. [CrossRef] [PubMed]

46. Cai, L.X.; Tanada, Y.; Bello, G.D.; Fleming, J.C.; Alkassis, F.F.; Ladd, T.; Golde, T.; Koh, J.; Chen, S.; Kasahara, H. Cardiac MLC2 Kinase Is Localized to the Z-Disc and Interacts with α-Actinin2. *Sci. Rep.* **2019**, *9*, 12580. [CrossRef] [PubMed]

International Journal of
Molecular Sciences

Review

Novel Approaches Used to Examine and Control Neurogenesis in Parkinson's Disease

Alla B. Salmina [1,2,*], Marina R. Kapkaeva [1], Anna S. Vetchinova [1] and Sergey N. Illarioshkin [1]

[1]   Research Center of Neurology, 125367 Moscow, Russia; mareenqa@yandex.ru (M.R.K.); annvet@mail.ru (A.S.V.); snillario@gmail.com (S.N.I.)

[2]   Research Institute of Molecular Medicine & Pathobiochemistry, Prof. V.F. Voino-Yasenetsky Krasnoyarsk State Medical University, 660022 Krasnoyarsk, Russia

*   Correspondence: allasalmina@mail.ru

Abstract: Neurogenesis is a key mechanism of brain development and plasticity, which is impaired in chronic neurodegeneration, including Parkinson's disease. The accumulation of aberrant α-synuclein is one of the features of PD. Being secreted, this protein produces a prominent neurotoxic effect, alters synaptic plasticity, deregulates intercellular communication, and supports the development of neuroinflammation, thereby providing propagation of pathological events leading to the establishment of a PD-specific phenotype. Multidirectional and ambiguous effects of α-synuclein on adult neurogenesis suggest that impaired neurogenesis should be considered as a target for the prevention of cell loss and restoration of neurological functions. Thus, stimulation of endogenous neurogenesis or cell-replacement therapy with stem cell-derived differentiated neurons raises new hopes for the development of effective and safe technologies for treating PD neurodegeneration. Given the rapid development of optogenetics, it is not surprising that this method has already been repeatedly tested in manipulating neurogenesis in vivo and in vitro via targeting stem or progenitor cells. However, niche astrocytes could also serve as promising candidates for controlling neuronal differentiation and improving the functional integration of newly formed neurons within the brain tissue. In this review, we mainly focus on current approaches to assess neurogenesis and prospects in the application of optogenetic protocols to restore the neurogenesis in Parkinson's disease.

Keywords: Parkinson's disease; α-synuclein; neurogenesis; neural stem cell; neural progenitor cell; astrocyte; optogenetics

Citation: Salmina, A.B.; Kapkaeva, M.R.; Vetchinova, A.S.; Illarioshkin, S.N. Novel Approaches Used to Examine and Control Neurogenesis in Parkinson's Disease. *Int. J. Mol. Sci.* **2021**, *22*, 9608. https://doi.org/10.3390/ijms22179608

Academic Editors: Piotr D. Bregestovski and Carlo Matera

Received: 29 July 2021
Accepted: 2 September 2021
Published: 4 September 2021

Publisher's Note: MDPI stays neutral with regard to jurisdictional claims in published maps and institutional affiliations.

1. Introduction: Neurogenesis in the Healthy and Parkinson's Disease-Affected Brains

1.1. Key Characteristics of Adult Neurogenesis

Neurogenesis is a mechanism of brain development and plasticity. Embryonic neurogenesis provides new neurons for brain growth, whereas adult neurogenesis is required for memory consolidation and tissue repair, mood regulation, and social recognition [1–4]. Functional competence of adult-born neurons results in their successful integration into pre-existing neural circuits, for instance, in the hippocampus, which is associated with activity-mediated acceleration of dendritic spines formation [5]. Therefore, the efficacy of neurogenesis in the embryonic brain corresponds to appropriate brain development and maturation. Adult neurogenesis could be considered as efficient if brain plasticity meets the current needs in cognition, social interactions, expression of emotions, memory consolidation and retrieval, forgetting, and experience-driven circuits remodeling.

The key pool of cells that could be activated to provide new neurons and astrocytes is represented by neural stem cells (NSCs) that are found in highly specialized neurogenic niches (subventricular zone, SVZ, and subgranular zone of the hippocampus, SGZ), as well in some other brain regions (cerebellum, substantia nigra, cortex), and in loci with the facilitated access to regulatory molecules, i.e., in the periventricular area with ischemia- or neuroinflammation-mediated compromised blood-brain barrier (BBB) [6,7]. NSCs exhibit two pivotal characteristics: (1) the ability to self-renew and to produce copies of themselves by symmetric or asymmetric division; (2) multipotency to produce neural progenitor cells (NPCs) that are able to differentiate into neurons, astrocytes, or oligodendroglia [8–10]. The proliferation of NSCs is under the tight control of the local microenvironment consisting of numerous soluble molecules: neurotransmitters (GABA, glutamate, dopamine, etc.), neuropeptides (oxytocin, angiotensin, etc.), cytokines (interleukins, chemokines, etc.), metabolites (lactate, NAD$^+$), extracellular matrix proteins, and accessory cells (astrocytes, brain microvessel endothelial cells) aimed to prevent non-reasonable utilization of the NSCs pool, to coordinate cell fate, and to drive cell proliferation, differentiation, and migration on demand (Figure 1). In addition to the above-mentioned factors, local hypoxia in the neurogenic niche serves as a signal to control the NSC's recruitment [11].

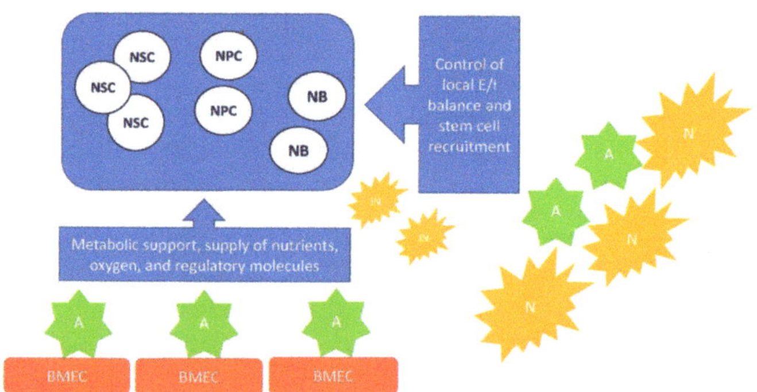

Figure 1. A simplified scheme of intercellular communications within the neurogenic niche. The local microenvironment is established due to the activity of neuronal, astroglial, and endothelial cells that are able to release various molecules (growth factors, neurotransmitters, cytokines, gliotransmitters, metabolites) affecting cell fate within the niche. Abbreviations used: NSC—neural stem cell, NPC—neural progenitor cell, NB—neuroblast, IN—immature neuron, N—neuron, A—astrocyte, BMEC—brain microvessel endothelial cell, E/I—excitation-inhibition balance.

Postnatal ontogenesis, neurodegeneration, and aging are associated with progressive loss of NSCs (radial glia cells, RGCs) in the rodent hippocampus [10], thereby suggesting that mechanisms preventing the depletion of the NSCs pool came to be less efficient in older brains. Indeed, in human SVZ, the density of RGCs reduces from mid-gestation until the perinatal period, and in the human SGZ, the decline of RGCs number is observed from early ages of development until 5 years old and then in adulthood [12].

The predominant view on neurogenic events in the adult brain states that enhanced neurogenesis is required for (re)cognition and memory, whereas reduced neurogenesis manifests aberrant brain plasticity [13]. However, recent data suggest that the general picture is not so simple, at least in some details. Firstly, even the addition of new neurons to the dentate gyrus of the hippocampus in vivo provides a fresh substrate for new memories, blocking adult neurogenesis in rats results in the elongation of long-term potentiation (LTP); therefore, newly-formed cells are required for the phenomenon of hippocampal clearance and consolidation of memory in extra-hippocampal brain regions [14]. It links elevated hippocampal neurogenesis to mechanisms of forgetting when newly-arrived young cells make a vacant position for new memories by eliminating recently-learned information, but not remotely acquired ones, which already exist in extra-hippocampal brain structures [15]. Secondly, adult-born neurons inhibit the dentate gyrus activity by recruiting local interneurons, and it seems to be important for preventing memory interference and engrams overlapping in subsequent learning episodes (so-called cognitive flexibility) [13]. This mechanism underlies the ability of young dentate gyrus cells to support pattern separation and the ability of old dentate gyrus cells to support rapid recall and pattern completion [16]. That is why hyperexcitable dentate gyrus results in cognitive deficits and the impairment of pattern separation in mice [17].

1.2. Aberrant Neurogenesis in Parkinson's Disease

Parkinson's disease (PD) is a chronic neurodegenerative disease primarily affecting dopaminergic neurons in the substantia nigra pars compacta (SN) and leading to prominent motor and cognitive dysfunction. Several hypotheses have been proposed to explain the progressive cell loss in SN, including mitochondrial alterations, genetic predisposition, accumulation of abnormal proteins, development of oxidative stress, chronic neuroinflammation, and aberrant neurogenesis [18–20]. PD belongs to the group of neurodegenerative diseases with Lewy body and Lewy neurite pathology that are associated with the accumulation of wild-type α-synuclein protein as intracellular neuronal and glial filamentous deposits (other examples are dementia with Lewy bodies, multiple system atrophy) [21].

Presynaptic protein α-synuclein encoded by the SNCA gene belongs to the group of so-called "natively unfolded proteins" that lack ordered structure, has high flexibility and the ability to get the conformation upon binding to ligands, and are prone to aggregation and deposition [22]. In physiological conditions, it may have two states: the unfolded state in the cytosol or the helical multimeric state at the cell membranes due to its interactions with lipid rafts and phospholipids [23]. It is interesting that binding to membranes with a larger diameter (~100 nm) produces an elongated helix conformation in α-synuclein, whereas binding to membranes with small and highly curved vesicles (i.e., synaptic vesicles) results in a broken helix conformation [23]. Being located in close vicinity to vesicles in the presynaptic terminal, α-synuclein significantly affects synaptic transmission via the regulation of vesicle formation and neurotransmitter release [24]. Neural activity triggers the dispersion of α-synuclein from synapses during exocytosis in a Ca^{2+} entry-dependent manner [24]. In PD, a mutated form of α-synuclein has a tendency to aggregate and not disperse from synaptic boutons, thereby leading to deposition of synuclein-containing protofibrils [25].

In familial autosomal dominant PD, several missense mutations and multiplications of SNCA have been reported even though they are a rare cause of the disease, but exonic duplication and deletion mutations in parkin (PRKN), protein deglycase (DJ-1), and PINK1 genes have been identified in the early-onset parkinsonism [26]. SNCA multiplications are also present in some cases of sporadic PD [27].

Tissue accumulation of defective α-synuclein is one of the key features of PD. Being present intracellularly or in the extracellular space, this protein produces prominent neurotoxic effects, alters synaptic plasticity, affects autophagy, induces mitochondrial dysfunction and endoplasmic reticulum stress, deregulates intercellular communication, and supports the development of neuroinflammation, thereby providing propagation of pathological events leading to the establishment of a PD-specific phenotype [21,26,28,29]. Moreover, there is a documented transneuronal propagation of abnormal α-synuclein aggregates in PD, leading to prion-like synuclein dissemination within the nervous tissue [30,31]. The distribution of α-synuclein in the tissue might depend on the connexin 32 (Cx32)-based activity of gap junctions between cells and within the astroglial syncytium [32]. As we and others have shown before, this is confirmed by extra-brain localization of α-synuclein in PD patients, thereby suggesting the retrograde dissemination of α-synuclein forms olfactory bulbs and intestinal autonomic neurons on the brainstem structures and determining the staging of synucleinopathy development [30,33–35].

The role of α-synuclein in the regulation of adult neurogenesis has been partly identified: (i) SGZ neurogenesis is increased in α-synuclein knock-out mice, whereas overexpression of wild type synuclein results in decreased dendritic growth [21]; (ii) injections of α-synuclein oligomers in mice produces a significant increase in the number of proliferating cells and immature neurons in the SGZ, corresponding to the loss of dopaminergic neurons in SN [36]; (iii) expression of wild type and mutant α-synuclein in embryonic stem cells results in their reduced proliferation and neuronal differentiation associated with lowered

Notch signaling in vitro [37]; (iv) in mice overexpressing wild type α-synuclein, the number of transcription factor paired box protein (Pax6)-expressing NPCs in SGZ increases, presumably, due to the deregulation of dopamine levels or altered excitation/inhibition balance in the hippocampus [38]; (v) in the experimental (MPTP) model of PD in mice, SVZ NPCs show a reduced capacity of proliferation in aged but not young animals, whereas in transgenic mice overexpressing mutant (A53T) α-synuclein and treated with MPTP, neurogenesis is reduced in olfactory bulbs and SN [39]; (vi) defects in neurogenesis seen in the olfactory bulbs and hippocampus of transgenic mice with the overexpression of wild type of α-synuclein have been attributed to the development of olfactory deficits in PD [21].

Figure 2 illustrates the current understandings of the role of α-synuclein in brain (patho)physiology.

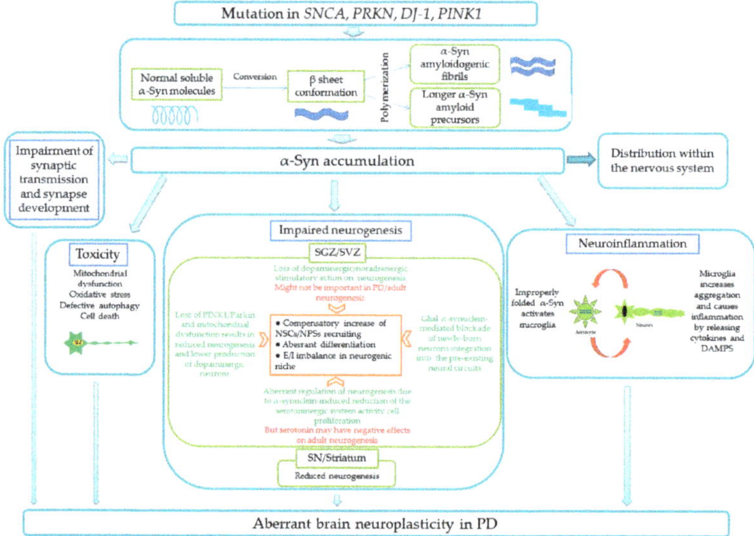

Figure 2. Impairment of brain plasticity in PD caused by the accumulation of α-synuclein in the brain tissue. Aberrant neurogenesis in SGZ/SVZ as well as in the substantia nigra /striatum is a result of numerous mechanisms triggered by improperly folded α-synuclein (α-syn) that are associated with neuroinflammation, direct cell toxicity, and synaptic dysfunction. Abbreviations used: *SNCA— synuclein* gene, *PRKN—parkin* gene, *DJ-1—deglycase* gene, PINK1—PTEN-induced kinase *1* gene, E/I—excitation/inhibition balance, DAMPs—damage-associated molecular patterns, SN—substantia nigra.

Neurogenesis in a PD-affected brain is believed to be altered by several mechanisms that are not fully understood and are even based on controversial experimental findings: (i) loss of dopaminergic and noradrenergic stimulatory action on SGZ neurogenesis [40] and SVZ neurogenesis [41], however, some studies suggest that this mechanism might not be important in PD and in the adult neurogenesis, in general [42,43], or dopaminergic neurodegeneration increases SVZ- and midbrain-derived progenitor cell proliferation [44]; (ii) aberrant regulation of neurogenesis in neurogenic niches due to the α-synuclein-induced reduction of the local serotoninergic system activity, which is required for SGZ cells proliferation [45,46], however, there are some data on the negative effect of serotonin on adult neurogenesis [47]; (iii) loss of PTEN-induced kinase 1 (PINK1) and parkin, as well as mitochondrial dysfunction results in reduced SGZ and SVZ neurogenesis or suppressed production of dopaminergic neurons [48,49]; (iv) glial α-synuclein-mediated blockade of newly-born neurons integration into the pre-existing neural circuits [50].

While analyzing all these data, including the controversial data, one should keep in mind that: (i) α-synuclein demonstrates physiological activity towards newly-formed neurons, and promotes dendrite and spine development and maturation depending on the expression level [51], therefore, impairment of neurogenesis could be caused by the action of supraphysiological concentrations or aggregates of this protein, which is specific for PD pathogenesis [21]; (ii) analysis of cell proliferation and NSCs/NPCs number might not be relevant in the assessment of neurogenesis efficacy, since preserved neurogenesis could be linked to predominant self-renewal and prevention of excessive recruitment of stem and progenitor cells; (iii) compensatory increase in striatal neurogenesis and intensive migration of SVZ-generated neuroblasts to SN might be evident at the initial stages of development of the neurodegenerative process as it was shown in 1-methyl-4-phenyl-1,2,3,6-tetrahydropyridine (MPTP)-induced and 6-hydroxydopamine (6-OHDA)-induced mouse models of PD, or in human brain samples [21,52–55]. Such a compensatory increase in neurogenesis seems to utilize Wnt1-dependent and growth factors-triggered signaling machinery [40]. The contribution of other neurogenic regions of the brain (i.e., periventricular parts of the aqueduct and the fourth ventricle) into the compensatory increase in neurogenesis is not evident in experimental PD [56].

In the α-synuclein transgenic rat PD model, the impairment of SGZ neurogenesis due to excessive cell loss is evident prior to the development of motor symptoms being associated with the serotonergic deficit in the hippocampus and anxiety-like phenotype [57]. It is interesting to note that SVZ neurogenic niches are under the control of dopaminergic neurons located in the substantia nigra [58]. Thus, one could speculate that altered neurogenesis at the early (pre-motor) stage of PD would result in the insufficient production of dopaminergic neurons, but later, when the number of these cells comes to be very low, the loss of dopaminergic stimulation of SVZ niche activity would lead to the secondary suppression of neurogenesis. In sum, the widely-accepted view on the neurogenesis alterations in PD states that survival, recruitment, and proliferation of NSCs/NPCs is greatly affected by the accumulation of improperly folded proteins or signaling pathways associated with neurodegeneration and neuroinflammation, thereby leading to abnormal brain plasticity and motor and cognitive impairments [21,59].

In addition to the numerous experimental data obtained in rodent PD models, aberrant neurogenesis was found in the brain of patients with PD. Particularly, in humans, the number of the RNA-binding protein Musashi-immunopositive cells (NSCs/NPCs) within the SVZ positively correlates with the extent of dopaminergic treatment, whereas disease duration shows a negative correlation; the number of the transcription factor Sox2-immunopositive cells (NSCs) in the SGZ is significantly decreased compared with a control group [59]. Since Sox2 inhibits paracrine and autocrine Wingless/Int-1 (Wnt) signaling and maintains the cells in the proliferative state [60], one may suggest that the recruitment of stem and progenitor cells in PD is diminished.

Another important mechanism of neurogenesis impairment is directly linked to the pathology of SN as a non-conventional neurogenic niche in the adult brain: the generation of dopaminergic neurons has been shown locally in the SN by means of tracing analysis revealing newly-generated neurons either in a normal or degenerated brain in PD [61]. The physiological rate of neurogenesis in SN is several orders of magnitude lower than in SGZ but contributes to the replacement of dopaminergic neurons during the lifespan in mice [62]; however, some other studies show no evidence for local neurogenesis in the SN [63]. Neurogenesis in the SN depends on the presence of angiogenic factors and the establishment of new microvessels, thereby resembling the situation in SVZ and SGZ [64]. Recent findings on dopamine-stimulated hippocampal SGZ and striatal neurogenesis [65] suggest that PD-associated impairment of neurogenesis might have links to insufficiency of dopaminergic mechanisms of neurogenesis regulation. Stimulation of dopaminergic receptors results in the induction of neurogenesis within the SN in rodents [66]; however, it was proposed that the local microenvironment in the midbrain supports gliogenesis, but not neurogenesis [40,67]. Data on the experimental transplantation of adult rat-derived

NPCs into the lesioned striatum demonstrate that special solutions should be found to drive the differentiation of grafted cells toward the desired phenotype (neuronal but not glial) [51].

2. iPSC-Based Platform for Studying PD-Affected Neurogenesis In Vitro

2.1. Generation of iPSC-Derived Dopaminergic Neurons

The development of protocols for induced pluripotent stem cells (iPSCs) generation had revolutionized the methodology of studying the brain. Particularly, the generation of neurons and glial cells from a patient became possible, thereby allowing the reconstruction of key processes of brain plasticity, development of brain tissue in vitro models, and establishment of isogenic platforms for cell-replacement therapy and cell transplantation [68]. As we have shown before, such an approach was effective in the generation of PD-derived iPSC lines with different mutations for studying defects in neurotrophic factors signaling affecting neuronal development [19], personalized modeling of PD pathogenesis [69], and screening of drug candidates [70,71]. The optimized protocols for getting dopaminergic differentiated neurons from iPSCs have been suggested [72,73], and they include the recruitment of stem cells with the transforming growth factor-beta (TGFβ) antagonists, activation of Hedgehog, Wnt, and fibroblast growth factor 8 (FGF8) signaling pathways or expression of Lmx1a, Foxa2, and Nurr1 and other midbrain-specific transcription factors for getting the midbrain floor-plate progenitors, followed by the application of neurotrophic factors (brain-derived neurotrophic factor BDNF, glia cell line-derived neurotrophic factor GDNF) and Notch receptor antagonists to induce the terminal differentiation of cells toward a dopaminergic phenotype. However, the final populations of cells are rather heterogeneous, consisting of post-mitotic neurons of different subtypes and immature cells; therefore, 3D cultures, including cerebral organoids, have been applied to improve the quality of the final cellular composition [72].

After the differentiation in vitro, dopaminergic neurons appeared to be not fully matured; therefore, acute progerin overexpression or co-culture with isogenic astrocytes is highly recommended [73–75]. There is growing evidence that co-culturing with astrocytes results in the promotion of neuronal differentiation and functional maturation of newly-formed iPSC-derived neurons in various models [75–78]. Thus, various local factors and types of intercellular communication affect the development of iPSC-derived dopaminergic neurons [79].

The overexpression of α-synuclein could be achieved in iPSC-derived neurons with SNCA multiplication, i.e., triplication of the gene leads to abnormally high expression and deposition of α-synuclein in differentiated cells [80], such as in PARK4 PD patients [81], hereby providing a relevant model of PD pathogenesis. In addition, severe changes in neuronal differentiation and maturation have been detected upon SNCA triplication, whereas the knock-down of SNCA mRNA in iPSC-derived cells prevents such abnormalities [82]. A53T point mutation in the SNCA gene in iPSC-derived neurons results in the development of pathological alterations in cell metabolism and defective proteostasis, early neurite degeneration, and down-regulation of some synaptic proteins [28]. Interneuronal spreading of α-synuclein within and between iPSCs cortical neurons was reproduced in the in vitro microfluidic systems allowing unidirectional axonal growth [83].

2.2. Generation of iPSC-Derived Midbrain Astrocytes

The establishment of a co-culture of midbrain neurons and astrocytes of the same origin would have several advantages in developing the platform for PD study and drug testing in vitro [73].

The differentiation of astrocytes from human iPSC-derived progenitors has been demonstrated in [84–86]. Basically, the protocols utilize the application of a medium with low fetal bovine serum (FBS) (1%–2%) and the replacement of half of the conditioned medium with the fresh one to induced astroglial phenotype acquisition within 1 month in vitro. iPSC-derived astrocytes express astroglial markers (S100 calcium-binding protein B

(S100β), gap junction protein connexin 43, and water channel aquaporin 4 (AQ4)) and other molecules whose pattern resembles quiescent astrocytes. Being functionally competent, these cells respond to inflammatory stimuli by cytokine release and demonstrate Ca^{2+} elevations in basal conditions or after the stimulation with ATP or glutamate.

Another compatible approach is based on embryonic stem cells converted into neural progenitors through the stage of embryoid body (EB) formation and adding transforming growth factor (TGF) and bone morphogenetic protein 4 (BMP4) inhibitors SB431542 and LDN193189 to suppress the production of cystic embryonic bodies, followed by differentiation supported by epidermal growth factor (EGF) and ciliary neurotrophic factor (CNTF). The astrocytes obtained express astroglial markers and appropriately responded to pro-inflammatory stimuli [87]. Direct conversion of embryonic and postnatal mouse and human fibroblasts into astrocytes in vitro was proposed in [88] with NFIA, NFIB, and SOX9 transcription factors. The astrocytes obtained in this protocol demonstrate gene expression pattern, K^+ and Ca^{2+} membrane permeability, glutamate transport, and response to cytokines stimulation similar to native brain astrocytes.

The generation of midbrain astrocytes from human iPSCs was demonstrated from small molecule-treated NPCs (smNPCs) that are able to differentiate by the withdrawal of the small molecules used for their expansion (TGF and BMP inhibitors SB431542 and dorsomorphin, Wnt stimulator and glycogen synthase kinase 3 (GSK3) inhibitor CHIR 99021, SHH stimulator purmorphamine) into midbrain dopaminergic neurons, midbrain astrocytes (that could be obtained in a medium with 4% fetal calf serum (FCS) and CNTF later replaced with dibutyryl cyclic AMP), and oligodendrocytes [89].

In another protocol, midbrain astrocytes have been obtained from SNCA-mutated iPSCs generated from PD patient's fibroblasts according to the following conditions: fibroblast growth factor 8 (FGF8) for getting the midbrain identity, epidermal growth factor (EGF), leukemia inhibitory factor (LIF), FGF2 + heparin for effective gliogenesis, and histone deacetylase (HDAC) inhibitor valproic acid for increased expression of glial cell line-derived neurotrophic factor (GDNF). The obtained cells express astrocyte markers, such as aldehyde dehydrogenase 1 family member L1 (ALDH1L1), Vimentin, Connexin 43 (Cx43), and aquaporin 4 (AQP4), as well as S100β, accumulate α-synuclein, and release the excess of Ca^{2+} into cytosol, but demonstrate pathological mitochondrial fragmentation and aberrant respiration [90].

PD patient-specific astrocytes derived from iPSCs with mutations in the LRKK2 gene and further run in the neuron-astrocyte co-culture system support the development of a neurodegeneration characteristic for PD (incl. morphological alterations, α-synuclein accumulation, shortened neurites, and reduced cell survival), whereas astrocytes themselves demonstrate signs of incomplete autophagy [91].

In summary, the establishment of differentiated astrocytes from PD patient-derived iPSCs allows studying the astroglial contribution to the local control of neurogenesis, promoting differentiation of dopaminergic neurons co-cultured with astroglia, and developing novel methodological approaches to modulate astroglial activity with optogenetic protocols (as discussed below).

2.3. Generation of iPSC-Derived Midbrain Cerebral Organoids

Cerebral organoids represent another type of brain tissue model with reconstituted processes of neurogenesis and brain development. Actually, cerebral organoids reproduce embryonic neurogenesis, and the data obtained cannot be extrapolated directly to mechanisms of adult neurogenesis [92]. The key characteristics of cerebral organoids are the ability of stem cells to produce self-organized structures resembling various brain regions. This methodology is based on the production of embryonic bodies and clusters of neuroepithelial cells followed by the establishment of apico-basally polarized neural tissue that is achieved by so-called un-guided or guided protocols to get spontaneous differentiation or specification of cell development, respectively [93]. The main advantage in using human iPSC-derived cerebral organoids is an opportunity to get the human-specific cell types and tissue developmental traits that could not be reproduced in the rodent tissue [94].

In the protocols of organoids generation, the application of growth factors and morphogens is rather limited because of a shortage of knowledge on their action in a stage-specific manner and general disorganization of cells positioning with the organoids. However, successful attempts to produce cortical organoids, hippocampal organoids, ventricular zone-like regions, and their assembloids resembling some periods of brain development with the specific transcriptomic and proteomic changes have been reported and analyzed [93–96].

Another methodological problem with the lack of microvasculature and microglia in neuroectodermal progenitors-derived organoids is currently getting some solution with the approaches to prevent organoid core hypoxia by co-culturing with brain microvessel endothelial cells [97–99], or to support normal neuronal development with microglia cells incorporated into organoids in vitro [100]. Moreover, data on the presence of ectodermal, mesodermal, and endodermal progenitors at the earliest stages of organoids development [101] suggest that mesodermal progenitors might be able to develop into microglial cells simultaneously with neurons and astrocytes or into BMECs to provide a vascular scaffold for developing and maturing cells.

Cerebral organoids contain various cells (radial glia, intermediate progenitors) whose self-organization results in establishing the structures resembling brain development during the first trimester of human gestation [93]; therefore, they are mainly applied in studying the molecular pathogenesis of neurodevelopmental disorders. However, neurodegeneration associated with impaired neurogenesis might be examined with the cerebral organoid methodology; even the trajectory of brain cells development and their diversity are quite different in the embryonic and adult brain.

Actually, it is hard to imagine that generation of region-specific cerebral organoids from patient-derived iPSCs would give the same phenotype of neuronal and glial cells that are seen in advanced neurodegeneration. However, it was confirmed that this in vitro model provides unique opportunities for analyzing the entire mechanisms of brain plasticity under the conditions of abnormal expression of genes and proteins in the particular type of neurodegeneration. For instance, organoids generated from Alzheimer's disease patients and aged in culture (up to 60–90 days) produce significantly higher levels of beta-amyloid and show sensitivity to inhibitors of gamma-secretase [102]. Cerebral organoids obtained from iPSCs from patients with frontotemporal dementia allow novel aspects of tau-mediated pathology to be revealed [103]. At present, the establishment of organoids correctly resembling aging- or neurodegeneration-associated changes in cell development is a great challenge for the current neurobiology and bioengineering, as was discussed recently in [104].

In modeling Parkinson's disease with cerebral midbrain organoids, the specification of cells induces the expression of transcription factors FOXA1/2, LMX1A, and LMX1B in midbrain dopaminergic progenitor cells that are able to express tyrosine hydroxylase and produce dopamine [105]. Numerous attempts have been applied to increase the yield of tyrosine hydroxylase-immunopositive cells in organoids and to produce earlier differentiated midbrain organoids in vitro [106,107]. In some cases, cerebral organoids have been used for in vitro modeling of Parkinson's disease, e.g., by the treatment with 1-methyl-4-phenyl-1,2,3,6-tetrahydropyridine (MPTP) [108], but patient-derived midbrain cerebral organoids appear to be more informative in studying the pathogenesis of Parkinson's disease. These midbrain organoids allowed demonstrating the characteristics specific for Parkinson's disease: impaired differentiation of progenitor cells, reduced number of differentiated dopaminergic neurons, higher number of progenitors, elevated expression of markers of mitophagy and autophagy, appearance of mitochondrial dysfunction, low viability of cells, and dysfunctional response to neuroinflammatory stimuli in LRKK2, DJ-1, or PRKN mutants [105,109–111].

The general principles of the current methodology used for the generation of iPSC-derived cells and organoids in PD are summarized in Figure 3. As we discussed above, numerous protocols have been applied to get differentiated neuronal and glial cells, or cerebral organoids from human iPSCs. All these protocols have their own strengths and limitations; therefore, the development of novel, probably unified, approaches are a big challenge in modern bioengineering and neurobiology.

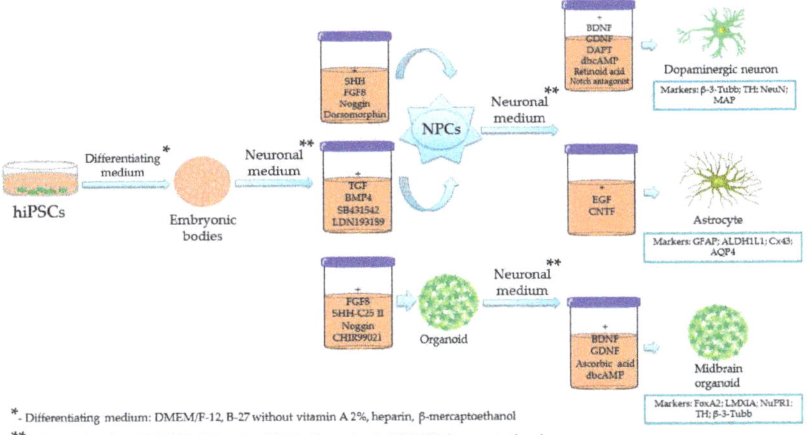

Figure 3. Summary of widely-used protocols for the generation of iPSC-derived midbrain cells and cerebral organoids. The scheme shows the main procedures aimed to establish the appropriate local microenvironment for the in vitro differentiation of cells toward midbrain neurons, glia, and multicellular structures (cerebral organoids).

Very recently, some new data on the establishment of iPSC-derived brain cells and multicellular ensembles suggest that we might have more efficient tools for deciphering cellular and molecular mechanisms of brain plasticity in (patho)physiological conditions. Some of these revolutionizing protocols are CRISPR-Cas9 generation of iPSC-derived cell lines that are suitable for live imaging and selective isolation of dopaminergic neurons in the culture [112], application of 3D organoids to study the idiopathic form of PD [113], development of in vitro BBB model from iPSCs for the assessment of BBB breakdown in PD [114], generation of brain-on-chips with the microfluidic technologies that are helpful in separating the cell-specific effects or studying the BBB integrity in vitro [115], and establishment of novel cell products matching the requirements for pre-clinical studies or even of clinical-grade quality [116].

3. Adult Neurogenesis as a Target for Therapy and Optogenetic Control

Neurogenesis is a well-known target for the pharmaceutical correction of brain plasticity and treatment of neurological and mental disorders [117,118]. Neurogenesis is affected not by drugs or small molecules only [119] but also by other various exogenous stimuli. For instance, restricted sleep results in reduced neurogenesis [120], social interactions promote neurogenesis in the post-ischemic brain [121], and an enriched (multi-stimuli) environment activates neurogenesis in the brain tissue in the postnatal period and leads to obvious effects in NSCs proliferation in physiological aging and Alzheimer's type neurodegeneration in vivo and in vitro [122,123]. Other factors that affect adult neurogenesis (nutrients, metabolites, hormones, cytokines, etc.) have also been tested as potential modulators of brain plasticity. As an example, lactate produced by niche astrocytes and stem cells, or transported by BMECs from the extra-niche compartment, stimulates adult neurogenesis and mediates the pro-neurogenic effects of physical exercise [124]. Potent regulators of glucose metabolism, such as insulin, insulin-like growth factor-I, glucagon-like peptide-1, and ghrelin, control NSC fate and stimulates SGZ neurogenesis [125]. Pro-inflammatory cytokine IL-6 supports NSCs self-renewal, but a transient surge of systemic IL-6 levels results in an increase in NPCs proliferation and long-lasting depletion of NSCs pools [126]. The stimulation of NAD^+ synthesis in niche cells leads to the restoration of adult neurogenesis affected in neurodegeneration, presumably due to the activity of NAD^+-consuming enzymes (NAD^+-glycohydrolases, poly (ADP-ribose)polymerase) or NAD^+-dependent sirtuins [127].

Targeting neurogenesis with drugs and compounds affecting some of the above-mentioned regulatory mechanisms is in the focus of neurobiologists and neuropharmacologists. At the same time, given the rapid development of optogenetics, it is not surprising that this method has already been repeatedly tested to modulate the adult neurogenesis with higher precision either in vivo or in vitro [128,129]. Indeed, neural stem or progenitor cells could be transfected with light-sensitive channelrhodopsin2 (ChR2) or other variants of chimeric opsins, for the induction of large photocurrents, either with viral vectors (i.e., lentivirus) or via a non-viral transfection system (i.e., piggyBac transposons). Photostimulation of these cells results in the production of a larger number of neuroblasts and functionally competent neurons in vitro [130], with up-regulated Wnt/β-catenin pathway, or induces differentiation of NPCs into oligodendrocytes and neurons, as well as the polarization of astrocytes to a pro-regenerative/anti-inflammatory phenotype [131]. Expression of ChR2 in human iPSC-derived neuronal cells under the calcium/calmodulin-dependent kinase II (CaMKII) and synapsin 1 (SYN1) promoters was effective in the detection of the differentiated status of the progeny and in the optical control of their growth in vitro [132].

Optogenetic protocols have also been tested in grafted NSCs to increase the expression of genes involved in neurotransmission, neuronal differentiation, axonal guidance, and synaptic plasticity [133]. Cre-lox strategy and piggyBac vectors have been applied for getting the optogenetic stem cell lines from human iPSCs that can switch on optogenetic expression via Cre-induction in vitro for further photomanipulations (activation or silencing) with the differentiated neurons [134]. Embryonic stem cell-derived NPCs stably expressing ChR2 can be efficiently transplanted into the mouse cortex where they show good integration capacity and differentiation toward GABAergic phenotype; however, photostimulation of such optogenetic cells in vivo produce rather controversial effects [135]. In some cases, optogenetics might be used for studying the response of host cells on the transplantation of iPSC-derived neurons: expression of ChR2 in host neurons allows detecting the development of host-to-graft synaptic afferents and establishment of ample output from host cells to the grafted ones [136].

Despite the fact that the role of neurogenesis in the adult brain of humans and non-human primates is still a controversial issue, most neuroscientists believe that the management of neurogenesis could enhance cognitive reserve and stimulate restoration of the brain tissue after injury or in chronic neurodegeneration [10,137–139]. Neurobiologists and neurologists are still rather optimistic about using NSCs as a substrate for brain tissue-replacement therapy or stimulating endogenous sources of adult-born neurons for brain tissue repair and facilitation of cognitive functions. As an example, recent experimental data demonstrate that the stimulation of even a small pool of NSCs "rejuvenates" the brain and reduces some age-associated manifestations of cognitive deficits [140], but at the same time, stimulation of neurogenesis may alter forgetting [15].

In PD, impairment of neurogenesis suggests that the effective and long-lasting treatment for PD motor symptoms might be replacing SN dopaminergic cells by means of improved endogenous neurogenesis or by cell-replacement therapy/cell transplantation [40,119]. At present, optogenetic photostimulation has been mainly tested for the modulation of SN neurons. Particularly, light-induced activation of ChR2 dopaminergic neurons in the genetic model Drosophila larva rescues PD symptoms caused by α-synuclein [141], light-dependent activation of mitochondrially expressed proton pump dR reinforces mitochondrial function and prevents α-synuclein-driven mitochondrial dysfunction in a Drosophila model of PD [142]. It should be mentioned that some additional optogenetics-based options became available: light-inducible protein aggregation system that allows photoinduced aggregation of α-synuclein in vitro and in vivo to model the Parkinson's type neurodegeneration and to find out the effects of abnormal protein accumulation in the brain [143]. Thus, even though there are some positive results of optogenetic manipulations with dopaminergic neurons in experimental PD, there are no examples of a similar approach applied for NSCs/NPCs in PD-neurogenic niches.

4. Optogenetic Activation of Niche Astrocytes for the Control of Cell Development in PD

4.1. Astrocytes as Potent Regulators of Neurogenesis and Parkinson's Type Neurodegeneration

The analysis of neurogenesis impairments in neurodegeneration leads to some critical questions: how is it possible to manage the fate of transplanted NSCs/NPCs in the SN since the local microenvironment there supports glial, but not neuronal phenotype acquisition? If so, would the transplantation of mature well-differentiated neurons be the only solution, or could another strategy for precise control of cell proliferation and differentiation within the SN be applied? Taking into consideration the above-mentioned issues, one could propose that for the efficient therapy of PD, two major approaches should be evaluated: (i) stimulation of endogenous neurogenesis in vivo by targeting SN NSCs/NPCs along with the prevention of their development toward the astroglial phenotype; or (ii) obtaining the pool of dopaminergic functionally competent neurons from iPSCs in vitro and their transplantation in SN. Recently, another approach based on the reprogramming of midbrain astrocytes into dopaminergic neurons has been suggested for the treatment of PD [144].

In all cases, astroglial cells could be considered key regulators of neurogenesis and maturation. Glial fibrillary acidic protein (GFAP)-immunopositive radial glial cells (RGCs) located within neurogenic niches are the NSCs that give rise to multipotent and dividing progenitors. In addition, RGS controls cell migration to ensure reparative neurogenesis. The activation of neurogenesis is always associated with an accumulation of astrocytes in neurogenic niches, and the establishment of a niche astroglial network is required for the local microenvironment supporting the proliferation of cell clusters in neurogenic niches [145]. The close contact of niche astrocytes with NSCs/NPCs and brain microvessel endothelial cells (BMECs) of the niche vascular scaffold, as well as secretory activity of astrocytes, control the promotion of neuronal differentiation of stem cells in the SGZ [145,146]. Some data suggest that astrocytes negatively affect neurogenesis and inhibit neuronal differentiation through direct cell-to-cell contacts with NSCs and the modulation of Notch/Jagged1 signaling pathways in an intermediate filament protein GFAP-dependent manner [147].

In PD, midbrain astrocytes play a dual role in the disease progression and tissue repair: (i) astroglial production of reactive oxygen species (ROS) and reactive nitrogen species (RNS) in activated astrocytes partially caused by aberrant expression of α-synuclein (SNCA), parkin (PARK2), protein deglycase DJ-1 (PARK7), and PINK1 genes lead to the damage of dopaminergic neurons and progression of neuroinflammation [148]; (ii) astroglial production of gliotransmitters and neurotrophic factors is important for governing cell proliferation, differentiation, and tissue repair in chronic neurodegeneration [149,150]. Particularly, astrocytes-derived Wnt contributes to dopaminergic neurons survival [151], stimulation of neurogenesis from SN stem cells [152], and tissue regeneration in PD [153] through canonical (Wnt/β-catenin) and non-canonical (Wnt/planar cell polarity and Wnt/Ca^{2+}) pathways that are involved in the differentiation of dopaminergic neurons [151].

The activity of astrocyte-derived Wnt is required for NSCs proliferation and differentiation in the SGZ [154], whereas Notch signaling regulates the maintenance of adult NSCs governing them out of cell cycle exit, thereby decreasing the pool of NPCs [155]. Thus, Notch signaling prevents excessive recruiting of NSCs, while Wnt signaling supports the proliferation and differentiation of NPCs and neuroblasts. Recent data reveal novel aspects of Notch and Wnt signaling in NSCs development: when iPSCs cortical spheroids are treated with Wnt and Notch modulators, they demonstrate a synergistic effect on neural regional patterning and occurrence of neurogenesis and gliogenesis (increase in Notch and Wnt activity results in the development of a larger number of glial cells), thus, repressing impact of Notch inhibitor on Wnt inhibition and the positive impact of Wnt activation on Notch signaling are proposed [156]. It should be noted that Wnt signaling is a well-known target for the activity of proteins involved in the pathogenesis of PD: in the healthy brain, LRKK2 (product of *PARK8* gene) serves as a scaffold protein and positive regulator of the canonical pathways, whereas parkin (product of PARK2 gene) induces β-catenin degradation and suppression of the canonical pathway [157]. Thus, in physiological conditions, parkin protects dopaminergic neurons from excessive activation of the Wnt/β-catenin pathway [158], but in PD, this mechanism is lost due to parkin alterations. Thus, the data on enhanced neurogenesis due to Wnt/β-catenin overactivation associated with impaired differentiation of dopaminergic neurons [159] are rather reasonable. The stimulatory effect of LRRK2 on the non-canonical Wnt/planar cell polarity pathway was reported as well [160].

Other than Wnt/β-catenin signaling, FGF8 plays a great role in the regulation of differentiation toward dopaminergic phenotype: FGF receptors (FGFRs) regulate the self-renewal and dopaminergic differentiation of NPCs in the developing midbrain [161–163]. Dysfunction of the FGF-driven mechanisms of midbrain development and control of midbrain neurons survival and metabolism is implicated in the pathogenesis of PD [163]. It was reported that the dopaminergic differentiation of embryonic stem cells in vitro could be facilitated by astrocytes providing FGF. Moreover, the optogenetic activation of astrocytes transplanted in SN in vivo results in elevated FGF release and promotion of appropriate differentiation of co-transplanted stem cells [161]. Novel optogenetics tools, such as optoFGFR based on the cryptochrome2 domain and cytoplasmic region of FGFR, enable light-guided activation and clustering of FGFRs for efficient analyzing of the downstream molecular events [164,165]. We suggest that a similar approach could be tested to modulate the FGF-driven regulation of stem cell development and dopaminergic differentiation in PD.

4.2. Optogenetic Targeting of GFAP⁺ Cells in the Neurogenic Niche: Established and Prospective Approaches to Cells Activation and Signal Propagation

The essence of astroglial activation is the elevation of intracellular Ca^{2+} levels due to Ca^{2+} influx through membrane channels, i.e., L-type voltage-operated calcium channels, VOCC [166], connexin 43 (Cx43) hemichannels [167], and transient receptor potential channels, TRP [168], or Ca^{2+} release from intracellular stores (endoplasmic reticulum, mitochondria, nucleus) via activation of inositol-3-phosphate receptors of cyclic ADP-ribose-sensitive ryanodine receptors [169,170]. The activation of Ca^{2+} release mechanisms is a result of stimulation of astroglial Gq, Gi/o, or Gs G-protein-coupled receptors (GPCRs) culminating in the synthesis of second messengers with Ca^{2+}-mobilizing activity, whereas the opening of VOCC is triggered by high extracellular concentrations of glutamate, K^+, and ATP, i.e., in active brain regions or in inflammatory loci [166,171]. Thus, "artificial" induction of Ca^{2+} rise in astroglial cells might mimic the activation achieved by ligands of GPCRs, K^+, ATP, or cytokines. As a result of activation, extracellular K+ concentrations transiently rise, astrocytes release gliotransmitters and change their mitochondrial activity and proliferative status [172–174]. Actually, this is a principle of optogenetic photostimulation of astroglial cells expressing ChR2 or optoGPCRs under the astroglial promoters (i.e., GFAP), which recently appeared as a new approach to control brain activity [175–178].

Midbrain astrocytes in PD with SNCA mutations demonstrate aberrant Ca^{2+} release from intracellular stores into cytosol, presumably, caused by mitochondrial dysfunction [90]. Thus, one may propose that optogenetic stimulation of PD-specific astrocytes with the mutant form of SNCA would result in an abnormal pattern of their activation.

It should be kept in mind that recent complex proteomic and transcriptomic analyses revealed interesting differences in the expression pattern of astrocytes in various brain regions. Particularly, hippocampal astrocytes and striatal astrocytes predominantly express GFAP or μ-crystalline, respectively, and they are different in gap-junctional coupling (lower in striatal astroglia) and GPCR-mediated Ca^{2+} signals (weaker response in hippocampal astrocytes) [179]. Thus, any, including optogenetic, manipulations with hippocampal and striatal astroglial cells would have a priori different efficacy and results: expression of light-sensitive molecules under the GFAP promoter would be higher in the hippocampus, but Ca^{2+}-driven activation of glial cells would be more evident in the striatum.

The expression of light-sensitive molecules under the control of the astroglial promoter (GFAP) raises the question of what type of cells within the neurogenic niche could be affected by photostimulation. Even though there is a heterogeneity of astroglial cells, the expression of GFAP could be easily detected in the majority of mature resting and reactive astrocytes throughout the brain [180]. Higher expression of GFAP in reactive astrocytes, particularly in neurodegeneration [181], makes it possible to increase the efficacy of photostimulation in the affected brain vs. the healthy brain. Indeed, optogenetic protocols targeting astrocytes provide precise manipulation with their functional status, secretory phenotype, and interactions with mature neurons in physiological conditions and in neurodegeneration [175,177,182,183].

Less is known about the application of optogenetics for controlling astroglia-driven regulation of adult neurogenesis. We have demonstrated before that optogenetic stimulation of niche astrocytes expressing channelrhodopsin-2 under the GFAP promoter was efficient in activating the neurogenic potential of NSCs/NPCs in the in vitro neurogenic niche model or in implanted intrahippocampal neurospheres ex vivo in experimental Alzheimer's disease [184,185]. Photostimulation of iPSC-derived ChR2-expressing astrocytes co-cultured with iPSC-derived neurons results in effective bidirectional signaling, which is important for supporting the maturation of neurons and the establishment of a functional synaptic network, even though the transcriptomic analysis confirms that iPSCs-originated astrocytes are relatively immature compared to adult cortical astrocytes [186].

Quiescent NSCs, as RGs, demonstrates the expression pattern as $GFAP^+Nestin^+PCNA^-$ $Pax6^+NeuroD1^-$. Type-1 NSCs, as slowly dividing, cells have the phenotype $GFAP^{+/-}$ $Nestin^+PCNA^+Pax6^+NeuroD1^-$ and express lower GFAP. NPCs, as amplifying progenitors, with the phenotype $GFAP^-Nestin^+PCNA^+Pax6^+NeuroD1^+$ do not express GFAP during neurogenesis [187]. Thus, the expression of light-activated molecules in NSCs under the GFAP promoter could regulate their activity. However, it might be impossible to use the same strategy to express a construct in post-mitotic astrocytes and NSCs: adeno-associated viruses (AAV) used as vectors are inefficient in transducing stem cells; therefore, engineered AAV variants or other delivery tools (i.e., polymer complexes containing plasmids, episomes, or retrovirus- and lentivirus-based vectors) should be applied [188–191].

The most attractive feature of optogenetic protocols is an opportunity to stimulate the particular cell precisely and in a controllable manner. While considering astroglial optogenetic stimulation, one should remember the existence of the so-called astroglial syncytium due to the activity of intercellular gap junctions [192]. It is well-known that Ca^{2+} waves in astrocytes propagate via gap junctions consisted of connexin 43 (Cx43) channels [193], thereby resulting in the activation of astroglia located distantly from the focus of primary activation [194] or via extracellular ATP-dependent mechanisms [193]. However, whether or not this mechanism is relevant in optogenetically-stimulated astrocytes remains to be evaluated. The photostimulation of astrocytes expressing light-gated glutamate receptor 6 (LiGluR) in vitro results in the activation of adjacent non-expressing LiGluR astrocytes; this effect was insensitive to the blockers of gap junctions but sensitive to inhibitors of ATP-driven purinergic signaling [195]. Taking into consideration that reactive astrocytes have permissive conditions for Ca^{2+}-dependent ATP release [196], one could suggest that in a neurodegeneration-affected brain, the propagation of light-induced signals from the particular astrocytes would be facilitated.

In a rat rotenone-induced model of PD, increased expression of Cx43 was detected in SN, striatum, and basal ganglia astrocytes [197], thus suggesting that metabolic and functional coupling of astroglial cells might be enhanced in Parkinson's type neurodegeneration. The same phenomenon is evident in brain ischemia [198] and Alzheimer's disease [199] and might reflect the neuroprotective potential of reactive astrocytes, or could be a consequence of Cx43 functional coupling with another protein, $CD38/NAD^+$-glycohydrolase [200], whose expression is elevated in neuroinflammation and neurodegeneration [199]. This suggestion turns us to the idea of NAD^+-dependent mechanisms in the molecular pathogenesis of PD [201]. Since Cx43 may act as an NAD^+-transporting molecule in the plasma membrane [202], higher expression of Cx43 might be beneficial for cell survival. Probably, this is a reason why NSCs/NPCs express up-regulated functional Cx43 in brain injury [203]. It is interesting that when NSCs are engrafted into the striatum, they express much higher levels of Cx43, and host cells do the same within the limited period of time [204]; thus, it could be utilized for the improvement of transplantation outcomes. Indeed, the positive effect of gap junction-mediated communication between host cells and NSCs was demonstrated in organotypic slice cultures [205]. Along with this idea, the optogenetic control of cell engraftment might be rather useful: photostimulation of transplanted neurons helps in the assessment of their functional integration into neuronal circuits and communication with other types of cells in the host brain tissue [206]. However, such an approach has never been tested for astroglial cells.

The elevated expression of Cx43 might not correspond to facilitated intercellular communication only: membrane Cx43 hemichannels serve as efflux-oriented transporters for NAD^+, lactate, or Ca^{2+} in astrocytes; therefore, they provide release of low molecular weight substances and ions into extracellular space with no apparent effect on direct cell-to-cell communication [207]. The activity of this machinery is abnormal in PD: α-synuclein induces the opening of Cx43 hemichannels, excessive Ca2+ rise in the cytosol, gliotransmitters, and cytokines release [208]. In sum, astroglial cells in PD might respond to photostimulation-induced Ca^{2+} intracellular elevations to a greater extent than normal cells due to higher expression of gap junction proteins of Cx43 hemichannels. What might be an outcome for such effects within the neurogenic niche or in the affected brain regions remains to be evaluated. However, experimental data on elevated expression and activity of Cx43 hemichannels driving better communication of host cells and transplanted NSCs [209] allow considering Cx43 hemichannels as a target for light-guided control of engrafting efficacy. Since the normalization of neuron-astroglial gap junction-mediated crosstalk by optogenetic manipulations with astrocytes was proposed in [210], a similar approach should be tested for niche astrocytes communicating with stem cell grafts.

We have proposed before [211] that Cx43 expression in different cells of the neurogenic niche (radial glia, mature astrocytes, endothelial cells) could be utilized to control neurogenesis. Indeed, the expression of Cx43 is indispensable for RGs proliferation in the adult hippocampus [212]. Neuroectodermal specialization of embryonic stem cells depends on the rate of Cx43 expression [213]. Deletion of Cx43 suppresses hippocampal adult neurogenesis due to the inhibition of NSCs proliferation and survival [212], whereas the absence of Cx43 expression in NPCs results in their predominant differentiation toward a neuronal, but not astroglial, phenotype, probably, due to increased β-catenin signaling and Wnt-driven expression of pro-neuronal genes [214]. Indeed, Cx43 in NPCs down-regulates β-catenin signaling, reduces the proliferation of progenitors, and promotes astroglial differentiation [215].

It should be noted that in resting astrocytes, Cx43 localizes in intracellular vesicles, but the activation of cells drives Cx43 expression at the plasma membrane contact sites [216]. Cx43 interacts with β-catenin directly, and the activation of Wnt signaling results in the re-dislocation of Cx43 in some cell lines leading to enhanced expression of Cx43 in the nucleus, but not at the plasma membrane or cytosol [217]. The same phenomenon has not been reproduced in NSCs/NPCs yet, but it might be tempting to speculate that the positive effects of Wnt on dopaminergic neuron generation and survival are disrupted in PD due to the abnormal activity of parkin and the overactivity of the Wnt/β-catenin signaling pathway could be modulated via Cx43-β-catenin interactions at the plasma membrane of NSCs/NPCs.

Since neuronal activity increases Cx43 expression in astrocytes [218], and excitatory (NMDA, or depolarizing action of GABA) stimuli directly promote differentiation of NPCs toward neuronal phenotypes [6,219], it is tempting to speculate that the establishment of an in vitro neurogenic niche model with mature neurons or with conditions mimicking excitation/inhibition balance specific for neurogenic niches, would give us new opportunities in increasing the efficacy of astroglial (photo)activation for the local control of neurogenesis.

Application of up-to-date protocols for optical mapping of gap junctions, for instance, PARIS, "pairing actuators and receivers to optically isolate gap junctions" [220] or optoGap, an optogenetics-based tool for the analysis of cell-to-cell connexin-driven coupling [221], would be helpful in further elucidating the Cx43 activity in NSCs/NPCs and niche astrocytes.

5. Alternative Approaches to Restoring Impaired Neurogenesis in PD

In the context of PD pathogenesis, motor dysfunction appears as a result of excessive GABAergic output in the striatum: normally, dopaminergic neurons of SN terminate at the striatum and release dopamine there to propagate signals to cholinergic and GABAergic neurons, resulting in the inhibition of the output from GABAergic neurons [222,223]. In the healthy striatum, the majority of cells are the GABAergic interneurons, whereas the role of striatal dopaminergic cells is not clear [224].

New dopaminergic neurons adjacent to the band of preserved nigral input and expressing tyrosine hydroxylase and dopamine transporter have been found in the striatum of PD patients [54,225]. Optogenetic activation of striatal tyrosine hydroxylase-expressing interneurons in mice in vivo produce strong GABAergic inhibition, but no evidence for dopamine production has been obtained [226]. Loss of nigrostriatal innervation results in morphological and functional changes in this population of cells aimed to compensate the GABAergic inhibition [227]. Probably, endogenous dopamine negatively controls the number of these cells [228]. Currently, optogenetic activation of striatal neurons was found to be an efficient tool for studying striatum-dependent neurological processes, i.e., reward behavioral encoding, reinforcement learning, and motivation, as it was reviewed in detail elsewhere [229].

Insulin receptors are expressed on midbrain dopamine neurons, so their stimulation controls dopaminergic transmission in the striatum [230]. Particularly, insulin may enhance dopamine release in the striatum through cholinergic interneurons [231]. Data obtained in Drosophila reveal that the induction of NSCs from glia, their proliferation and limited neurogenesis are regulated by insulin signaling [232]. Neurogenesis in conventional neurogenic niches (SVZ and SGZ) also depends on insulin and insulin-like growth factors (IGF) [233]; thus, cerebral insulin resistance evident in chronic neurodegeneration (Alzheimer's disease, Parkinson's disease) negatively affects neurogenesis [234], whereas peptides facilitating insulin effects promote the development of new dopaminergic neurons in SN in a model of PD [235]. Thus, the modulation of insulin signaling in SN and striatum might be important for restoring neurogenesis in these non-conventional neurogenic niches. In this context, the photodynamic reversible opening of the blood-brain barrier (BBB) [236] might be useful for driving insulin or insulin-like growth factors transport into the particular brain region, as it was shown for IGF-I in the active brain [237].

Neuroblasts differentiating into mature GABAergic interneurons have been found in the striatum close to the SVZ in the adult brain in humans, and local neurogenic events here are diminished in Huntington's disease [238]. Since dopaminergic activity is required for stimulating the striatal neurogenesis [65], the loss of SN neurons in PD would result in the suppression of striatal neurogenesis. However, alternative hypotheses on the origin of adult-born striatal interneurons have been proposed, including differentiation of local NPCs [239,240] or conversion of striatal astrocytes into mature neurons by blocking Notch signaling [241].

The latter approach has attracted a lot of attention in recent years because the direct reprogramming of adult post-mitotic cells might be quite useful in the replacement of lost neurons with new cells in brain regions (i.e., SN and striatum) with very limited neurogenic capacity in adults [242]. However, the conversion of SN GFAP$^+$ cells into neurons was found, thereby providing an alternative neurogenic mechanism within SN and striatum:

GFAP$^+$/s100β^+ astrocytes could transdifferentiate into dividing cells (neuroblasts) or even dedifferentiate back to NSCs [64].

In the adult mouse brain, transcription factor SOX2 can induce the transformation of astrocytes into neuroblasts that can be further driven to mature neurons with BDNF or valproic acid as a histone deacetylase inhibitor [243]. The induction of expression of Ascl1 in cortex astrocytes results in the formation of GABAergic neurons, while Neurog2 expression is responsible for the glutamatergic phenotype, but NeuroD4 is capable of reprogramming astrocytes into neurons that cannot complete synaptic maturation [244]. The combination of several transcription factors, NEUROD1, ASCL1, and LMX1A, and the microRNA miR218 is helpful in the in vivo and in vitro reprogramming of striatal astrocytes of mouse and human origin into functional dopaminergic neurons [245]. Moreover, when small molecules that promote chromatin remodeling and activate the TGFβ, Shh, and Wnt signaling pathways, such as ascorbic acid, valproic acid, or 5-aza-2′-deoxycytidine, TGF and BMP4 inhibitors SB431542 and LDN193189, sonic hedgehog (SHH) and the GSK3β inhibitor CT99021, dual-Smad inhibitors SB431542 and LDN193189, and midbrain patterning signals CT99021 and purmorphamine, have been applied, the number of tyrosine hydroxylase-expressing neurons was increased [245]. It is important to note that the in vivo reprogramming of striatal astrocytes into dopaminergic neurons results in the improvement of behavioral characteristics of 6-OHDA-treated mice with PD [245]. The reconstruction of nigrostriatal circuits, replenishment of dopaminergic neurons and reduction of neurological deficits were achieved in mice with a 6-OHDA model of PD by the reprogramming of astrocytes to functional neurons via depletion of RNA-binding protein PTB (PTBP1), and the functional characteristics of newly-developed neurons were confirmed with the chemogenetic protocols [144]. Astrocytes of different origins might demonstrate various abilities to be reprogrammed into neurons: adult human astrocytes could be reprogrammed to neuroblasts with miRNAs (miR-302/367), but mouse astrocytes required valproic acid for successful conversion [246].

When mouse embryonic bodies (EBs) were transplanted into SN of rats and mice, stimulation of neurogenesis was observed, but the establishment of fully differentiated dopaminergic neurons failed; however, previously non-dividing resident GFAP+/S100b+ cells acquired neuroblast markers after EBs transplantation [64]. Similarly, in the 6-OHDA rat model of PD, chronic (10 days) infusion of platelet-derived growth factor (PDGF-BB) and brain-derived neurotrophic factor (BDNF) results in the generation of newly-formed cells in the striatum and SN, but these cells do not demonstrate the expression pattern of striatal mature projection neurons or dopaminergic neurons in SN [247]. The combination of 6-OHDA-lesion of SN dopaminergic neurons and infusions of transforming growth factor α (TGFα) into forebrain structures results in a massive migration of neural progenitors from the SVZ to the striatum, their differentiation to dopaminergic neurons, and the improvement of rotational behavior in rats [248]. However, data obtained in humans and in rodents with PD models seem to be controversial: the striatum of PD patients was found to contain six times fewer tyrosine hydroxylase-expressing cells [249]. Moreover, as it was resumed in [250], there is no confirmation that the enhancement of striatal neurogenesis would result in the improvement of behavioral effects in PD in a similar way, as it was shown in the striatal transplantation of dopaminergic neurons, but the stimulation of striatal neurogenesis and reinnervation of local interneurons is a promising strategy in PD.

Attempts to produce dopaminergic neurons from SVZ stem cells have shown that adult NSC-derived cells co-express Nestin and tyrosine hydroxylase and demonstrated a low survival rate, but embryonic stem cell-derived neurons have characteristics of mature cells with strong dopamine release upon the action of depolarizing stimuli [251]. Thus, the use of stem cells close to embryonic parameters (iPSCs) should have obvious advantages in cell-replacement therapy.

Undoubtedly, all the attempts aimed to reduce α-synuclein-induced alterations (including prevention of its aggregation and dissemination or enhancing degradation of α-synuclein aggregates) would be efficient in restoring the neurogenesis in PD-affected brains, thereby contributing to functional recovery [29].

6. Conclusion and Perspectives

The current attempts to establish reliable and safe therapeutic platforms for the restoration of impaired brain plasticity in neurodegeneration are facing the complexity of adult neurogenesis. Recent achievements in understanding the key molecular mechanisms of NSCs/NPCs maintenance and development, the role of intercellular communications in the adjustment of neurogenesis to the actual needs of the active brain, and application of up-to-date tools for getting the desired cellular phenotypes (e.g., in iPSC-based protocols) and precise activation of target cells (e.g., in optogenetic protocols) suggest new opportunities in the cell-replacement therapy, either via the stimulation of endogenous neurogenesis or the generation of cells for efficient engrafting.

In the case of synucleinopathies, this approach should be based on the molecular mechanisms of impaired brain plasticity caused by abnormal accumulation and distribution of α-synuclein in various brain regions, including conventional and non-conventional neurogenic niches. There are no doubts that aberrant neurogenesis is ultimately involved in the pathogenesis of PD from the very early, even pre-motor and pre-manifesting, stages. Correct analysis of neurogenesis impairments, as well as the development of novel approaches to manipulate the neurogenic capacity of NSCs/NPCs, would give progress in the early diagnostics, effective prevention, and treatment of PD.

In this context, various niche cellular components (NSCs, NPCs, astrocytes, BMECs, and mature neurons) serve as promising targets for the optogenetic control of the local microenvironment. Modulating the functional activity of niche cells might be helpful in the control of cell proliferation, reprogramming, and differentiation either in vitro or in vivo. The same approach is rather prospective for improving the outcomes of cells transplantation and their functional integration in the affected brain regions.

Thus, the application of novel optogenetic/chemogenetic tools and advanced in vitro models, including those based on iPSC-derived cells, organoids, or utilizing 3D brain-on-chip platforms, are of great importance for the development of new therapeutic options and assessment of aberrant neurogenesis in Parkinson's type neurodegeneration.

Author Contributions: A.B.S. and S.N.I., writing: initial hypothesis, writing the manuscript; M.R.K. and A.S.V., writing: drawing the figures, editing; S.N.I., writing: editing and supervision. All authors have read and agree to the published version of the manuscript.

Funding: This research was supported by the grant of the Russian Science Foundation, project № 19-15-00320 (MRK, ASV, SNI).

Abbreviations

AV	adeno-associated virus
ALDH1L1	aldehyde dehydrogenase 1 family member L1
AMP	adenosine monophosphate
ADP	adenosine diphosphate
AQ4	aquaporin 4
ASCL1	Achaete-Scute homolog 1
ATP	adenosine triphosphate
BBB	blood-brain barrier
BDNF	brain-derived neurotrophic factor
BMECs	brain microvessel endothelial cells
BMP4	bone morphogenetic protein 4
CaMKII	calcium/calmodulin-dependent kinase II

CD	cluster of differentiation
ChR2	channelrhodopsin 2
CNTF	ciliary neurotrophic factor
CRISPR-Cas9	clustered regularly interspaced short palindromic repeats-associated protein 9
Cx	connexin
3D	three-dimensional
DAMPs	damage-associated molecular patterns
DJ-1	deglycase-1
EB	embryonic body
EGF	epidermal growth factor
E/I	excitation/inhibition balance
FBS	fetal bovine serum
FCS	fetal calf serum
FGF	fibroblast growth factor
Foxa2	forkhead box protein A2
GABA	gamma-aminobutyric acid
GDNF	glia cell line-derived neurotrophic factor
GFAP	glial fibrillary acidic protein
GPCRs	G-protein-coupled receptors
GSK3	glycogen synthase kinase 3
HDAC	histone deacetylase
IGF	insulin-like growth factor
IL	interleukin
iPSCs	induced pluripotent stem cells
LIF	leukemia inhibitory factor
LiGluR	light-gated glutamate receptor 6
Lmx1a	LIM homeobox transcription factor 1 alpha
LRKK2	leucine-rich repeat kinase 2
MPTP	1-methyl-4-phenyl-1,2,3,6-tetrahydropyridine
mRNA	messenger ribonucleic acid
NAD^+	nicotinamide adenine dinucleotide
NeuroD1	neurogenic differentiation 1 transcription factor
Neurog2	neurogenin 2
NFIA	nuclear factor I A
NFIB	nuclear factor I B
NMDA	N-methyl-D-aspartate
NPCs	neural progenitor cells
NSCs	neural stem cells
Nurr1	nuclear receptor related 1
6-OHDA	6-hydroxydopamine
optoFGFR	light-activatable fibroblast growth factor receptor
optoGap	light-activatable gap junction
optoGPCR	light-activatable G-protein coupled receptor
PARK2	parkin ubiquitin protein ligase
Pax6	transcription factor paired box protein
PCNA	proliferating cell nuclear antigen
PD	Parkinson's disease
PDGF-BB	platelet-derived growth factor
PINK-1	PTEN (phosphatase and tensin homolog deleted)-induced kinase 1
PRKN	parkin
RGCs	radial glia cells
PTBP1	RNA-binding protein PTB
RNS	reactive nitrogen species
ROS	reactive oxygen species
S100β	S100 calcium-binding protein B
SGZ	subgranular zone

SHH	sonic hedgehog
smNPCs	small molecule-treated neural progenitor cells
SN	substantia nigra
SNCA	synuclein
Sox6	SRY-Box transcription factor 6
Sox9	SRY-Box transcription factor 9
SVZ	subventricular zone
SYN1	synapsin 1
TGFβ	transforming growth factor-beta
TRP	transient receptor potential channel
VOCC	voltage-operated calcium channel
Wnt	Wingless/Int-1

References

1. Saxe, M.D.; Battaglia, F.; Wang, J.-W.; Malleret, G.; David, D.J.; Monckton, J.E.; Garcia, A.D.R.; Sofroniew, M.V.; Kandel, E.R.; Santarelli, L.; et al. Ablation of hippocampal neurogenesis impairs contextual fear conditioning and synaptic plasticity in the dentate gyrus. *Proc. Natl. Acad. Sci. USA* **2006**, *103*, 17501–17506. [CrossRef]
2. Gonçalves, J.T.; Schafer, S.T.; Gage, F.H. Adult Neurogenesis in the Hippocampus: From Stem Cells to Behavior. *Cell* **2016**, *167*, 897–914. [CrossRef]
3. Toda, T.; Parylak, S.L.; Linker, S.B.; Gage, F.H. The role of adult hippocampal neurogenesis in brain health and disease. *Mol. Psychiatry* **2019**, *24*, 67–87. [CrossRef] [PubMed]
4. Pereira-Caixeta, A.R.; Guarnieri, L.O.; Medeiros, D.C.; Mendes, E.; Ladeira, L.C.D.; Pereira, M.T.; Moraes, M.F.D.; Pereira, G.S. Inhibiting constitutive neurogenesis compromises long-term social recognition memory. *Neurobiol. Learn. Mem.* **2018**, *155*, 92–103. [CrossRef]
5. Petsophonsakul, P.; Richetin, K.; Andraini, T.; Roybon, L.; Rampon, C. Memory formation orchestrates the wiring of adult-born hippocampal neurons into brain circuits. *Brain Struct. Funct.* **2017**, *222*, 2585–2601. [CrossRef] [PubMed]
6. Lopatina, O.L.; Malinovskaya, N.A.; Komleva, Y.K.; Gorina, Y.V.; Shuvaev, A.N.; Olovyannikova, R.Y.; Belozor, O.S.; Belova, O.A.; Higashida, H.; Salmina, A.B. Excitation/inhibition imbalance and impaired neurogenesis in neurodevelopmental and neurodegenerative disorders. *Rev. Neurosci.* **2019**, *30*, 807–820. [CrossRef] [PubMed]
7. Pozhilenkova, E.A.; Lopatina, O.L.; Komleva, Y.K.; Salmin, V.V.; Salmina, A.B. Blood-brain barrier-supported neurogenesis in healthy and diseased brain. *Rev. Neurosci.* **2017**, *28*, 397–415. [CrossRef]
8. Wu, Y.; Chen, X.; Xi, G.; Zhou, X.; Pan, S.; Ying, Q.-L. Long-term self-renewal of naïve neural stem cells in a defined condition. *Biochim. Biophys. Acta (BBA)-Bioenerg.* **2019**, *1866*, 971–977. [CrossRef]
9. Alexanian, A.R.; Kurpad, S.N. Quiescent neural cells regain multipotent stem cell characteristics influenced by adult neural stem cells in co-culture. *Exp. Neurol.* **2005**, *191*, 193–197. [CrossRef]
10. Semënov, M.V. Adult Hippocampal Neurogenesis Is a Developmental Process Involved in Cognitive Development. *Front. Neurosci.* **2019**, *13*, 159. [CrossRef]
11. Santilli, G.; Lamorte, G.; Carlessi, L.; Ferrari, D.; Rota Nodari, L.; Binda, E.; Delia, D.; Vescovi, A.L.; De Filippis, L. Mild Hypoxia Enhances Proliferation and Multipotency of Human Neural Stem Cells. *PLoS ONE* **2010**, *5*, e8575. [CrossRef]
12. Cipriani, S.; Ferrer, I.; Aronica, E.; Kovacs, G.G.; Verney, C.; Nardelli, J.; Khung, S.; Delezoide, A.-L.; Milenkovic, I.; Rasika, S.; et al. Hippocampal Radial Glial Subtypes and Their Neurogenic Potential in Human Fetuses and Healthy and Alzheimer's Disease Adults. *Cereb. Cortex* **2018**, *28*, 2458–2478. [CrossRef]
13. Anacker, C.; Hen, R. Adult hippocampal neurogenesis and cognitive flexibility—Linking memory and mood. *Nat. Rev. Neurosci.* **2017**, *18*, 335–346. [CrossRef] [PubMed]
14. Kitamura, T.; Saitoh, Y.; Takashima, N.; Murayama, A.; Niibori, Y.; Ageta, H.; Sekiguchi, M.; Sugiyama, H.; Inokuchi, K. Adult Neurogenesis Modulates the Hippocampus-Dependent Period of Associative Fear Memory. *Cell* **2009**, *139*, 814–827. [CrossRef] [PubMed]
15. Gao, A.; Xia, F.; Guskjolen, A.J.; Ramsaran, A.I.; Santoro, A.; Josselyn, S.A.; Frankland, P.W. Elevation of Hippocampal Neurogenesis Induces a Temporally Graded Pattern of Forgetting of Contextual Fear Memories. *J. Neurosci.* **2018**, *38*, 3190–3198. [CrossRef] [PubMed]
16. Nakashiba, T.; Cushman, J.D.; Pelkey, K.A.; Renaudineau, S.; Buhl, D.L.; McHugh, T.J.; Rodriguez Barrera, V.; Chittajallu, R.; Iwamoto, K.S.; McBain, C.J.; et al. Young dentate granule cells mediate pattern separation, whereas old granule cells facilitate pattern completion. *Cell* **2012**, *149*, 188–201. [CrossRef] [PubMed]
17. Kim, K.R.; Kim, Y.; Jeong, H.J.; Kang, J.S.; Lee, S.H.; Ho, W.K. Impaired pattern separation in Tg2576 mice is associated with hyperexcitable dentate gyrus caused by Kv4.1 downregulation. *Mol. Brain* **2021**, *14*, 62. [CrossRef]
18. Shulman, J.M.; Jager, P.L.D.; Feany, M.B. Parkinson's Disease: Genetics and Pathogenesis. *Annu. Rev. Pathol. Mech. Dis.* **2011**, *6*, 193–222. [CrossRef]
19. Novosadova, E.V.; Nenasheva, V.V.; Makarova, I.V.; Dolotov, O.V.; Inozemtseva, L.S.; Arsenyeva, E.L.; Chernyshenko, S.V.; Sultanov, R.I.; Illarioshkin, S.N.; Grivennikov, I.A.; et al. Parkinson's Disease-Associated Changes in the Expression of Neurotrophic Factors and their Receptors upon Neuronal Differentiation of Human Induced Pluripotent Stem Cells. *J. Mol. Neurosci.* **2020**, *70*, 514–521. [CrossRef]

20. Trifonova, O.P.; Maslov, D.L.; Balashova, E.E.; Urazgildeeva, G.R.; Abaimov, D.A.; Fedotova, E.Y.; Poleschuk, V.V.; Illarioshkin, S.N.; Lokhov, P.G. Parkinson's Disease: Available Clinical and Promising Omics Tests for Diagnostics, Disease Risk Assessment, and Pharmacotherapy Personalization. *Diagnostics* **2020**, *10*, 339. [CrossRef]

21. Winner, B.; Lie, D.C.; Rockenstein, E.; Aigner, R.; Aigner, L.; Masliah, E.; Kuhn, H.G.; Winkler, J. Human Wild-Type α-Synuclein Impairs Neurogenesis. *J. Neuropathol. Exp. Neurol.* **2004**, *63*, 1155–1166. [CrossRef] [PubMed]

22. Uversky, V.N. What does it mean to be natively unfolded? *Eur. J. Biochem.* **2002**, *269*, 2–12. [CrossRef]

23. Burré, J. The Synaptic Function of α-Synuclein. *J. Park. Dis.* **2015**, *5*, 699–713. [CrossRef]

24. Fortin, D.L.; Nemani, V.M.; Voglmaier, S.M.; Anthony, M.D.; Ryan, T.A.; Edwards, R.H. Neural activity controls the synaptic accumulation of alpha-synuclein. *J. Neurosci.* **2005**, *25*, 10913–10921. [CrossRef]

25. Guhathakurta, S.; Bok, E.; Evangelista, B.A.; Kim, Y.-S. Deregulation of α-synuclein in Parkinson's disease: Insight from epigenetic structure and transcriptional regulation of SNCA. *Prog. Neurobiol.* **2017**, *154*, 21–36. [CrossRef]

26. Hope, A.D.; Myhre, R.; Kachergus, J.; Lincoln, S.; Bisceglio, G.; Hulihan, M.; Farrer, M.J. α-Synuclein missense and multiplication mutations in autosomal dominant Parkinson's disease. *Neurosci. Lett.* **2004**, *367*, 97–100. [CrossRef] [PubMed]

27. Ahn, T.B.; Kim, S.Y.; Kim, J.Y.; Park, S.S.; Lee, D.S.; Min, H.J.; Kim, Y.K.; Kim, S.E.; Kim, J.M.; Kim, H.J.; et al. alpha-Synuclein gene duplication is present in sporadic Parkinson disease. *Neurology* **2008**, *70*, 43–49. [CrossRef]

28. Baena-Montes, J.M.; Avazzadeh, S.; Quinlan, L.R. α-synuclein pathogenesis in hiPSC models of Parkinson's disease. *Neuronal Signal.* **2021**, *5*, NS20210021. [CrossRef]

29. Fields, C.R.; Bengoa-Vergniory, N.; Wade-Martins, R. Targeting Alpha-Synuclein as a Therapy for Parkinson's Disease. *Front. Mol. Neurosci.* **2019**, *12*, 299. [CrossRef] [PubMed]

30. Braak, H.; Del Tredici, K. Potential Pathways of Abnormal Tau and α-Synuclein Dissemination in Sporadic Alzheimer's and Parkinson's Diseases. *Cold Spring Harb. Perspect. Biol.* **2016**, *8*, a023630. [CrossRef] [PubMed]

31. Okuzumi, A.; Kurosawa, M.; Hatano, T.; Takanashi, M.; Nojiri, S.; Fukuhara, T.; Yamanaka, T.; Miyazaki, H.; Yoshinaga, S.; Furukawa, Y.; et al. Rapid dissemination of alpha-synuclein seeds through neural circuits in an in-vivo prion-like seeding experiment. *Acta Neuropathol. Commun.* **2018**, *6*, 96. [CrossRef] [PubMed]

32. Reyes, J.F.; Sackmann, C.; Hoffmann, A.; Svenningsson, P.; Winkler, J.; Ingelsson, M.; Hallbeck, M. Binding of α-synuclein oligomers to Cx32 facilitates protein uptake and transfer in neurons and oligodendrocytes. *Acta Neuropathol.* **2019**, *138*, 23–47. [CrossRef]

33. Salkov, V.N.; Voronkov, D.V.; Khacheva, K.K.; Fedotova, E.Y.; Khudoerkov, R.M.; Illarioshkin, S.N. Clinical and morphological analysis of a caseof Parkinson's disease. *Arch. Patol.* **2020**, *82*, 52–56. [CrossRef]

34. Schaser, A.J.; Stackhouse, T.L.; Weston, L.J.; Kerstein, P.C.; Osterberg, V.R.; López, C.S.; Dickson, D.W.; Luk, K.C.; Meshul, C.K.; Woltjer, R.L.; et al. Trans-synaptic and retrograde axonal spread of Lewy pathology following pre-formed fibril injection in an in vivo A53T alpha-synuclein mouse model of synucleinopathy. *Acta Neuropathol. Commun.* **2020**, *8*, 150. [CrossRef]

35. Stavrovskaya, A.V.; Voronkov, D.N.; Kutukova, K.A.; Ivanov, M.V.; Gushchina, A.S.; Illarioshkin, S.N. Paraquat-induced model of Parkinson's disease and detection of phosphorylated A-synuclein in the enteric nervous system of rats. *Vestn. RGMU* **2019**, *6*, 63–69. [CrossRef]

36. Sherstnev, V.V.; Kedrov, A.V.; Solov'eva, O.A.; Gruden', M.A.; Konovalova, E.V.; Kalinin, I.A.; Proshin, A.T. The effects of α-synuclein oligomers on neurogenesis in the hippocampus and the behavior of aged mice. *Neurochem. J.* **2017**, *11*, 282–289. [CrossRef]

37. Crews, L.; Mizuno, H.; Desplats, P.; Rockenstein, E.; Adame, A.; Patrick, C.; Winner, B.; Winkler, J.; Masliah, E. α-Synuclein Alters Notch-1 Expression and Neurogenesis in Mouse Embryonic Stem Cells and in the Hippocampus of Transgenic Mice. *J. Neurosci.* **2008**, *28*, 4250–4260. [CrossRef] [PubMed]

38. Bender, H.; Fietz, S.A.; Richter, F.; Stanojlovic, M. Alpha-Synuclein Pathology Coincides with Increased Number of Early Stage Neural Progenitors in the Adult Hippocampus. *Front. Cell Dev. Biol.* **2021**, *9*, 691560. [CrossRef] [PubMed]

39. Peng, J.; Andersen, J.K. Mutant α-synuclein and aging reduce neurogenesis in the acute 1-methyl-4-phenyl-1,2,3,6-tetrahydropyridine model of Parkinson's disease. *Aging Cell* **2011**, *10*, 255–262. [CrossRef] [PubMed]

40. Farzanehfar, P. Towards a Better Treatment Option for Parkinson's Disease: A Review of Adult Neurogenesis. *Neurochem. Res.* **2016**, *41*, 3161–3170. [CrossRef]

41. Baker, S.A.; Baker, K.A.; Hagg, T. Dopaminergic nigrostriatal projections regulate neural precursor proliferation in the adult mouse subventricular zone. *Eur. J. Neurosci.* **2004**, *20*, 575–579. [CrossRef]

42. Ermine, C.M.; Wright, J.L.; Frausin, S.; Kauhausen, J.A.; Parish, C.L.; Stanic, D.; Thompson, L.H. Modelling the dopamine and noradrenergic cell loss that occurs in Parkinson's disease and the impact on hippocampal neurogenesis. *Hippocampus* **2018**, *28*, 327–337. [CrossRef] [PubMed]

43. Brandt, M.D.; Krüger-Gerlach, D.; Hermann, A.; Meyer, A.K.; Kim, K.-S.; Storch, A. Early Postnatal but Not Late Adult Neurogenesis Is Impaired in the Pitx3-Mutant Animal Model of Parkinson's Disease. *Front. Neurosci.* **2017**, *11*. [CrossRef] [PubMed]

44. Aponso, P.M.; Faull, R.L.; Connor, B. Increased progenitor cell proliferation and astrogenesis in the partial progressive 6-hydroxydopamine model of Parkinson's disease. *Neuroscience* **2008**, *151*, 1142–1153. [CrossRef] [PubMed]

45. Arnold, S.A.; Hagg, T. Serotonin 1A receptor agonist increases species- and region-selective adult CNS proliferation, but not through CNTF. *Neuropharmacology* **2012**, *63*, 1238–1247. [CrossRef]

46. Blum, R.; Lesch, K.-P. Parkinson's disease, anxious depression and serotonin—Zooming in on hippocampal neurogenesis. *J. Neurochem.* **2015**, *135*, 441–444. [CrossRef]

47. Song, N.-N.; Jia, Y.-F.; Zhang, L.; Zhang, Q.; Huang, Y.; Liu, X.-Z.; Hu, L.; Lan, W.; Chen, L.; Lesch, K.-P.; et al. Reducing central serotonin in adulthood promotes hippocampal neurogenesis. *Sci. Rep.* **2016**, *6*, 20338. [CrossRef]

48. Brown, S.J.; Boussaad, I.; Jarazo, J.; Fitzgerald, J.C.; Antony, P.; Keatinge, M.; Blechman, J.; Schwamborn, J.C.; Krüger, R.; Placzek, M.; et al. PINK1 deficiency impairs adult neurogenesis of dopaminergic neurons. *Sci. Rep.* **2021**, *11*, 6617. [CrossRef]

49. Agnihotri, S.K.; Shen, R.; Li, J.; Gao, X.; Büeler, H. Loss of PINK1 leads to metabolic deficits in adult neural stem cells and impedes differentiation of newborn neurons in the mouse hippocampus. *FASEB J.* **2017**, *31*, 2839–2853. [CrossRef]

50. Marxreiter, F.; Ettle, B.; May, V.E.L.; Esmer, H.; Patrick, C.; Kragh, C.L.; Klucken, J.; Winner, B.; Riess, O.; Winkler, J.; et al. Glial A30P alpha-synuclein pathology segregates neurogenesis from anxiety-related behavior in conditional transgenic mice. *Neurobiol. Dis.* **2013**, *59*, 38–51. [CrossRef] [PubMed]

51. Dziewczapolski, G.; Lie, D.C.; Ray, J.; Gage, F.H.; Shults, C.W. Survival and differentiation of adult rat-derived neural progenitor cells transplanted to the striatum of hemiparkinsonian rats. *Exp. Neurol.* **2003**, *183*, 653–664. [CrossRef]

52. Tattersfield, A.S.; Croon, R.J.; Liu, Y.W.; Kells, A.P.; Faull, R.L.; Connor, B. Neurogenesis in the striatum of the quinolinic acid lesion model of Huntington's disease. *Neuroscience* **2004**, *127*, 319–332. [CrossRef]

53. Kay, J.N.; Blum, M. Differential Response of Ventral Midbrain and Striatal Progenitor Cells to Lesions of the Nigrostriatal Dopaminergic Projection. *Dev. Neurosci.* **2000**, *22*, 56–67. [CrossRef] [PubMed]

54. Porritt, M.J.; Batchelor, P.E.; Hughes, A.J.; Kalnins, R.; Donnan, G.A.; Howells, D.W. New dopaminergic neurons in Parkinson's disease striatum. *Lancet* **2000**, *356*, 44–45. [CrossRef]

55. Shan, X.; Chi, L.; Bishop, M.; Luo, C.; Lien, L.; Zhang, Z.; Liu, R. Enhanced de novo neurogenesis and dopaminergic neurogenesis in the substantia nigra of 1-methyl-4-phenyl-1,2,3,6-tetrahydropyridine-induced Parkinson's disease-like mice. *Stem Cells* **2006**, *24*, 1280–1287. [CrossRef]

56. Fauser, M.; Pan-Montojo, F.; Richter, C.; Kahle, P.J.; Schwarz, S.C.; Schwarz, J.; Storch, A.; Hermann, A. Chronic-Progressive Dopaminergic Deficiency Does Not Induce Midbrain Neurogenesis. *Cells* **2021**, *10*, 775. [CrossRef] [PubMed]

57. Kohl, Z.; Ben Abdallah, N.; Vogelgsang, J.; Tischer, L.; Deusser, J.; Amato, D.; Anderson, S.; Müller, C.P.; Riess, O.; Masliah, E.; et al. Severely impaired hippocampal neurogenesis associates with an early serotonergic deficit in a BAC α-synuclein transgenic rat model of Parkinson's disease. *Neurobiol. Dis.* **2016**, *85*, 206–217. [CrossRef]

58. Mori, K.; Kaneko, Y.S.; Nakashima, A.; Nagasaki, H.; Nagatsu, T.; Nagatsu, I.; Ota, A. Subventricular zone under the neuroinflammatory stress and Parkinson's disease. *Cell. Mol. Neurobiol.* **2012**, *32*, 777–785. [CrossRef]

59. Regensburger, M.; Prots, I.; Winner, B. Adult Hippocampal Neurogenesis in Parkinson's Disease: Impact on Neuronal Survival and Plasticity. *Neural Plast.* **2014**, *2014*, 454696. [CrossRef] [PubMed]

60. Jurkowski, M.P.; Bettio, L.; Woo, E.K.; Patten, A.; Yau, S.-Y.; Gil-Mohapel, J. Beyond the Hippocampus and the SVZ: Adult Neurogenesis Throughout the Brain. *Front. Cell. Neurosci.* **2020**, *14*, 576444. [CrossRef]

61. Mourtzi, T.; Dimitrakopoulos, D.; Kakogiannis, D.; Salodimitris, C.; Botsakis, K.; Meri, D.K.; Anesti, M.; Dimopoulou, A.; Charalampopoulos, I.; Gravanis, A.; et al. Characterization of substantia nigra neurogenesis in homeostasis and dopaminergic degeneration: Beneficial effects of the microneurotrophin BNN-20. *Stem Cell Res. Ther.* **2021**, *12*, 335. [CrossRef]

62. Zhao, M.; Momma, S.; Delfani, K.; Carlen, M.; Cassidy, R.M.; Johansson, C.B.; Brismar, H.; Shupliakov, O.; Frisen, J.; Janson, A.M. Evidence for neurogenesis in the adult mammalian substantia nigra. *Proc. Natl. Acad. Sci. USA* **2003**, *100*, 7925–7930. [CrossRef]

63. Frielingsdorf, H.; Schwarz, K.; Brundin, P.; Mohapel, P. No evidence for new dopaminergic neurons in the adult mammalian substantia nigra. *Proc. Natl. Acad. Sci. USA* **2004**, *101*, 10177–10182. [CrossRef]

64. Arzate, D.M.; Guerra-Crespo, M.; Covarrubias, L. Induction of typical and atypical neurogenesis in the adult substantia nigra after mouse embryonic stem cells transplantation. *Neuroscience* **2019**, *408*, 308–326. [CrossRef] [PubMed]

65. Salvi, R.; Steigleder, T.; Schlachetzki, J.C.M.; Waldmann, E.; Schwab, S.; Winner, B.; Winkler, J.; Kohl, Z. Distinct Effects of Chronic Dopaminergic Stimulation on Hippocampal Neurogenesis and Striatal Doublecortin Expression in Adult Mice. *Front. Neurosci.* **2016**, *10*, 77. [CrossRef] [PubMed]

66. Van Kampen, J.M.; Robertson, H.A. A possible role for dopamine D3 receptor stimulation in the induction of neurogenesis in the adult rat substantia nigra. *Neuroscience* **2005**, *136*, 381–386. [CrossRef]

67. Lie, D.C.; Dziewczapolski, G.; Willhoite, A.R.; Kaspar, B.K.; Shults, C.W.; Gage, F.H. The adult substantia nigra contains progenitor cells with neurogenic potential. *J. Neurosci.* **2002**, *22*, 6639–6649. [CrossRef] [PubMed]

68. Seah, Y.F.; El Farran, C.A.; Warrier, T.; Xu, J.; Loh, Y.H. Induced Pluripotency and Gene Editing in Disease Modelling: Perspectives and Challenges. *Int. J. Mol. Sci.* **2015**, *16*, 28614–28634. [CrossRef]

69. Shuvalova, L.D.; Eremeev, A.V.; Bogomazova, A.N.; Novosadova, E.V.; Zerkalenkova, E.A.; Olshanskaya, Y.V.; Fedotova, E.Y.; Glagoleva, E.S.; Illarioshkin, S.N.; Lebedeva, O.S.; et al. Generation of induced pluripotent stem cell line RCPCMi004-A derived from patient with Parkinson's disease with deletion of the exon 2 in PARK2 gene. *Stem Cell Res.* **2020**, *44*, 101733. [CrossRef]

70. Novosadova, E.; Antonov, S.; Arsenyeva, E.; Kobylanskiy, A.; Vanyushina, Y.; Malova, T.; Khaspekov, L.; Bobrov, M.; Bezuglov, V.; Tarantul, V.; et al. Neuroprotective and neurotoxic effects of endocannabinoid-like compounds, N-arachidonoyl dopamine and N-docosahexaenoyl dopamine in differentiated cultures of induced pluripotent stem cells derived from patients with Parkinson's disease. *Neurotoxicology* **2021**, *82*, 108–118. [CrossRef]

71. Novosadova, E.V.; Arsenyeva, E.L.; Manuilova, E.S.; Khaspekov, L.G.; Bobrov, M.Y.; Bezuglov, V.V.; Illarioshkin, S.N.; Grivennikov, I.A. Neuroprotective Properties of Endocannabinoids N-Arachidonoyl Dopamine and N-Docosahexaenoyl Dopamine Examined in Neuronal Precursors Derived from Human Pluripotent Stem Cells. *Biochemistry* **2017**, *82*, 1367–1372. [CrossRef] [PubMed]
72. Antonov, S.A.; Novosadova, E.V. Current State-of-the-Art and Unresolved Problems in Using Human Induced Pluripotent Stem Cell-Derived Dopamine Neurons for Parkinson's Disease Drug Development. *Int. J. Mol. Sci.* **2021**, *22*, 3381. [CrossRef] [PubMed]
73. de Rus Jacquet, A. Preparation and Co-Culture of iPSC-Derived Dopaminergic Neurons and Astrocytes. *Curr. Protoc. Cell Biol.* **2019**, *85*, e98. [CrossRef] [PubMed]
74. Miller, J.D.; Ganat, Y.M.; Kishinevsky, S.; Bowman, R.L.; Liu, B.; Tu, E.Y.; Mandal, P.; Vera, E.; Shim, J.-W.; Kriks, S.; et al. Human iPSC-Based Modeling of Late-Onset Disease via Progerin-Induced Aging. *Cell Stem Cell* **2013**, *13*, 691–705. [CrossRef]
75. Tang, X.; Zhou, L.; Wagner, A.M.; Marchetto, M.C.N.; Muotri, A.R.; Gage, F.H.; Chen, G. Astroglial cells regulate the developmental timeline of human neurons differentiated from induced pluripotent stem cells. *Stem Cell Res.* **2013**, *11*, 743–757. [CrossRef]
76. Nadadhur, A.G.; Alsaqati, M.; Gasparotto, L.; Cornelissen-Steijger, P.; van Hugte, E.; Dooves, S.; Harwood, A.J.; Heine, V.M. Neuron-Glia Interactions Increase Neuronal Phenotypes in Tuberous Sclerosis Complex Patient iPSC-Derived Models. *Stem Cell Rep.* **2019**, *12*, 42–56. [CrossRef]
77. Taga, A.; Dastgheyb, R.; Habela, C.; Joseph, J.; Richard, J.-P.; Gross, S.K.; Lauria, G.; Lee, G.; Haughey, N.; Maragakis, N.J. Role of Human-Induced Pluripotent Stem Cell-Derived Spinal Cord Astrocytes in the Functional Maturation of Motor Neurons in a Multielectrode Array System. *STEM CELLS Transl. Med.* **2019**, *8*, 1272–1285. [CrossRef] [PubMed]
78. Klapper, S.D.; Garg, P.; Dagar, S.; Lenk, K.; Gottmann, K.; Nieweg, K. Astrocyte lineage cells are essential for functional neuronal differentiation and synapse maturation in human iPSC-derived neural networks. *Glia* **2019**, *67*, 1893–1909. [CrossRef]
79. Coccia, E.; Ahfeldt, T. Towards physiologically relevant human pluripotent stem cell (hPSC) models of Parkinson's disease. *Stem Cell Res. Ther.* **2021**, *12*, 253. [CrossRef]
80. Tagliafierro, L.; Chiba-Falek, O. Up-regulation of SNCA gene expression: Implications to synucleinopathies. *Neurogenetics* **2016**, *17*, 145–157. [CrossRef]
81. Fukusumi, H.; Togo, K.; Sumida, M.; Nakamori, M.; Obika, S.; Baba, K.; Shofuda, T.; Ito, D.; Okano, H.; Mochizuki, H.; et al. Alpha-synuclein dynamics in induced pluripotent stem cell-derived dopaminergic neurons from a Parkinson's disease patient (PARK4) with SNCA triplication. *FEBS Open Bio* **2021**, *11*, 354–366. [CrossRef] [PubMed]
82. Oliveira, L.M.; Falomir-Lockhart, L.J.; Botelho, M.G.; Lin, K.H.; Wales, P.; Koch, J.C.; Gerhardt, E.; Taschenberger, H.; Outeiro, T.F.; Lingor, P.; et al. Elevated α-synuclein caused by SNCA gene triplication impairs neuronal differentiation and maturation in Parkinson's patient-derived induced pluripotent stem cells. *Cell Death Dis.* **2015**, *6*, e1994. [CrossRef]
83. Gribaudo, S.; Tixador, P.; Bousset, L.; Fenyi, A.; Lino, P.; Melki, R.; Peyrin, J.-M.; Perrier, A.L. Propagation of α-Synuclein Strains within Human Reconstructed Neuronal Network. *Stem Cell Rep.* **2019**, *12*, 230–244. [CrossRef] [PubMed]
84. Soubannier, V.; Maussion, G.; Chaineau, M.; Sigutova, V.; Rouleau, G.; Durcan, T.M.; Stifani, S. Characterization of human iPSC-derived astrocytes with potential for disease modeling and drug discovery. *Neurosci. Lett.* **2020**, *731*, 135028. [CrossRef]
85. Tcw, J.; Wang, M.; Pimenova, A.A.; Bowles, K.R.; Hartley, B.J.; Lacin, E.; Machlovi, S.I.; Abdelaal, R.; Karch, C.M.; Phatnani, H.; et al. An Efficient Platform for Astrocyte Differentiation from Human Induced Pluripotent Stem Cells. *Stem Cell Rep.* **2017**, *9*, 600–614. [CrossRef]
86. Serio, A.; Bilican, B.; Barmada, S.J.; Ando, D.M.; Zhao, C.; Siller, R.; Burr, K.; Haghi, G.; Story, D.; Nishimura, A.L.; et al. Astrocyte pathology and the absence of non-cell autonomy in an induced pluripotent stem cell model of TDP-43 proteinopathy. *Proc. Natl. Acad. Sci. USA* **2013**, *110*, 4697–4702. [CrossRef] [PubMed]
87. Byun, J.S.; Lee, C.O.; Oh, M.; Cha, D.; Kim, W.-K.; Oh, K.-J.; Bae, K.-H.; Lee, S.-C.; Han, B.-S. Rapid differentiation of astrocytes from human embryonic stem cells. *Neurosci. Lett.* **2020**, *716*, 134681. [CrossRef]
88. Caiazzo, M.; Giannelli, S.; Valente, P.; Lignani, G.; Carissimo, A.; Sessa, A.; Colasante, G.; Bartolomeo, R.; Massimino, L.; Ferroni, S.; et al. Direct Conversion of Fibroblasts into Functional Astrocytes by Defined Transcription Factors. *Stem Cell Rep.* **2015**, *4*, 25–36. [CrossRef]
89. Reinhardt, P.; Glatza, M.; Hemmer, K.; Tsytsyura, Y.; Thiel, C.S.; Höing, S.; Moritz, S.; Parga, J.A.; Wagner, L.; Bruder, J.M.; et al. Derivation and Expansion Using Only Small Molecules of Human Neural Progenitors for Neurodegenerative Disease Modeling. *PLoS ONE* **2013**, *8*, e59252. [CrossRef]
90. Barbuti, P.A.; Antony, P.; Novak, G.; Larsen, S.B.; Berenguer-Escuder, C.; Santos, B.F.; Massart, F.; Grossmann, D.; Shiga, T.; Ishikawa, K.-I.; et al. IPSC-derived midbrain astrocytes from Parkinson's disease patients carrying pathogenic *SNCA* mutations exhibit alpha-synuclein aggregation, mitochondrial fragmentation and excess calcium release. *bioRxiv* **2020**. [CrossRef]
91. di Domenico, A.; Carola, G.; Calatayud, C.; Pons-Espinal, M.; Muñoz, J.P.; Richaud-Patin, Y.; Fernandez-Carasa, I.; Gut, M.; Faella, A.; Parameswaran, J.; et al. Patient-Specific iPSC-Derived Astrocytes Contribute to Non-Cell-Autonomous Neurodegeneration in Parkinson's Disease. *Stem Cell Rep.* **2019**, *12*, 213–229. [CrossRef] [PubMed]
92. Chan, W.K.; Griffiths, R.; Price, D.J.; Mason, J.O. Cerebral organoids as tools to identify the developmental roots of autism. *Mol. Autism* **2020**, *11*, 58. [CrossRef]
93. Khakipoor, S.; Crouch, E.E.; Mayer, S. Human organoids to model the developing human neocortex in health and disease. *Brain Res.* **2020**, *1742*, 146803. [CrossRef]
94. Di Lullo, E.; Kriegstein, A.R. The use of brain organoids to investigate neural development and disease. *Nat. Rev. Neurosci.* **2017**, *18*, 573–584. [CrossRef]

95. Karzbrun, E.; Reiner, O. Brain Organoids-A Bottom-Up Approach for Studying Human Neurodevelopment. *Bioengineering* **2019**, *6*, 9. [CrossRef]
96. Nascimento, J.M.; Saia-Cereda, V.M.; Sartore, R.C.; da Costa, R.M.; Schitine, C.S.; Freitas, H.R.; Murgu, M.; de Melo Reis, R.A.; Rehen, S.K.; Martins-de-Souza, D. Human Cerebral Organoids and Fetal Brain Tissue Share Proteomic Similarities. *Front. Cell Dev. Biol.* **2019**, *7*, 303. [CrossRef]
97. Vargas-Valderrama, A.; Messina, A.; Mitjavila-Garcia, M.T.; Guenou, H. The endothelium, a key actor in organ development and hPSC-derived organoid vascularization. *J. Biomed. Sci.* **2020**, *27*, 67. [CrossRef] [PubMed]
98. Cakir, B.; Xiang, Y.; Tanaka, Y.; Kural, M.H.; Parent, M.; Kang, Y.-J.; Chapeton, K.; Patterson, B.; Yuan, Y.; He, C.-S.; et al. Engineering of human brain organoids with a functional vascular-like system. *Nat. Methods* **2019**, *16*, 1169–1175. [CrossRef] [PubMed]
99. Pham, M.T.; Pollock, K.M.; Rose, M.D.; Cary, W.A.; Stewart, H.R.; Zhou, P.; Nolta, J.A.; Waldau, B. Generation of human vascularized brain organoids. *Neuroreport* **2018**, *29*, 588–593. [CrossRef]
100. Fagerlund, I.; Dougalis, A.; Shakirzyanova, A.; Gómez-Budia, M.; Konttinen, H.; Ohtonen, S.; Feroze, F.; Koskuvi, M.; Kuusisto, J.; Hernández, D.; et al. Microglia orchestrate neuronal activity in brain organoids. *bioRxiv* **2020**. [CrossRef]
101. Ormel, P.R.; Vieira de Sá, R.; van Bodegraven, E.J.; Karst, H.; Harschnitz, O.; Sneeboer, M.A.M.; Johansen, L.E.; van Dijk, R.E.; Scheefhals, N.; Berdenis van Berlekom, A.; et al. Microglia innately develop within cerebral organoids. *Nat. Commun.* **2018**, *9*, 4167. [CrossRef] [PubMed]
102. Raja, W.K.; Mungenast, A.E.; Lin, Y.-T.; Ko, T.; Abdurrob, F.; Seo, J.; Tsai, L.-H. Self-Organizing 3D Human Neural Tissue Derived from Induced Pluripotent Stem Cells Recapitulate Alzheimer's Disease Phenotypes. *PLoS ONE* **2016**, *11*, e0161969. [CrossRef] [PubMed]
103. Seo, J.; Kritskiy, O.; Watson, L.A.; Barker, S.J.; Dey, D.; Raja, W.K.; Lin, Y.-T.; Ko, T.; Cho, S.; Penney, J.; et al. Inhibition of p25/Cdk5 Attenuates Tauopathy in Mouse and iPSC Models of Frontotemporal Dementia. *J. Neurosci.* **2017**, *37*, 9917–9924. [CrossRef] [PubMed]
104. Grenier, K.; Kao, J.; Diamandis, P. Three-dimensional modeling of human neurodegeneration: Brain organoids coming of age. *Mol. Psychiatry* **2020**, *25*, 254–274. [CrossRef]
105. Galet, B.; Cheval, H.; Ravassard, P. Patient-Derived Midbrain Organoids to Explore the Molecular Basis of Parkinson's Disease. *Front. Neurol.* **2020**, *11*, 1005. [CrossRef] [PubMed]
106. Nickels, S.L.; Modamio, J.; Mendes-Pinheiro, B.; Monzel, A.S.; Betsou, F.; Schwamborn, J.C. Reproducible generation of human midbrain organoids for in vitro modeling of Parkinson's disease. *Stem Cell Res.* **2020**, *46*, 101870. [CrossRef]
107. Sánchez-Danés, A.; Consiglio, A.; Richaud, Y.; Rodríguez-Pizà, I.; Dehay, B.; Edel, M.; Bové, J.; Memo, M.; Vila, M.; Raya, A.; et al. Efficient generation of A9 midbrain dopaminergic neurons by lentiviral delivery of LMX1A in human embryonic stem cells and induced pluripotent stem cells. *Hum. Gene Ther.* **2012**, *23*, 56–69. [CrossRef]
108. Kwak, T.H.; Kang, J.H.; Hali, S.; Kim, J.; Kim, K.P.; Park, C.; Lee, J.H.; Ryu, H.K.; Na, J.E.; Jo, J.; et al. Generation of homogeneous midbrain organoids with in vivo-like cellular composition facilitates neurotoxin-based Parkinson's disease modeling. *Stem Cells* **2020**, *38*, 727–740. [CrossRef]
109. Zanetti, C.; Spitz, S.; Berger, E.; Bolognin, S.; Smits, L.M.; Crepaz, P.; Rothbauer, M.; Rosser, J.M.; Marchetti-Deschmann, M.; Schwarmborn, J.C.; et al. Monitoring the neurotransmitter release of human midbrain organoids using a redox cycling microsensor as a novel tool for personalized Parkinson's disease modelling and drug screening. *Analyst.* **2021**, *146*, 2358–2367. [CrossRef]
110. Schultz, E.M.; Jones, T.J.; Xu, S.; Dean, D.D.; Zechmann, B.; Barr, K.L. Cerebral Organoids Derived from a Parkinson's Patient Exhibit Unique Pathogenesis from Chikungunya Virus Infection When Compared to a Non-Parkinson's Patient. *Pathogens* **2021**, *10*, 913. [CrossRef]
111. Jarazo, J.; Barmpa, K.; Rosety, I.; Smits, L.M.; Arias-Fuenzalida, J.; Walter, J.; Gomez-Giro, G.; Monzel, A.S.; Qing, X.; Cruciani, G.; et al. Parkinson's disease phenotypes in patient specific brain organoids are improved by HP-β-CD treatment. *bioRxiv* **2019**, 813089. [CrossRef]
112. Calatayud, C.; Carola, G.; Fernández-Carasa, I.; Valtorta, M.; Jiménez-Delgado, S.; Díaz, M.; Soriano-Fradera, J.; Cappelletti, G.; García-Sancho, J.; Raya, Á.; et al. CRISPR/Cas9-mediated generation of a tyrosine hydroxylase reporter iPSC line for live imaging and isolation of dopaminergic neurons. *Sci. Rep.* **2019**, *9*, 6811. [CrossRef]
113. Chlebanowska, P.; Tejchman, A.; Sułkowski, M.; Skrzypek, K.; Majka, M. Use of 3D Organoids as a Model to Study Idiopathic Form of Parkinson's Disease. *Int. J. Mol. Sci.* **2020**, *21*, 694. [CrossRef]
114. Katt, M.E.; Mayo, L.N.; Ellis, S.E.; Mahairaki, V.; Rothstein, J.D.; Cheng, L.; Searson, P.C. The role of mutations associated with familial neurodegenerative disorders on blood-brain barrier function in an iPSC model. *Fluids Barriers CNS* **2019**, *16*, 20. [CrossRef]
115. Kane, K.I.W.; Jarazo, J.; Moreno, E.L.; Fleming, R.M.T.; Schwamborn, J.C. Passive controlled flow for Parkinson's disease neuronal cell culture in 3D microfluidic devices. *Organs-on-a-Chip* **2020**, *2*, 100005. [CrossRef]
116. Doi, D.; Magotani, H.; Kikuchi, T.; Ikeda, M.; Hiramatsu, S.; Yoshida, K.; Amano, N.; Nomura, M.; Umekage, M.; Morizane, A.; et al. Pre-clinical study of induced pluripotent stem cell-derived dopaminergic progenitor cells for Parkinson's disease. *Nat. Commun.* **2020**, *11*, 3369. [CrossRef] [PubMed]

117. DeCarolis, N.A.; Eisch, A.J. Hippocampal neurogenesis as a target for the treatment of mental illness: A critical evaluation. *Neuropharmacology* **2010**, *58*, 884–893. [CrossRef] [PubMed]

118. Bortolasci, C.C.; Spolding, B.; Kidnapillai, S.; Connor, T.; Truong, T.T.T.; Liu, Z.S.J.; Panizzutti, B.; Richardson, M.F.; Gray, L.; Berk, M.; et al. Transcriptional Effects of Psychoactive Drugs on Genes Involved in Neurogenesis. *Int. J. Mol. Sci.* **2020**, *21*, 8333. [CrossRef]

119. Latchney, S.E.; Eisch, A.J. Therapeutic application of neural stem cells and adult neurogenesis for neurodegenerative disorders: Regeneration and beyond. *Eur. J. Neurodegener. Dis.* **2012**, *1*, 335–351. [PubMed]

120. Kreutzmann, J.C.; Havekes, R.; Abel, T.; Meerlo, P. Sleep deprivation and hippocampal vulnerability: Changes in neuronal plasticity, neurogenesis and cognitive function. *Neuroscience* **2015**, *309*, 173–190. [CrossRef]

121. Venna, V.R.; Xu, Y.; Doran, S.J.; Patrizz, A.; McCullough, L.D. Social interaction plays a critical role in neurogenesis and recovery after stroke. *Transl. Psychiatry* **2014**, *4*, e351. [CrossRef]

122. Salmin, V.V.; Komleva, Y.K.; Kuvacheva, N.V.; Morgun, A.V.; Khilazheva, E.D.; Lopatina, O.L.; Pozhilenkova, E.A.; Shapovalov, K.A.; Uspenskaya, Y.A.; Salmina, A.B. Differential Roles of Environmental Enrichment in Alzheimer's Type of Neurodegeneration and Physiological Aging. *Front. Aging Neurosci.* **2017**, *9*, 245. [CrossRef]

123. Salmina, A.B.; Komleva, Y.K.; Salmin, V.V.; Lopatina, O.L.; Belova, O.A. Environmental enrichment and physiological aging. In *The Neuroscience of Aging*, 1st ed.; Martin, C., Preedy, V., Rajendram, R., Eds.; Elsevier: Amsterdam, The Netherlands; Academic Press: Cambridge, MA, USA, 2021.

124. Lev-Vachnish, Y.; Cadury, S.; Rotter-Maskowitz, A.; Feldman, N.; Roichman, A.; Illouz, T.; Varvak, A.; Nicola, R.; Madar, R.; Okun, E. L-Lactate Promotes Adult Hippocampal Neurogenesis. *Front. Neurosci.* **2019**, *13*, 403. [CrossRef]

125. Mainardi, M.; Fusco, S.; Grassi, C. Modulation of Hippocampal Neural Plasticity by Glucose-Related Signaling. *Neural Plast.* **2015**, *2015*, 657928. [CrossRef]

126. Storer, M.A.; Gallagher, D.; Fatt, M.P.; Simonetta, J.V.; Kaplan, D.R.; Miller, F.D. Interleukin-6 Regulates Adult Neural Stem Cell Numbers during Normal and Abnormal Post-natal Development. *Stem Cell Rep.* **2018**, *10*, 1464–1480. [CrossRef]

127. Wang, S.N.; Xu, T.Y.; Li, W.L.; Miao, C.Y. Targeting Nicotinamide Phosphoribosyltransferase as a Potential Therapeutic Strategy to Restore Adult Neurogenesis. *CNS Neurosci. Ther.* **2016**, *22*, 431–439. [CrossRef]

128. Asano, T.; Teh, D.B.L.; Yawo, H. Application of Optogenetics for Muscle Cells and Stem Cells. *Adv. Exp. Med. Biol.* **2021**, *1293*, 359–375. [CrossRef] [PubMed]

129. Boltze, J.; Stroh, A. Optogenetics in Stem Cell Research: Focus on the Central Nervous System. In *Optogenetics: A Roadmap*; Stroh, A., Ed.; Springer: New York, NY, USA, 2018; pp. 75–87.

130. Teh, D.B.L.; Prasad, A.; Jiang, W.; Zhang, N.; Wu, Y.; Yang, H.; Han, S.; Yi, Z.; Yeo, Y.; Ishizuka, T.; et al. Driving Neurogenesis in Neural Stem Cells with High Sensitivity Optogenetics. *NeuroMolecular Med.* **2020**, *22*, 139–149. [CrossRef] [PubMed]

131. Giraldo, E.; Palmero-Canton, D.; Martinez-Rojas, B.; Sanchez-Martin, M.D.M.; Moreno-Manzano, V. Optogenetic Modulation of Neural Progenitor Cells Improves Neuroregenerative Potential. *Int. J. Mol. Sci.* **2020**, *22*, 365. [CrossRef] [PubMed]

132. Lee, S.Y.; George, J.H.; Nagel, D.A.; Ye, H.; Kueberuwa, G.; Seymour, L.W. Optogenetic control of iPS cell-derived neurons in 2D and 3D culture systems using channelrhodopsin-2 expression driven by the synapsin-1 and calcium-calmodulin kinase II promoters. *J. Tissue Eng. Regen. Med.* **2019**, *13*, 369–384. [CrossRef] [PubMed]

133. Daadi, M.M.; Klausner, J.Q.; Bajar, B.; Goshen, I.; Lee-Messer, C.; Lee, S.Y.; Winge, M.C.; Ramakrishnan, C.; Lo, M.; Sun, G.; et al. Optogenetic Stimulation of Neural Grafts Enhances Neurotransmission and Downregulates the Inflammatory Response in Experimental Stroke Model. *Cell Transpl.* **2016**, *25*, 1371–1380. [CrossRef]

134. Klapper, S.D.; Sauter, E.J.; Swiersy, A.; Hyman, M.A.E.; Bamann, C.; Bamberg, E.; Busskamp, V. On-demand optogenetic activation of human stem-cell-derived neurons. *Sci. Rep.* **2017**, *7*, 14450. [CrossRef]

135. Ryu, J.; Vincent, P.F.Y.; Ziogas, N.K.; Xu, L.; Sadeghpour, S.; Curtin, J.; Alexandris, A.S.; Stewart, N.; Sima, R.; du Lac, S.; et al. Optogenetically transduced human ES cell-derived neural progenitors and their neuronal progenies: Phenotypic characterization and responses to optical stimulation. *PLoS ONE* **2019**, *14*, e0224846. [CrossRef]

136. Avaliani, N.; Sørensen, A.T.; Ledri, M.; Bengzon, J.; Koch, P.; Brüstle, O.; Deisseroth, K.; Andersson, M.; Kokaia, M. Optogenetics reveal delayed afferent synaptogenesis on grafted human-induced pluripotent stem cell-derived neural progenitors. *Stem Cells* **2014**, *32*, 3088–3098. [CrossRef]

137. Shohayeb, B.; Diab, M.; Ahmed, M.; Ng, D.C.H. Factors that influence adult neurogenesis as potential therapy. *Transl. Neurodegener.* **2018**, *7*, 4. [CrossRef]

138. Isaev, N.K.; Stelmashook, E.V.; Genrikhs, E.E. Neurogenesis and brain aging. *Rev. Neurosci.* **2019**, *30*, 573–580. [CrossRef] [PubMed]

139. Kempermann, G. Activity Dependency and Aging in the Regulation of Adult Neurogenesis. *Cold Spring Harb. Perspect. Biol.* **2015**, *7*, a018929. [CrossRef]

140. Berdugo-Vega, G.; Arias-Gil, G.; López-Fernández, A.; Artegiani, B.; Wasielewska, J.M.; Lee, C.-C.; Lippert, M.T.; Kempermann, G.; Takagaki, K.; Calegari, F. Increasing neurogenesis refines hippocampal activity rejuvenating navigational learning strategies and contextual memory throughout life. *Nat. Commun.* **2020**, *11*, 135. [CrossRef]

141. Qi, C.; Varga, S.; Oh, S.-J.; Lee, C.J.; Lee, D. Optogenetic Rescue of Locomotor Dysfunction and Dopaminergic Degeneration Caused by Alpha-Synuclein and EKO Genes. *Exp. Neurobiol.* **2017**, *26*, 97–103. [CrossRef] [PubMed]

142. Imai, Y.; Inoshita, T.; Meng, H.; Shiba-Fukushima, K.; Hara, K.Y.; Sawamura, N.; Hattori, N. Light-driven activation of mitochondrial proton-motive force improves motor behaviors in a Drosophila model of Parkinson's disease. *Commun. Biol.* **2019**, *2*, 424. [CrossRef] [PubMed]

143. Bérard, M.; Sheta, R.; Malvaut, S.; Turmel, R.; Alpaugh, M.; Dubois, M.; Dahmene, M.; Salesse, C.; Profes, M.; Lamontagne-Proulx, J.; et al. Optogenetic-Mediated Spatiotemporal Control of α-Synuclein Aggregation Disrupts Nigrostriatal Transmission and Precipitates Neurodegeneration. *SSRN Electron. J.* **2019**. [CrossRef]

144. Qian, H.; Kang, X.; Hu, J.; Zhang, D.; Liang, Z.; Meng, F.; Zhang, X.; Xue, Y.; Maimon, R.; Dowdy, S.F.; et al. Reversing a model of Parkinson's disease with in situ converted nigral neurons. *Nature* **2020**, *582*, 550–556. [CrossRef]

145. Cassé, F.; Richetin, K.; Toni, N. Astrocytes' Contribution to Adult Neurogenesis in Physiology and Alzheimer's Disease. *Front. Cell. Neurosci.* **2018**, *12*, 432. [CrossRef]

146. Araki, T.; Ikegaya, Y.; Koyama, R. The effects of microglia- and astrocyte-derived factors on neurogenesis in health and disease. *Eur. J. Neurosci.* **2020**, *54*, 5880–5901. [CrossRef] [PubMed]

147. Wilhelmsson, U.; Faiz, M.; de Pablo, Y.; Sjöqvist, M.; Andersson, D.; Widestrand, Å.; Potokar, M.; Stenovec, M.; Smith, P.L.P.; Shinjyo, N.; et al. Astrocytes Negatively Regulate Neurogenesis Through the Jagged1-Mediated Notch Pathway. *Stem Cells* **2012**, *30*, 2320–2329. [CrossRef]

148. Rizor, A.; Pajarillo, E.; Johnson, J.; Aschner, M.; Lee, E. Astrocytic Oxidative/Nitrosative Stress Contributes to Parkinson's Disease Pathogenesis: The Dual Role of Reactive Astrocytes. *Antioxidants* **2019**, *8*, 265. [CrossRef]

149. Hindeya Gebreyesus, H.; Gebrehiwot Gebremichael, T. The Potential Role of Astrocytes in Parkinson's Disease (PD). *Med. Sci.* **2020**, *8*, 7. [CrossRef]

150. Mosher, K.I.; Schaffer, D.V. Influence of hippocampal niche signals on neural stem cell functions during aging. *Cell Tissue Res.* **2018**, *371*, 115–124. [CrossRef] [PubMed]

151. Marchetti, B.; L'Episcopo, F.; Morale, M.C.; Tirolo, C.; Testa, N.; Caniglia, S.; Serapide, M.F.; Pluchino, S. Uncovering novel actors in astrocyte-neuron crosstalk in Parkinson's disease: The Wnt/β-catenin signaling cascade as the common final pathway for neuroprotection and self-repair. *Eur. J. Neurosci.* **2013**, *37*, 1550–1563. [CrossRef]

152. L'Episcopo, F.; Tirolo, C.; Testa, N.; Caniglia, S.; Morale, M.C.; Cossetti, C.; D'Adamo, P.; Zardini, E.; Andreoni, L.; Ihekwaba, A.E.; et al. Reactive astrocytes and Wnt/β-catenin signaling link nigrostriatal injury to repair in 1-methyl-4-phenyl-1,2,3,6-tetrahydropyridine model of Parkinson's disease. *Neurobiol. Dis.* **2011**, *41*, 508–527. [CrossRef]

153. Marchetti, B. Wnt/β-Catenin Signaling Pathway Governs a Full Program for Dopaminergic Neuron Survival, Neurorescue and Regeneration in the MPTP Mouse Model of Parkinson's Disease. *Int. J. Mol. Sci.* **2018**, *19*, 3743. [CrossRef]

154. Faigle, R.; Song, H. Signaling mechanisms regulating adult neural stem cells and neurogenesis. *Biochim. Biophys. Acta* **2013**, *1830*, 2435–2448. [CrossRef] [PubMed]

155. Hitoshi, S.; Alexson, T.; Tropepe, V.; Donoviel, D.; Elia, A.J.; Nye, J.S.; Conlon, R.A.; Mak, T.W.; Bernstein, A.; van der Kooy, D. Notch pathway molecules are essential for the maintenance, but not the generation, of mammalian neural stem cells. *Genes Dev.* **2002**, *16*, 846–858. [CrossRef]

156. Bejoy, J.; Bijonowski, B.; Marzano, M.; Jeske, R.; Ma, T.; Li, Y. Wnt-Notch Signaling Interactions During Neural and Astroglial Patterning of Human Stem Cells. *Tissue Eng. Part A* **2020**, *26*, 419–431. [CrossRef] [PubMed]

157. Berwick, D.C.; Harvey, K. The regulation and deregulation of Wnt signaling by PARK genes in health and disease. *J. Mol. Cell Biol.* **2014**, *6*, 3–12. [CrossRef] [PubMed]

158. Rawal, N.; Corti, O.; Sacchetti, P.; Ardilla-Osorio, H.; Sehat, B.; Brice, A.; Arenas, E. Parkin protects dopaminergic neurons from excessive Wnt/beta-catenin signaling. *Biochem. Biophys. Res. Commun.* **2009**, *388*, 473–478. [CrossRef] [PubMed]

159. Nouri, N.; Patel, M.J.; Joksimovic, M.; Poulin, J.F.; Anderegg, A.; Taketo, M.M.; Ma, Y.C.; Awatramani, R. Excessive Wnt/beta-catenin signaling promotes midbrain floor plate neurogenesis, but results in vacillating dopamine progenitors. *Mol. Cell. Neurosci.* **2015**, *68*, 131–142. [CrossRef] [PubMed]

160. Salašová, A.; Yokota, C.; Potěšil, D.; Zdráhal, Z.; Bryja, V.; Arenas, E. A proteomic analysis of LRRK2 binding partners reveals interactions with multiple signaling components of the WNT/PCP pathway. *Mol. Neurodegener.* **2017**, *12*, 54. [CrossRef]

161. Yang, F.; Liu, Y.; Tu, J.; Wan, J.; Zhang, J.; Wu, B.; Chen, S.; Zhou, J.; Mu, Y.; Wang, L. Activated astrocytes enhance the dopaminergic differentiation of stem cells and promote brain repair through bFGF. *Nat. Commun.* **2014**, *5*, 5627. [CrossRef]

162. Saarimäki-Vire, J.; Peltopuro, P.; Lahti, L.; Naserke, T.; Blak, A.A.; Vogt Weisenhorn, D.M.; Yu, K.; Ornitz, D.M.; Wurst, W.; Partanen, J. Fibroblast growth factor receptors cooperate to regulate neural progenitor properties in the developing midbrain and hindbrain. *J. Neurosci.* **2007**, *27*, 8581–8592. [CrossRef] [PubMed]

163. Puelles, E.; Annino, A.; Tuorto, F.; Usiello, A.; Acampora, D.; Czerny, T.; Brodski, C.; Ang, S.L.; Wurst, W.; Simeone, A. Otx2 regulates the extent, identity and fate of neuronal progenitor domains in the ventral midbrain. *Development* **2004**, *131*, 2037–2048. [CrossRef] [PubMed]

164. Kim, N.; Kim, J.M.; Heo, W.D. Optogenetic Control of Fibroblast Growth Factor Receptor Signaling. *Methods Mol. Biol.* **2016**, *1408*, 345–362. [CrossRef] [PubMed]

165. Kim, N.; Kim, J.M.; Lee, M.; Kim, C.Y.; Chang, K.-Y.; Heo, W.D. Spatiotemporal Control of Fibroblast Growth Factor Receptor Signals by Blue Light. *Chem. Biol.* **2014**, *21*, 903–912. [CrossRef]

166. Cheli, V.T.; Santiago González, D.A.; Smith, J.; Spreuer, V.; Murphy, G.G.; Paez, P.M. L-type voltage-operated calcium channels contribute to astrocyte activation in vitro. *Glia* **2016**, *64*, 1396–1415. [CrossRef] [PubMed]

167. De Bock, M.; Wang, N.; Bol, M.; Decrock, E.; Ponsaerts, R.; Bultynck, G.; Dupont, G.; Leybaert, L. Connexin 43 hemichannels contribute to cytoplasmic Ca2+ oscillations by providing a bimodal Ca2+-dependent Ca2+ entry pathway. *J. Biol. Chem.* **2012**, *287*, 12250–12266. [CrossRef]

168. Shigetomi, E.; Tong, X.; Kwan, K.Y.; Corey, D.P.; Khakh, B.S. TRPA1 channels regulate astrocyte resting calcium and inhibitory synapse efficacy through GAT-3. *Nat. Neurosci.* **2012**, *15*, 70–80. [CrossRef]

169. Banerjee, S.; Walseth, T.F.; Borgmann, K.; Wu, L.; Bidasee, K.R.; Kannan, M.S.; Ghorpade, A. CD38/cyclic ADP-ribose regulates astrocyte calcium signaling: Implications for neuroinflammation and HIV-1-associated dementia. *J. Neuroimmune Pharm.* **2008**, *3*, 154–164. [CrossRef] [PubMed]

170. Navarrete, M.; Perea, G.; de Sevilla, D.F.; Gómez-Gonzalo, M.; Núñez, A.; Martín, E.D.; Araque, A. Astrocytes Mediate In Vivo Cholinergic-Induced Synaptic Plasticity. *PLOS Biol.* **2012**, *10*, e1001259. [CrossRef]

171. Guerra-Gomes, S.; Sousa, N.; Pinto, L.; Oliveira, J.F. Functional Roles of Astrocyte Calcium Elevations: From Synapses to Behavior. *Front. Cell. Neurosci.* **2018**, *11*, 427. [CrossRef]

172. Li, Y.; Li, L.; Wu, J.; Zhu, Z.; Feng, X.; Qin, L.; Zhu, Y.; Sun, L.; Liu, Y.; Qiu, Z.; et al. Activation of astrocytes in hippocampus decreases fear memory through adenosine A(1) receptors. *Elife* **2020**, *9*, 9. [CrossRef]

173. Ono, K.; Suzuki, H.; Higa, M.; Tabata, K.; Sawada, M. Glutamate release from astrocyte cell-line GL261 via alterations in the intracellular ion environment. *J. Neural Transm.* **2014**, *121*, 245–257. [CrossRef] [PubMed]

174. Okubo, Y.; Iino, M. Visualization of astrocytic intracellular Ca(2+) mobilization. *J. Physiol.* **2020**, *598*, 1671–1681. [CrossRef] [PubMed]

175. Figueiredo, M.; Lane, S.; Tang, F.; Liu, B.H.; Hewinson, J.; Marina, N.; Kasymov, V.; Souslova, E.A.; Chudakov, D.M.; Gourine, A.V.; et al. Optogenetic experimentation on astrocytes. *Exp. Physiol.* **2011**, *96*, 40–50. [CrossRef] [PubMed]

176. Salmina, A.B.; Gorina, Y.V.; Erofeev, A.I.; Balaban, P.M.; Bezprozvanny, I.B.; Vlasova, O.L. Optogenetic and chemogenetic modulation of astroglial secretory phenotype. *Rev. Neurosci.* **2021**, *32*, 459–479. [CrossRef] [PubMed]

177. Borodinova, A.A.; Balaban, P.M.; Bezprozvanny, I.B.; Salmina, A.B.; Vlasova, O.L. Genetic Constructs for the Control of Astrocytes' Activity. *Cells* **2021**, *10*, 1600. [CrossRef]

178. Bang, J.; Kim, H.Y.; Lee, H. Optogenetic and Chemogenetic Approaches for Studying Astrocytes and Gliotransmitters. *Exp. Neurobiol.* **2016**, *25*, 205–221. [CrossRef]

179. Chai, H.; Diaz-Castro, B.; Shigetomi, E.; Monte, E.; Octeau, J.C.; Yu, X.; Cohn, W.; Rajendran, P.S.; Vondriska, T.M.; Whitelegge, J.P.; et al. Neural Circuit-Specialized Astrocytes: Transcriptomic, Proteomic, Morphological, and Functional Evidence. *Neuron* **2017**, *95*, 531–549.e539. [CrossRef]

180. Morgun, A.V.; Malinovskaya, N.A.; Komleva, Y.K.; Lopatina, O.L.; Kuvacheva, N.V.; Panina, Y.A.; Taranushenko, T.Y.; Solonchuk, Y.R.; Salmina, A.B. Structural and functional heterogeneity of astrocytes in the brain: Role in neurodegeneration and neuroinflammation. *Bull. Sib. Med.* **2014**, *13*, 138–148. [CrossRef]

181. Li, K.; Li, J.; Zheng, J.; Qin, S. Reactive Astrocytes in Neurodegenerative Diseases. *Aging Dis.* **2019**, *10*, 664–675. [CrossRef]

182. Figueiredo, M.; Lane, S.; Stout, R.F.J.; Liu, B.; Parpura, V.; Teschemacher, A.G.; Kasparov, S. Comparative analysis of optogenetic actuators in cultured astrocytes. *Cell Calcium* **2014**, *56*, 208–214. [CrossRef]

183. Shuvaev, A.N.; Belozor, O.S.; Mozhei, O.; Yakovleva, D.A.; Potapenko, I.V.; Shuvaev, A.N.; Smolnikova, M.V.; Salmin, V.V.; Salmina, A.B.; Hirai, H.; et al. Chronic optogenetic stimulation of Bergman glia leads to dysfunction of EAAT1 and Purkinje cell death, mimicking the events caused by expression of pathogenic ataxin-1. *Neurobiol. Dis.* **2021**, *154*, 105340. [CrossRef]

184. Morgun, A.V.; Osipova, E.D.; Boytsova, E.B.; Shuvaev, A.N.; Komleva, Y.K.; Trufanova, L.V.; Vais, E.F.; Salmina, A.B. Astroglia-mediated regulation of cell development in the model of neurogenic niche in vitro treated with Aβ1-42. *Biomed. Khim.* **2019**, *65*, 366–373. [CrossRef]

185. Morgun, A.V.; Osipova, E.D.; Boitsova, E.B.; Shuvaev, A.N.; Malinovskaya, N.A.; Mosiagina, A.I.; Salmina, A.B. Neurogenic Potential of Implanted Neurospheres Is Regulated by Optogenetic Stimulation of Hippocampal Astrocytes Ex Vivo. *Bull. Exp. Biol. Med.* **2021**, *170*, 693–698. [CrossRef]

186. Hedegaard, A.; Monzón-Sandoval, J.; Newey, S.E.; Whiteley, E.S.; Webber, C.; Akerman, C.J. Pro-maturational Effects of Human iPSC-Derived Cortical Astrocytes upon iPSC-Derived Cortical Neurons. *Stem Cell Rep.* **2020**, *15*, 38–51. [CrossRef] [PubMed]

187. Zhang, J.; Jiao, J. Molecular Biomarkers for Embryonic and Adult Neural Stem Cell and Neurogenesis. *Biomed. Res. Int.* **2015**, *2015*, 727542. [CrossRef] [PubMed]

188. Kotterman, M.A.; Vazin, T.; Schaffer, D.V. Enhanced selective gene delivery to neural stem cells in vivo by an adeno-associated viral variant. *Development* **2015**, *142*, 1885–1892. [CrossRef] [PubMed]

189. Parr-Brownlie, L.C.; Bosch-Bouju, C.; Schoderboeck, L.; Sizemore, R.J.; Abraham, W.C.; Hughes, S.M. Lentiviral vectors as tools to understand central nervous system biology in mammalian model organisms. *Front. Mol. Neurosci.* **2015**, *8*, 14. [CrossRef]

190. Abdolahi, S.; Khodakaram-Tafti, A.; Aligholi, H.; Ziaei, S.; Khaleghi Ghadiri, M.; Stummer, W.; Gorji, A. Lentiviral vector-mediated transduction of adult neural stem/progenitor cells isolated from the temporal tissues of epileptic patients. *Iran. J. Basic Med. Sci.* **2020**, *23*, 354–361. [CrossRef]

191. Jandial, R.; Singec, I.; Ames, C.P.; Snyder, E.Y. Genetic Modification of Neural Stem Cells. *Mol. Ther.* **2008**, *16*, 450–457. [CrossRef]

192. Mayorquin, L.C.; Rodriguez, A.V.; Sutachan, J.-J.; Albarracín, S.L. Connexin-Mediated Functional and Metabolic Coupling between Astrocytes and Neurons. *Front. Mol. Neurosci.* **2018**, *11*, 118. [CrossRef]

193. Fujii, Y.; Maekawa, S.; Morita, M. Astrocyte calcium waves propagate proximally by gap junction and distally by extracellular diffusion of ATP released from volume-regulated anion channels. *Sci. Rep.* **2017**, *7*, 13115. [CrossRef]
194. De Bock, M.; Decrock, E.; Wang, N.; Bol, M.; Vinken, M.; Bultynck, G.; Leybaert, L. The dual face of connexin-based astroglial Ca2+ communication: A key player in brain physiology and a prime target in pathology. *Biochim. Biophys. Acta (BBA)-Mol. Cell Res.* **2014**, *1843*, 2211–2232. [CrossRef]
195. Li, D.; Hérault, K.; Isacoff, E.Y.; Oheim, M.; Ropert, N. Optogenetic activation of LiGluR-expressing astrocytes evokes anion channel-mediated glutamate release. *J. Physiol.* **2012**, *590*, 855–873. [CrossRef] [PubMed]
196. Agulhon, C.; Sun, M.-Y.; Murphy, T.; Myers, T.; Lauderdale, K.; Fiacco, T.A. Calcium Signaling and Gliotransmission in Normal vs. Reactive Astrocytes. *Front. Pharmacol.* **2012**, *3*, 139. [CrossRef] [PubMed]
197. Kawasaki, A.; Hayashi, T.; Nakachi, K.; Trosko, J.E.; Sugihara, K.; Kotake, Y.; Ohta, S. Modulation of connexin 43 in rotenone-induced model of Parkinson's disease. *Neuroscience* **2009**, *160*, 61–68. [CrossRef] [PubMed]
198. Nakase, T.; Fushiki, S.; Naus, C.C. Astrocytic gap junctions composed of connexin 43 reduce apoptotic neuronal damage in cerebral ischemia. *Stroke* **2003**, *34*, 1987–1993. [CrossRef]
199. Salmina, A.B.; Komleva, Y.K.; Lopatina, O.L.; Gorina, Y.V.; Malinovskaya, N.A.; Pozhilenkova, E.A.; Panina, Y.A.; Zhukov, E.L.; Medvedeva, N.N. CD38 and CD157 expression: Glial control of neurodegeneration and neuroinflammation. *Messenger* **2014**, *3*, 78–85. [CrossRef]
200. Malavasi, F.; Deaglio, S.; Funaro, A.; Ferrero, E.; Horenstein, A.L.; Ortolan, E.; Vaisitti, T.; Aydin, S. Evolution and Function of the ADP Ribosyl Cyclase/CD38 Gene Family in Physiology and Pathology. *Physiol. Rev.* **2008**, *88*, 841–886. [CrossRef] [PubMed]
201. Schöndorf, D.C.; Ivanyuk, D.; Baden, P.; Sanchez-Martinez, A.; De Cicco, S.; Yu, C.; Giunta, I.; Schwarz, L.K.; Di Napoli, G.; Panagiotakopoulou, V.; et al. The NAD+ Precursor Nicotinamide Riboside Rescues Mitochondrial Defects and Neuronal Loss in iPSC and Fly Models of Parkinson's Disease. *Cell Rep.* **2018**, *23*, 2976–2988. [CrossRef]
202. Song, E.K.; Rah, S.Y.; Lee, Y.R.; Yoo, C.H.; Kim, Y.R.; Yeom, J.H.; Park, K.H.; Kim, J.S.; Kim, U.H.; Han, M.K. Connexin-43 hemichannels mediate cyclic ADP-ribose generation and its Ca2+-mobilizing activity by NAD+/cyclic ADP-ribose transport. *J. Biol. Chem.* **2011**, *286*, 44480–44490. [CrossRef]
203. Greer, K.; Chen, J.; Brickler, T.; Gourdie, R.; Theus, M.H. Modulation of gap junction-associated Cx43 in neural stem/progenitor cells following traumatic brain injury. *Brain Res. Bull.* **2017**, *134*, 38–46. [CrossRef] [PubMed]
204. Jäderstad, J.; Jäderstad, L.M.; Herlenius, E. Dynamic changes in connexin expression following engraftment of neural stem cells to striatal tissue. *Exp. Cell Res.* **2011**, *317*, 70–81. [CrossRef]
205. Jäderstad, J.; Jäderstad, L.M.; Li, J.; Chintawar, S.; Salto, C.; Pandolfo, M.; Ourednik, V.; Teng, Y.D.; Sidman, R.L.; Arenas, E.; et al. Communication via gap junctions underlies early functional and beneficial interactions between grafted neural stem cells and the host. *Proc. Natl. Acad. Sci. USA* **2010**, *107*, 5184–5189. [CrossRef]
206. Habibey, R.; Sharma, K.; Swiersy, A.; Busskamp, V. Optogenetics for neural transplant manipulation and functional analysis. *Biochem. Biophys. Res. Commun.* **2020**, *527*, 343–349. [CrossRef] [PubMed]
207. Bruzzone, S.; Guida, L.; Zocchi, E.; Franco, L.; De Flora, A. Connexin 43 hemi channels mediate Ca2+-regulated transmembrane NAD+ fluxes in intact cells. *FASEB J.* **2001**, *15*, 10–12. [CrossRef]
208. Díaz, E.F.; Labra, V.C.; Alvear, T.F.; Mellado, L.A.; Inostroza, C.A.; Oyarzún, J.E.; Salgado, N.; Quintanilla, R.A.; Orellana, J.A. Connexin 43 hemichannels and pannexin-1 channels contribute to the α-synuclein-induced dysfunction and death of astrocytes. *Glia* **2019**, *67*, 1598–1619. [CrossRef] [PubMed]
209. Jäderstad, J.; Brismar, H.; Herlenius, E. Hypoxic preconditioning increases gap-junctional graft and host communication. *Neuroreport* **2010**, *21*, 1126–1132. [CrossRef]
210. Xie, Z.; Yang, Q.; Song, D.; Quan, Z.; Qing, H. Optogenetic manipulation of astrocytes from synapses to neuronal networks: A potential therapeutic strategy for neurodegenerative diseases. *Glia* **2020**, *68*, 215–226. [CrossRef]
211. Salmina, A.B.; Morgun, A.V.; Kuvacheva, N.V.; Lopatina, O.L.; Komleva, Y.K.; Malinovskaya, N.A.; Pozhilenkova, E.A. Establishment of neurogenic microenvironment in the neurovascular unit: The connexin 43 story. *Rev. Neurosci.* **2014**, *25*, 97–111. [CrossRef]
212. Kunze, A.; Congreso, M.R.; Hartmann, C.; Wallraff-Beck, A.; Hüttmann, K.; Bedner, P.; Requardt, R.; Seifert, G.; Redecker, C.; Willecke, K.; et al. Connexin expression by radial glia-like cells is required for neurogenesis in the adult dentate gyrus. *Proc. Natl. Acad. Sci. USA* **2009**, *106*, 11336–11341. [CrossRef] [PubMed]
213. Parekkadan, B.; Berdichevsky, Y.; Irimia, D.; Leeder, A.; Yarmush, G.; Toner, M.; Levine, J.B.; Yarmush, M.L. Cell-cell interaction modulates neuroectodermal specification of embryonic stem cells. *Neurosci. Lett.* **2008**, *438*, 190–195. [CrossRef]
214. Rinaldi, F.; Hartfield, E.M.; Crompton, L.A.; Badger, J.L.; Glover, C.P.; Kelly, C.M.; Rosser, A.E.; Uney, J.B.; Caldwell, M.A. Cross-regulation of Connexin43 and β-catenin influences differentiation of human neural progenitor cells. *Cell Death Dis.* **2014**, *5*, e1017. [CrossRef]
215. Talaverón, R.; Matarredona, E.R.; Herrera, A.; Medina, J.M.; Tabernero, A. Connexin43 Region 266–283, via Src Inhibition, Reduces Neural Progenitor Cell Proliferation Promoted by EGF and FGF-2 and Increases Astrocytic Differentiation. *Int. J. Mol. Sci.* **2020**, *21*, 8852. [CrossRef]
216. Lagos-Cabré, R.; Brenet, M.; Díaz, J.; Pérez, R.D.; Pérez, L.A.; Herrera-Molina, R.; Quest, A.F.G.; Leyton, L. Intracellular Ca(2+) Increases and Connexin 43 Hemichannel Opening Are Necessary but Not Sufficient for Thy-1-Induced Astrocyte Migration. *Int. J. Mol. Sci.* **2018**, *19*, 2179. [CrossRef]

217. Hou, X.; Khan, M.R.A.; Turmaine, M.; Thrasivoulou, C.; Becker, D.L.; Ahmed, A. Wnt signaling regulates cytosolic translocation of connexin 43. *Am. J. Physiol.-Regul. Integr. Comp. Physiol.* **2019**, *317*, R248–R261. [CrossRef]
218. Koulakoff, A.; Ezan, P.; Giaume, C. Neurons control the expression of connexin 30 and connexin 43 in mouse cortical astrocytes. *Glia* **2008**, *56*, 1299–1311. [CrossRef] [PubMed]
219. Deisseroth, K.; Singla, S.; Toda, H.; Monje, M.; Palmer, T.D.; Malenka, R.C. Excitation-Neurogenesis Coupling in Adult Neural Stem/Progenitor Cells. *Neuron* **2004**, *42*, 535–552. [CrossRef]
220. Wu, L.; Dong, A.; Dong, L.; Wang, S.-Q.; Li, Y. PARIS, an optogenetic method for functionally mapping gap junctions. *bioRxiv* **2018**, 465781. [CrossRef]
221. Boyle, P.M.; Yu, J.; Klimas, A.; Williams, J.C.; Trayanova, N.A.; Entcheva, E. OptoGap is an optogenetics-enabled assay for quantification of cell–cell coupling in multicellular cardiac tissue. *Sci. Rep.* **2021**, *11*, 9310. [CrossRef] [PubMed]
222. Jones, M.B.; Siderovski, D.P.; Hooks, S.B. The G betagamma dimer as a novel source of selectivity in G-protein signaling: GGL-ing at convention. *Mol. Interv.* **2004**, *4*, 200–214. [CrossRef] [PubMed]
223. McKinley, J.W.; Shi, Z.; Kawikova, I.; Hur, M.; Bamford, I.J.; Sudarsana Devi, S.P.; Vahedipour, A.; Darvas, M.; Bamford, N.S. Dopamine Deficiency Reduces Striatal Cholinergic Interneuron Function in Models of Parkinson's Disease. *Neuron* **2019**, *103*, 1056–1072.e6. [CrossRef]
224. Bernácer, J.; Prensa, L.; Giménez-Amaya, J.M. Distribution of GABAergic interneurons and dopaminergic cells in the functional territories of the human striatum. *PLoS ONE* **2012**, *7*, e30504. [CrossRef]
225. Lane, E.M.; Handley, O.J.; Rosser, A.E.; Dunnett, S.B. Potential cellular and regenerative approaches for the treatment of Parkinson's disease. *Neuropsychiatr. Dis. Treat.* **2008**, *4*, 835–845. [CrossRef]
226. Xenias, H.S.; Ibáñez-Sandoval, O.; Koós, T.; Tepper, J.M. Are striatal tyrosine hydroxylase interneurons dopaminergic? *J. Neurosci.* **2015**, *35*, 6584–6599. [CrossRef] [PubMed]
227. Ünal, B.; Shah, F.; Kothari, J.; Tepper, J.M. Anatomical and electrophysiological changes in striatal TH interneurons after loss of the nigrostriatal dopaminergic pathway. *Brain Struct. Funct.* **2015**, *220*, 331–349. [CrossRef]
228. Busceti, C.L.; Bucci, D.; Molinaro, G.; Di Pietro, P.; Zangrandi, L.; Gradini, R.; Moratalla, R.; Battaglia, G.; Bruno, V.; Nicoletti, F.; et al. Lack or inhibition of dopaminergic stimulation induces a development increase of striatal tyrosine hydroxylase-positive interneurons. *PLoS ONE* **2012**, *7*, e44025. [CrossRef]
229. Lenz, J.D.; Lobo, M.K. Optogenetic insights into striatal function and behavior. *Behav. Brain Res.* **2013**, *255*, 44–54. [CrossRef] [PubMed]
230. Caravaggio, F.; Hahn, M.; Nakajima, S.; Gerretsen, P.; Remington, G.; Graff-Guerrero, A. Reduced insulin-receptor mediated modulation of striatal dopamine release by basal insulin as a possible contributing factor to hyperdopaminergia in schizophrenia. *Med. Hypotheses* **2015**, *85*, 391–396. [CrossRef]
231. Stouffer, M.A.; Woods, C.A.; Patel, J.C.; Lee, C.R.; Witkovsky, P.; Bao, L.; Machold, R.P.; Jones, K.T.; de Vaca, S.C.; Reith, M.E.A.; et al. Insulin enhances striatal dopamine release by activating cholinergic interneurons and thereby signals reward. *Nat. Commun.* **2015**, *6*, 8543. [CrossRef]
232. Harrison, N.J.; Connolly, E.; Gascón Gubieda, A.; Yang, Z.; Altenhein, B.; Losada Perez, M.; Moreira, M.; Sun, J.; Hidalgo, A. Regenerative neurogenic response from glia requires insulin-driven neuron-glia communication. *Elife* **2021**, *10*, e58756. [CrossRef]
233. Ziegler, A.N.; Levison, S.W.; Wood, T.L. Insulin and IGF receptor signalling in neural-stem-cell homeostasis. *Nat. Rev. Endocrinol.* **2015**, *11*, 161–170. [CrossRef]
234. Spinelli, M.; Fusco, S.; Grassi, C. Brain Insulin Resistance and Hippocampal Plasticity: Mechanisms and Biomarkers of Cognitive Decline. *Front. Neurosci.* **2019**, *13*, 788. [CrossRef]
235. Fiory, F.; Perruolo, G.; Cimmino, I.; Cabaro, S.; Pignalosa, F.C.; Miele, C.; Beguinot, F.; Formisano, P.; Oriente, F. The Relevance of Insulin Action in the Dopaminergic System. *Front. Neurosci.* **2019**, *13*, 868. [CrossRef]
236. Zhang, C.; Feng, W.; Vodovozova, E.; Tretiakova, D.; Boldyrevd, I.; Li, Y.; Kürths, J.; Yu, T.; Semyachkina-Glushkovskaya, O.; Zhu, D. Photodynamic opening of the blood-brain barrier to high weight molecules and liposomes through an optical clearing skull window. *Biomed. Opt. Express* **2018**, *9*, 4850–4862. [CrossRef]
237. Nishijima, T.; Piriz, J.; Duflot, S.; Fernandez, A.M.; Gaitan, G.; Gomez-Pinedo, U.; Verdugo, J.M.G.; Leroy, F.; Soya, H.; Nuñez, A.; et al. Neuronal Activity Drives Localized Blood-Brain-Barrier Transport of Serum Insulin-like Growth Factor-I into the CNS. *Neuron* **2010**, *67*, 834–846. [CrossRef] [PubMed]
238. Ernst, A.; Alkass, K.; Bernard, S.; Salehpour, M.; Perl, S.; Tisdale, J.; Possnert, G.; Druid, H.; Frisén, J. Neurogenesis in the Striatum of the Adult Human Brain. *Cell* **2014**, *156*, 1072–1083. [CrossRef] [PubMed]
239. Farzanehfar, P. Comparative review of adult midbrain and striatum neurogenesis with classical neurogenesis. *Neurosci. Res.* **2018**, *134*, 1–9. [CrossRef] [PubMed]
240. Inta, D.; Cameron, H.A.; Gass, P. New neurons in the adult striatum: From rodents to humans. *Trends Neurosci.* **2015**, *38*, 517–523. [CrossRef]
241. Magnusson, J.P.; Göritz, C.; Tatarishvili, J.; Dias, D.O.; Smith, E.M.; Lindvall, O.; Kokaia, Z.; Frisén, J. A latent neurogenic program in astrocytes regulated by Notch signaling in the mouse. *Science* **2014**, *346*, 237–241. [CrossRef] [PubMed]
242. Wei, Z.-Y.D.; Shetty, A.K. Treating Parkinson's disease by astrocyte reprogramming: Progress and challenges. *Sci. Adv.* **2021**, *7*, eabg3198. [CrossRef]

243. Niu, W.; Zang, T.; Zou, Y.; Fang, S.; Smith, D.K.; Bachoo, R.; Zhang, C.-L. In vivo reprogramming of astrocytes to neuroblasts in the adult brain. *Nat. Cell Biol.* **2013**, *15*, 1164–1175. [CrossRef] [PubMed]
244. Masserdotti, G.; Gillotin, S.; Sutor, B.; Drechsel, D.; Irmler, M.; Jørgensen, H.; Sass, S.; Theis, F.; Beckers, J.; Berninger, B.; et al. Transcriptional Mechanisms of Proneural Factors and REST in Regulating Neuronal Reprogramming of Astrocytes. *Cell Stem Cell* **2015**, *17*, 74–88. [CrossRef] [PubMed]
245. Rivetti di Val Cervo, P.; Romanov, R.A.; Spigolon, G.; Masini, D.; Martín-Montañez, E.; Toledo, E.M.; La Manno, G.; Feyder, M.; Pifl, C.; Ng, Y.-H.; et al. Induction of functional dopamine neurons from human astrocytes in vitro and mouse astrocytes in a Parkinson's disease model. *Nat. Biotechnol.* **2017**, *35*, 444–452. [CrossRef]
246. Ghasemi-Kasman, M.; Hajikaram, M.; Baharvand, H.; Javan, M. MicroRNA-Mediated In Vitro and In Vivo Direct Conversion of Astrocytes to Neuroblasts. *PLoS ONE* **2015**, *10*, e0127878. [CrossRef]
247. Mohapel, P.; Frielingsdorf, H.; Häggblad, J.; Zachrisson, O.; Brundin, P. Platelet-Derived Growth Factor (PDGF-BB) and Brain-Derived Neurotrophic Factor (BDNF) induce striatal neurogenesis in adult rats with 6-hydroxydopamine lesions. *Neuroscience* **2005**, *132*, 767–776. [CrossRef] [PubMed]
248. Fallon, J.; Reid, S.; Kinyamu, R.; Opole, I.; Opole, R.; Baratta, J.; Korc, M.; Endo, T.L.; Duong, A.; Nguyen, G.; et al. In vivo induction of massive proliferation, directed migration, and differentiation of neural cells in the adult mammalian brain. *Proc. Natl. Acad. Sci. USA* **2000**, *97*, 14686–14691. [CrossRef] [PubMed]
249. Huot, P.; Lévesque, M.; Parent, A. The fate of striatal dopaminergic neurons in Parkinson's disease and Huntington's chorea. *Brain* **2006**, *130*, 222–232. [CrossRef]
250. Hermann, A.; Storch, A. Endogenous regeneration in Parkinson's disease: Do we need orthotopic dopaminergic neurogenesis? *Stem Cells* **2008**, *26*, 2749–2752. [CrossRef]
251. Papanikolaou, T.; Lennington, J.B.; Betz, A.; Figueiredo, C.; Salamone, J.D.; Conover, J.C. In vitro generation of dopaminergic neurons from adult subventricular zone neural progenitor cells. *Stem Cells Dev.* **2008**, *17*, 157–172. [CrossRef]

International Journal of
Molecular Sciences

Article

Imaging of C-fos Activity in Neurons of the Mouse Parietal Association Cortex during Acquisition and Retrieval of Associative Fear Memory

Olga I. Ivashkina [1,2,3,*,†] , Anna M. Gruzdeva [2,†] , Marina A. Roshchina [4], Ksenia A. Toropova [1,2,3] and Konstantin V. Anokhin [1,3]

1   Institute for Advanced Brain Studies, Lomonosov Moscow State University, 119991 Moscow, Russia; xen.alexander@gmail.com (K.A.T.); k.anokhin@gmail.com (K.V.A.)
2   National Research Center "Kurchatov Institute", 123182 Moscow, Russia; annadronova@mail.ru
3   Laboratory for Neurobiology of Memory, P.K. Anokhin Institute of Normal Physiology, 125315 Moscow, Russia
4   Institute of Higher Nervous Activity and Neurophysiology of RAS, 117485 Moscow, Russia; marina.zots@gmail.com
*   Correspondence: oivashkina@gmail.com; Tel.: +7-9264289555
†   These authors contributed equally to this work.

Abstract: The parietal cortex of rodents participates in sensory and spatial processing, movement planning, and decision-making, but much less is known about its role in associative learning and memory formation. The present study aims to examine the involvement of the parietal association cortex (PtA) in associative fear memory acquisition and retrieval in mice. Using ex vivo c-Fos immunohistochemical mapping and in vivo Fos-EGFP two-photon imaging, we show that PtA neurons were specifically activated both during acquisition and retrieval of cued fear memory. Fos immunohistochemistry revealed specific activation of the PtA neurons during retrieval of the 1-day-old fear memory. In vivo two-photon Fos-EGFP imaging confirmed this result and in addition detected specific c-Fos responses of the PtA neurons during acquisition of cued fear memory. To allow a more detailed study of the long-term activity of such PtA engram neurons, we generated a Fos-Cre-GCaMP transgenic mouse line that employs the Targeted Recombination in Active Populations (TRAP) technique to detect calcium events specifically in cells that were Fos-active during conditioning. We show that gradual accumulation of GCaMP3 in the PtA neurons of Fos-Cre-GCaMP mice peaks at the 4th day after fear learning. We also describe calcium transients in the cell bodies and dendrites of the TRAPed neurons. This provides a proof-of-principle for TRAP-based calcium imaging of PtA functions during memory processes as well as in experimental models of fear- and anxiety-related psychiatric disorders and their specific therapies.

Keywords: parietal association cortex; fear memory; c-fos; calcium activity; transgenic mice; two-photon imaging

Citation: Ivashkina, O.I.; Gruzdeva, A.M.; Roshchina, M.A.; Toropova, K.A.; Anokhin, K.V. Imaging of C-fos Activity in Neurons of the Mouse Parietal Association Cortex during Acquisition and Retrieval of Associative Fear Memory. *Int. J. Mol. Sci.* **2021**, *22*, 8244. https://doi.org/10.3390/ijms22158244

Academic Editors: Piotr D. Bregestovski and Carlo Matera

Received: 30 June 2021
Accepted: 28 July 2021
Published: 31 July 2021

Publisher's Note: MDPI stays neutral with regard to jurisdictional claims in published maps and institutional affiliations.

1. Introduction

The parietal cortex is an associative cortical area that participates in various integrative brain functions, including multisensory processing, decision-making, motion planning, navigation, attention, and working memory [1]. Though this area has been extensively studied in cognitive tasks in primates [2], only more recently has it become the subject of corresponding analyses in rodents [3,4].

The parietal region in the rodent brain is defined as an area between visual and somatosensory cortices [3,5,6]. It is connected with diverse brain areas, including other associative cortical regions, such as orbitofrontal, retrosplenial and anterior cingulate cortices [4,7]. Its neurons respond to modality-specific (auditory, visual, or somatosensory) as well as complex stimuli [8–10]. In rodents, this area is also involved in spatial navigation [11] and decision-making [12,13].

According to Franklin and Paxinos (2007), the mouse parietal cortex consists of anterior and posterior parts, and the anterior part corresponds to the parietal association cortex (PtA). Though most rodent studies address the functions of the posterior parietal cortex, network analysis of mouse cortical connectivity revealed that the PtA has a high level of betweenness centrality that makes it a strong hub region within the cortical network [14]. Therefore, in this study, we focus on the role of mouse PtA in associative memory processes.

Retrosplenial, cingulate, and frontal associative cortical areas, which send projections to the PtA, are involved in the coding and retrieval of different types of memory [15–19]. Interestingly, there is no evidence about direct connections between the PtA and the prelimbic prefrontal cortex, which is known to be involved in the formation and retrieval of associative memory including conditioned fear memory in rats in mice [20–22]. The rat PtA is known to participate in the retrieval of recent and remote spatial memory [23,24]. However, the contribution of PtA to associative memory is still poorly understood. A direct way to address this question is to examine the specific expression of immediate-early genes (IEGs) involved in experience-dependent neuronal plasticity.

Expression of IEGs such as *c-fos* is commonly used to identify neurons activated by learning and involved in memory encoding [25–27]. *c-fos* encodes the transcription factor that regulates the activity of effector genes and the following long-term plasticity in neurons [28]. Different Fos-based methods are used to investigate the experience-induced changes of brain neuronal circuits in various learning and memory paradigms. C-Fos immunostaining is commonly used to access neuronal activity only at a single time point. In contrast, in vivo Fos-imaging allows observing the activation of the same neuronal population in different behavioral episodes. In this case, Fos-EGFP transgenic mice allow repeated imaging of Fos-positive neurons and comparison of neuronal populations activated during learning and memory retrieval [18,29–31]. Similarly, the method of targeted recombination in active populations (TRAP) was used to capture the candidate engram neurons [25,32]. TRAP is an approach to obtain permanent genetic access to distributed neuronal ensembles that are activated by experiences within a limited time window [32,33]. In Fos-TRAP transgenic mice, the tamoxifen-dependent recombinase CreERT2 is expressed in an activity-dependent manner under the control of the *c-fos* promoter. Active cells that express CreERT2 undergo recombination only in the presence of tamoxifen (TM). This allows genetic access to neurons that were active during a time window less than 24 h after injection. Nonactive cells do not express CreERT2 and do not undergo recombination, even if TM was injected [32]. Previously the TRAP approach was used to assess the involvement of neurons genetically captured during context fear conditioning (FC) in subsequent memory retrieval [33,34]. Optogenetic silencing of TRAP-labeled neuronal populations in CA3 or DG prevented the expression of the corresponding memory [33]. In the present study, we used c-Fos immunostaining, Fos-EGFP in vivo imaging, and Fos-TRAP to investigate the involvement of PtA cortex in the encoding and retrieval of associative fear memory in mice. Using c-Fos-immunohistochemistry, we found that PtA neurons were specifically activated during cued fear memory retrieval. In vivo Fos-EGFP imaging showed, in addition, the specific changes of activity during both fear conditioning and memory retrieval. Finally, we applied the TRAP technique for calcium imaging specifically in PtA engram cells. We showed no specific activity in such cells during conditioned stimulus (CS) presentation during the 1-day-old memory retrieval.

2. Results

2.1. Immunohistochemical Analysis of Fos Expression in the PtA

First, we analyzed c-Fos expression in the PtA during cued fear memory acquisition and retrieval. Wild-type mice were trained to associate CSs with foot shocks. We used three groups of mice: Paired (CS was paired with foot shock), CS-only (mice received only CS), and Home Cage (HC, mice were not tested in FC assay). Only Paired group but not CS-only group developed freezing behavior as the number of pairings increased (two-way ANOVA, $p < 0.0001$) (Figure 1a). Freezing response to the CS during memory retrieval was higher in the Paired group compared with CS-only group (two-way ANOVA, $p < 0.0001$) (Figure 1b). Freezing level during novel context exploration, however, was the same in the Paired and CS-only groups (two-way ANOVA, $p = 0.5341$), (Figure 1b). These results demonstrate that mice learned association between conditioned and unconditioned stimuli (CS-US association) and did not exhibit memory generalization to a novel context.

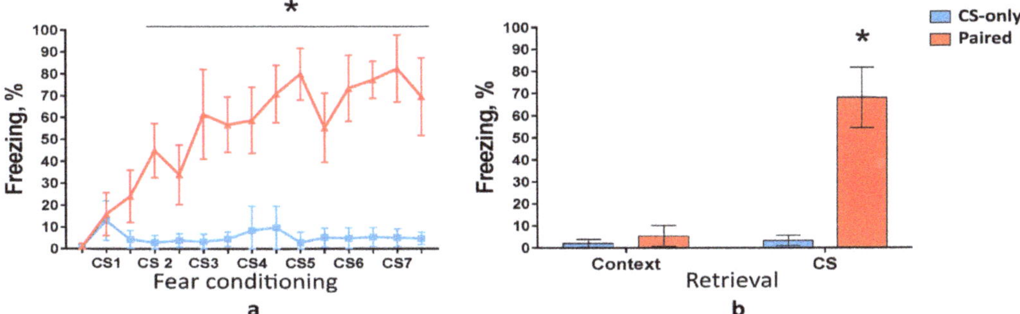

Figure 1. Freezing during (**a**) fear conditioning (FC) and (**b**) memory retrieval 24 h after the FC (mean, 95% confidence interval (CI)). Fear response increased during training only in the Paired group (* $p < 0.0001$, two-way ANOVA). Trained animals froze more than CS-only group during conditioned stimulus (CS) memory retrieval (* $p < 0.0001$, two-way ANOVA with post hoc Tukey test).

For the next step, we sacrificed the animals from all groups one hour after FC or memory retrieval and performed immunostaining to detect c-Fos in the PtA. The density of the PtA c-Fos-positive cells was higher in the Paired and CS-only groups than in the HC group during FC or memory retrieval (one-way ANOVA and post hoc Tukey test, $p < 0.001$) (Figure 2). However, there was no significant difference between Paired and CS-only groups during FC (one-way ANOVA and post hoc Tukey test, $p = 0.9532$) (Figure 2a). On the contrary, c-Fos-positive cell density was higher in the Paired group than in the CS-only group during memory retrieval (one-way ANOVA and post hoc Tukey test, $p = 0.0364$) (Figure 2b). This data suggest that PtA neurons are active during new experience acquisition regardless of the associative or nonassociative nature of this experience. At the same time, PtA is specifically activated during CS memory retrieval, suggesting its role in the recent fear memory storage.

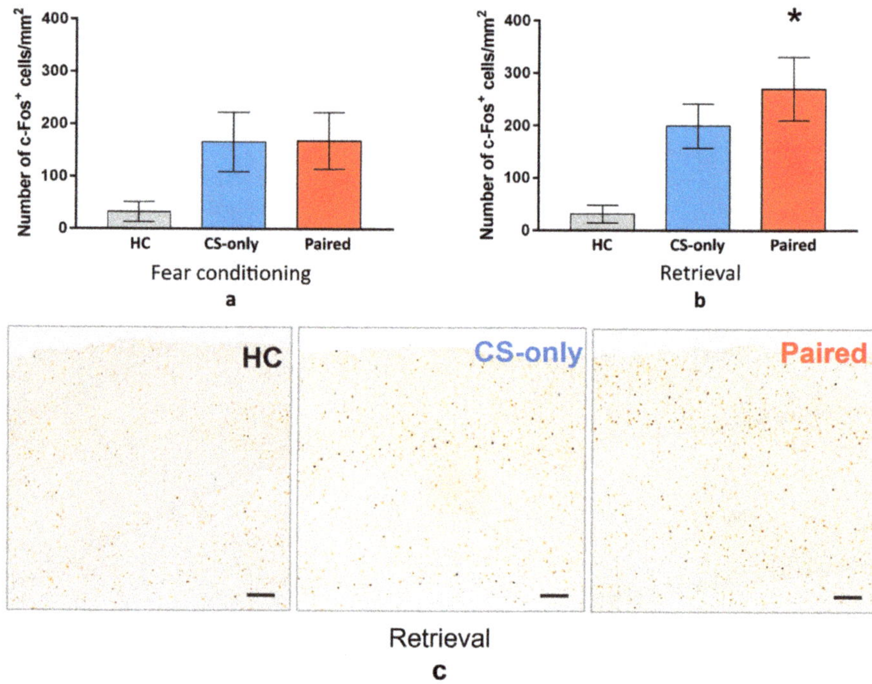

Figure 2. The density of c-Fos-positive cells in the PtA 1 h after (**a**) fear learning and (**b**) memory retrieval on the next day (mean, 95% CI). The density of the PtA c-Fos-positive cells was higher in the Paired and CS-only groups than in the Home cage (HC) group during FC or memory retrieval ($p < 0.001$, one-way ANOVA and post hoc Tukey test). During memory retrieval, c-Fos-positive cells density was higher in the Paired group compared with the CS-only group (* $p = 0.0364$, one-way ANOVA and post hoc Tukey test). (**c**) Image of Fos-positive cells in PtA in HC, CS-only and Paired groups (scale bar is 100 μm).

2.2. In Vivo Investigation of C-Fos Expression in the PtA of Fos-EGFP Mice

In the next experiment, we analyzed c-Fos activation of individual PtA neurons during cued memory acquisition and retrieval using in vivo two-photon imaging in Fos-EGFP transgenic mice. Imaging was performed three days before FC to access basal level of c-Fos expression in PtA neurons in a home cage, 90 min after FC session, and 90 min after memory test (Figure 3a). This approach allowed us to address and compare c-Fos activity of a specific neuronal population in PtA after different behavioral procedures in the same animal.

Overall, we identified 9325 neurons in all the mice during all sessions of two-photon visualization. In the control imaging session (home cage condition, 3 days before fear conditioning) 520 ± 160 Fos-EGFP positive neurons per mouse (mean $\pm$ 95% CI) in 16 mice were identified. 17% of all identified neurons changed their activity (i.e., were activated or inactivated) at least in one imaging session. We normalized the number of Fos-EGFP positive neurons in both conditioning and retrieval sessions to the number of neurons that were active during the control imaging session and found that the number of Fos-EGFP positive neurons increased during the FC training and memory recall in the Paired group but not in the CS-only and HC groups (t-test, compared with 1, $p = 0.0156$) (Figure 3b). Moreover, we found that during learning, but not during memory retrieval, the number of Fos-EGFP positive PtA cells was significantly higher in the Paired group than in the CS-only or HC groups (two-way ANOVA and post hoc Tukey test, $p < 0.05$).

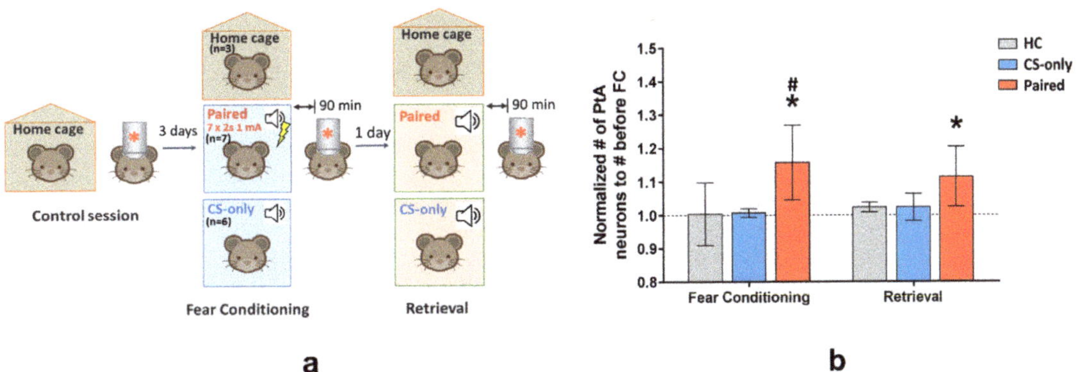

a

b

Figure 3. (**a**) Design and timeline of the in vivo imaging experiment with Fos-EGFP mice. (**b**) Number of PtA Fos-EGFP positive neurons normalized to the number of neurons active during the control session (mean, 95% CI). The number of Fos-EGFP positive neurons was increased during learning and memory retrieval on the next day in the Paired group ($n = 7$) but not in the control groups (* $p = 0.0156$, compared with 1, *t*-test; # $p < 0.05$ compared with CS-only ($n = 3$) and HC group ($n = 3$) groups, two-way ANOVA and post hoc Tukey test).

Next, we compared the number of neurons that were inactive during the control imaging session, but were activated during FC or memory test. The number of neurons activated during fear conditioning normalized to the total number of neurons for each group was higher in the Paired group than in the CS-only and HC groups (64 ($n = 7$ mice), 40 ($n = 6$), and 32 ($n = 3$) neurons in average, respectively, one-way ANOVA and post hoc Tukey test, $p < 0.05$) (Figure 4a). The number of neurons activated by memory retrieval was similar in all groups (64 ($n = 7$), 49 ($n = 6$), and 38 ($n = 3$ mice) neurons on average, respectively, one-way ANOVA, $p = 0.2185$) (Figure 4b).

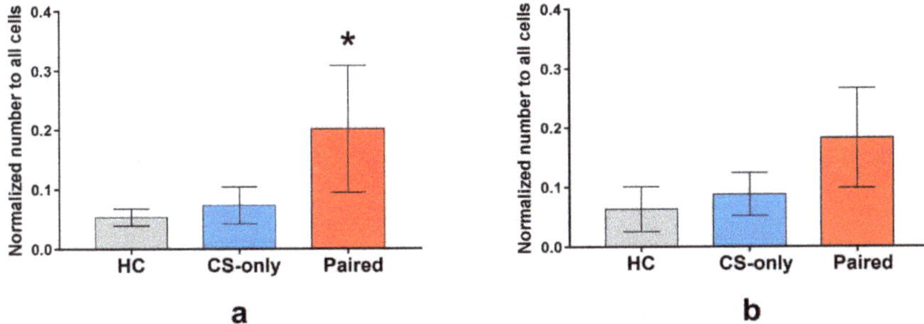

a

b

Figure 4. Number of the PtA neurons that were inactive before FC and active during (**a**) fear learning and (**b**) memory retrieval on the next day normalized to all identified neurons in Fos-EGFP mice (mean, 95% CI). The number of neurons that were inactive before FC and were active during FC was higher in the Paired group ($n = 7$) than in CS-only ($n = 6$) and HC ($n = 3$) groups (* $p < 0.05$, one-way ANOVA and post hoc Tukey test).

Altogether, these results indicate that, as sampled by in vivo Fos-EGFP activity, PtA neurons are involved in associative fear memory formation and retrieval in the mouse brain.

However, there are certain restrictions with the use of c-Fos as an indicator of conditioned neuronal activity. Long duration of c-Fos protein synthesis or of Fos-triggered EGFP accumulation does not allow precise matching of these biochemical responses to the stimuli that induced them. Furthermore, not all neuronal responses are accompanied by induction of *c-fos* transcription. To overcome these limitations and to address the causal links between learning and PtA activity in a more precise manner we used TRAP to image calcium activity specifically in neurons that expressed c-Fos during fear conditioning. The next sections describe the Fos-Cre-GCaMP transgenic mice used for this purpose as a proof-of-principle to capture and image calcium transients in learning-activated PtA neurons.

2.3. Characterization of Fos-Cre-GCaMP Transgenic Mice

First, Fos^{CreER} and Ai38 (RCL-GCaMP3) transgenic mice were crossed to obtain a double transgenic line Fos-Cre-GCaMP. Next, we performed the genotyping of Fos-Cre-GCaMP offspring mice. As was expected, we found specific sites for GCaMP (226 bp) and Fos-Cre (293 bp) compared to 128 bp and 215 bp from DNA of wild-type mice.

To determine the time course of GCaMP accumulation in TRAPed neurons we performed repeated two-photon imaging of the PtA area at different time points after the session of FC-triggered Cre-recombination. The first visualization was performed two hours after FC training to investigate the possibility of early spontaneous recombination. We found no GCaMP-expressing neurons in PtA at this time point. Thus, no background or spontaneous recombination events appear in a short time after tamoxifen injection. One day after FC training the number of GCaMP-expressing neurons was at a 15% level of the maximal number of all identified neurons (Figure 5a). We found 70% GCaMP-expressing neurons two days after the recombination event. The total number of Fos-TRAPed neurons reached 15, 35, 82, and 134 for 4 mice by the 4th day after the recombination event and remained at the same level during the following imaging sessions (Figure 5a). Thus, the maximum level of GCaMP accumulation occurs on the 4th day after the Cre-recombination and persists thereafter. Based on this data, we started calcium imaging sessions in the Fos-TRAPed neurons from the 4th day after the FC-induced recombination.

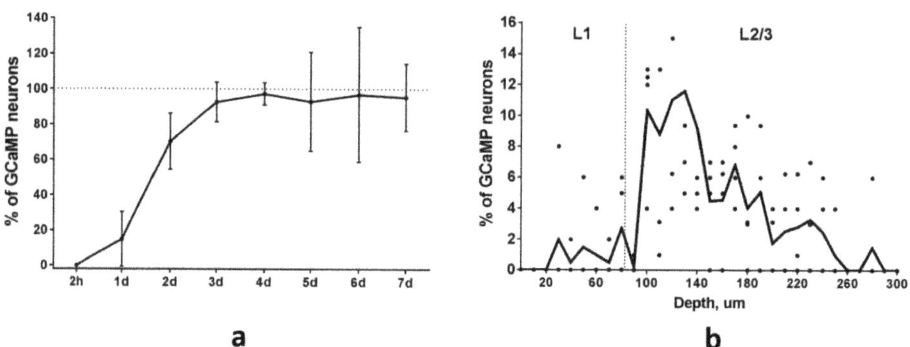

Figure 5. (**a**) GCaMP accumulation in the PtA neurons of Fos-Cre-GCaMP transgenic mice after cued fear learning. The total number of the GCaMP-positive neurons reached a maximum on the 4th day after training and remained at maximum level up to the 7th day (n = 4, mean, 95% CI). (**b**) Depth distribution of the GCaMP-expressing neurons in the PtA. Dots show the individual values for each mouse.

2.4. Two-Photon Imaging of TRAPed Neurons in the PtA

Volumetric two-photon reconstruction of a field of view (FOV) within the PtA was performed before calcium activity imaging in five Fos-Cre-GCaMP mice. The number of detected Fos-TRAPed neurons varied in the examined animals: 20, 30, 60, 70, and 86 neurons per mouse in 0.08 mm^3 volume. Neurons were visible at depths up to 450 μm. The maximum density of Fos-TRAPed GCaMP-expressing neurons was detected at 100–180 μm from the brain surface, a depth that matches the position of layer 2/3 of the PtA cortex (Figure 5b). Fos-TRAPed neurons showed fluorescence in the neuronal soma (nearly 15 μm in diameter), as well as in the processes (Figure 6A,B). Also, we visualized dendritic shafts (1–2 μm in diameter) with spines at depth of 5–20 μm under the brain surface (Figure 6B). Most of the Fos-TRAPed cells were pyramidal neurons according to the observed morphology (Figure 6C–E). This result is consistent with our data on the types of FC-induced Fos-TRAPed neurons in the mouse neocortex [30].

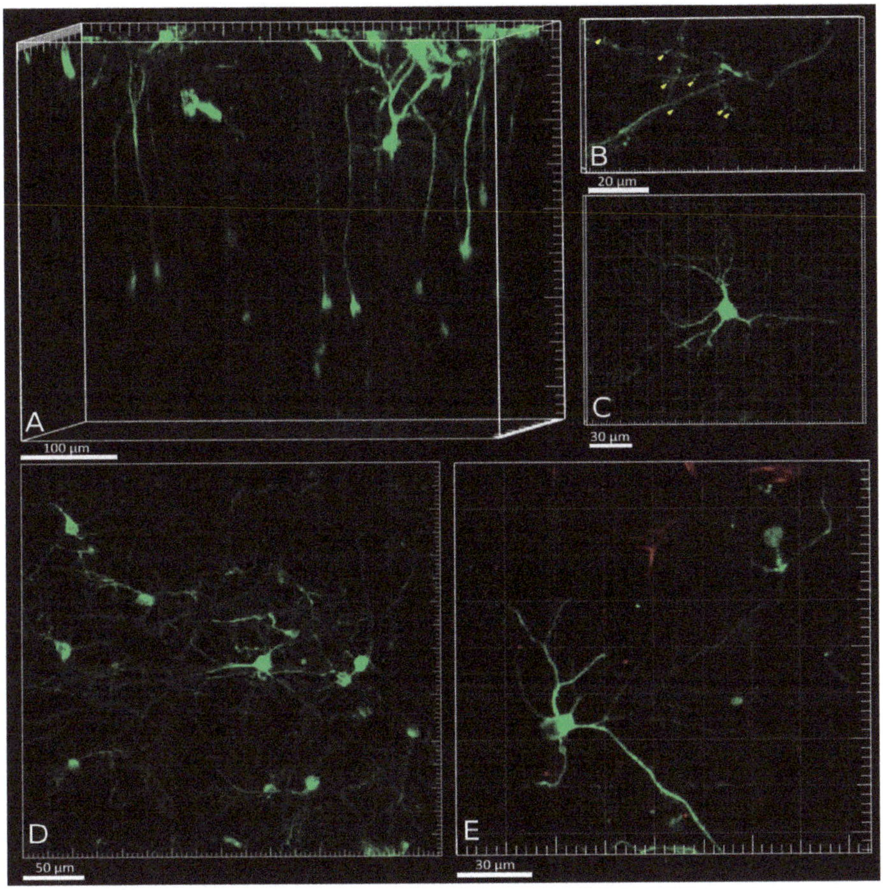

Figure 6. (**A**) 3D reconstruction of GCaMP-expressing neurons in the PtA. (**B**) Dendritic shafts and spines (tagging by yellow arrows) at cortical layer 1 of the PtA. (**C–E**) Examples of GCaMP-positive neurons in cortical layer 2 of the PtA. Green channel: GCaMP signal, red channel: autofluorescence of collagen at the brain surface.

To examine calcium activity in the Fos-TRAPed PtA neurons during fear memory recall we presented head-fixed mice with the CS and simultaneously recorded GCaMP fluorescence on the fourth day after FC training. In total, we recorded calcium activity in the somas of 28 neurons in layer 2/3 of PtA (Figure 7a). We found 11 unique calcium events in 6 neurons. Most of the neurons (80%) showed no calcium spikes during imaging sessions (Figure 7b). Surprisingly, no specific increase of calcium activity (i.e., reliably repetitive increase) during the presentation of the CS was observed (Figure 7c).

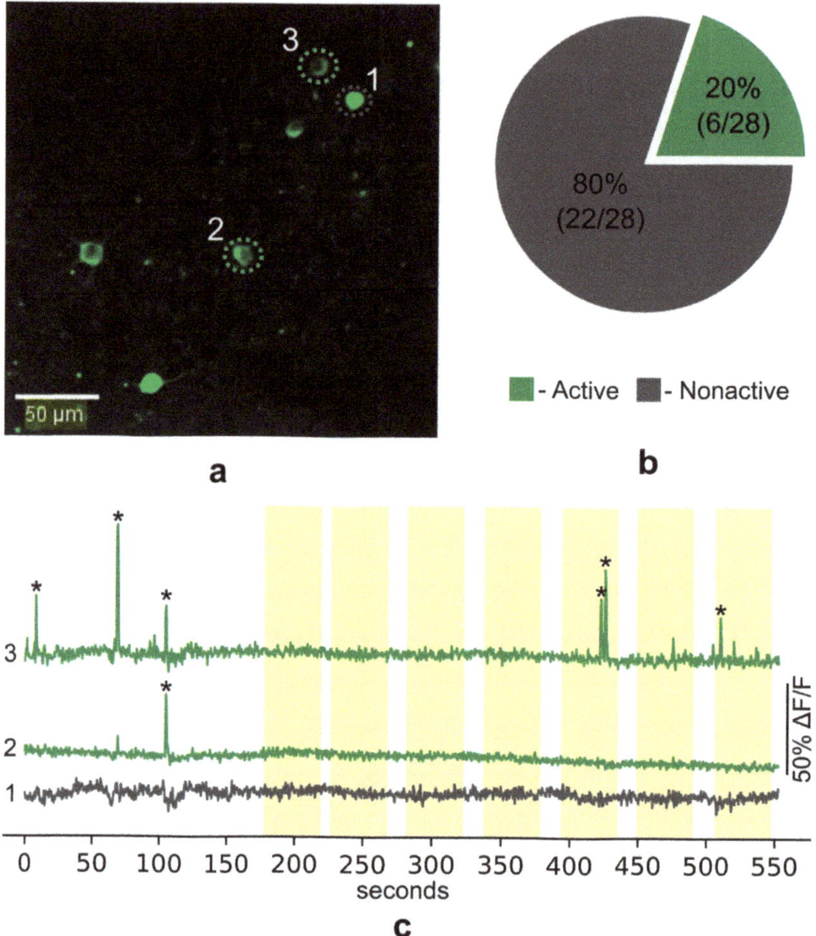

Figure 7. (**a**) Example field of view during in vivo imaging of Fos-Cre-GCaMP neurons; green-colored ROI—example of active cells, grey-colored ROI—example of nonactive cells. (**b**) Percentage of active and silent neurons (*n* = 3 mice). (**c**) Example calcium traces of Fos-TRAPed PtA neurons during presentation of the auditory CS. Asterisks * mark calcium events.

Additionally, spontaneous calcium transients were registered in the dendritic shaft and spines of the Fos-TRAPed neurons. Figure 8 shows an increase of GCaMP fluorescence in one area of the shaft and the following increase of fluorescence in the neighboring areas of the dendrite.

Taken together, these results suggest that the Fos-Cre-GCaMP mice are suitable for the investigation of calcium activity in the neurons, which were specifically activated during a particular learning episode.

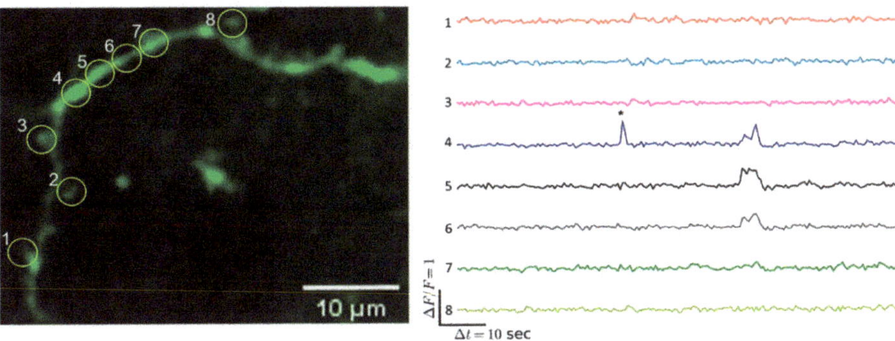

Figure 8. Example calcium traces in the dendritic spine of Fos-TRAPed PtA neuron. Yellow circles are regions of interest. We registered calcium event in ROI 4 and the following fluorescence increase in the neighboring ROIs 5 and 6 of the dendritic shaft. Asterisks * mark calcium events.

3. Discussion

Although it is known that the rodent parietal cortex is involved in various forms of sensory processing and decision-making tasks, it is less clear how this area contributes to associative memory encoding and retrieval [3,4]. Our results suggest that the anterior part of the parietal cortex (the parietal association cortex) is specifically activated during episodes of the cued associative fear memory encoding and retrieval. This conclusion is supported both by immunohistochemical analysis of c-Fos expression and by in-vivo activity imaging in Fos-EGFP mice.

A possible caveat in this conclusion appears whenever it is not possible to differentiate between sensory-induced and experience-dependent processes. According to previous studies, the parietal cortex participates in multisensory processing and receives diverse projections from other sensory brain regions [3,4]. Therefore, the increased number of c-Fos-expressing cells in PtA after the presentation of the auditory stimulus during FC training or fear memory retrieval may potentially reflect this sensory processing function of the parietal cortex. To exclude this explanation, we compared changes of the PtA activity in the Paired group with the CS-only control group, which received the same set of auditory stimuli without the subsequent foot shock exposure. Using such comparison in c-Fos-immunohistochemistry experiment, we found that PtA neurons were specifically activated in the Paired group during memory retrieval. In vivo Fos-EGFP imaging showed, in addition, the specific changes of activity during both FC training and memory retrieval. These results are consistent with the hypothesis that PtA is specifically involved in encoding and retrieval of associative memory. The diverging results of the two imaging approaches can be explained by different methods used for estimation of cell populations: by ex vivo c-Fos immunohistochemical analysis we compared the whole populations of activated neurons, while using in vivo analysis we compared cells, which were inactive before and changed their activity specifically during FC or memory test. Noticeably, in Fos-EGFP mice, we showed that the predominant proportion of identified neurons expressed Fos-EGFP during all two-photon imaging sessions. This observation is consistent with previous reports on Fos-EGFP expression in the mouse cerebral cortex [18,30,31].

Taken together our ex vivo and in vivo Fos imaging data suggest that the PtA neurons are actively engaged in fear conditioning processes. This highlights a potential role of PtA in associative memory functions as well as in fear- and anxiety-related psychiatric disorders.

In addition to Fos activity imaging, we used Fos-associated transgenic mouse approach as a proof-of-principle for long-term investigation of calcium activity in neurons specifically tagged during a cognitive episode. Currently, different IEG-based methods such as compartment analysis of temporal activity by fluorescent in-situ hybridization (catFISH) or TetTag and TRAP transgenic systems are used to label histochemically neurons that were activated during two behavioral episodes [25,32,33,35,36]. However, the catFISH technique has a limited tagging window that allows comparing populations of neurons that were activated in two episodes with only about 30 min time window [36]. Also, catFISH is not suitable for the investigation of the dynamic activity of labeled cells. Other approaches like TetTag and TRAP histochemical strategies allow comparing neuronal populations which were activated in two episodes spaced for at least 72 h [32,33]. In our study, we used the TRAP strategy to introduce genetically encoded calcium indicator GCaMP3 into neurons, that expressed *c-fos* during FC training. We showed gradual accumulation of GCaMP3 in the PtA neurons of Fos-Cre-GCaMP transgenic mice peaking on the 4th day after fear learning. We also detected calcium transients in such Fos-Cre-GCaMP cells and localized them both to cell bodies and dendrites of the TRAPed neurons.

However, to our surprise, we could not detect reliable calcium responses to the auditory CS. At least two potential reasons can account for such a result. One possibility relates to the laminar heterogeneity of neocortex involvement in memory storage [24,37]. In our experiments we used memory retrieval test 24 h after training, a period which qualifies as a recent memory [24,25,36]. Notably, consolidation from the recent to remote long-term memory was shown to be accompanied by a laminar reorganization of neuronal activity in the mouse parietal cortex [24]. This shift in the pattern of neuronal activation occurred from deep cortical layers at earlier times of memory storage to superficial cortical layers at later times. Due to limitations of two-photon imaging, here, we were restricted to sampling only superficial layers of the PtA. Neurons from these layers might be less involved in the recent memory retrieval compared to deeper layer cortical neurons as was previously shown for other cortical areas [24,36,37]. This hypothesis can be tested experimentally by using GRIN(Gradient-Index)-lens two-photon imaging, three-photon imaging, or miniscopes to examine the laminar distribution of Fos-Cre-GCaMP neurons responsive to CS presentation. A second possibility is that responses of all PtA engram neurons to CS mature during systems consolidation of memory, therefore these neurons would not yet be involved in CS encoding 1 day after training. This hypothesis is supported by the recent finding that 7d- and 14d-TRAPed neurons of PrL were significantly more likely to be reactivated during remote memory retrieval when compared to 1d testing [38]. Whether the PtA neurons have a response maturation profile similar to PrL is an important question that requires a specific study.

Altogether our results indicate the impication of the parietal association cortex in associative fear learning and emphasize the potential role of PtA in fear- and anxiety-related psychiatric disorders.

4. Materials and Methods

4.1. Animals

Male C57Bl/6J mice (2–3 months old) were used for ex vivo study of *c-fos* expression in the PtA after cued fear conditioning or memory retrieval. Transgenic Fos-EGFP male and female mice (B6.Cg-Tg(Fos/EGFP)1–3Brth/J, JAX Stock No: 014135, The Jackson Laboratory) were used for in vivo two-photon imaging of PtA neurons during FC or memory retrieval. Transgenic Fos-Cre-GCaMP male and female mice were obtained by crossing two transgenic mouse lines Ai38 (RCL-GCaMP3) (B6;129S-*Gt(ROSA)26Sor*$^{tm38(CAG\text{-}GCaMP3)Hze}$/J, JAX Stock No: 014538, The Jackson Laboratory) and Fos^{CreER} (B6.129(Cg)-*Fos*$^{tm1.1(cre/ERT2)Luo}$/J, JAX Stock No: 021882, The Jackson Laboratory).

Wild-type mice were group-housed 5–6 per cage. Transgenic mice were housed individually. All animals were kept under a 12 h light/dark cycle. All experiments were performed during the light phase of the cycle. All methods for animal care and all experiments were approved by the National Research Center "Kurchatov Institute" Committee on Animal Care (protocol code NG-1/109PR, date of approval 13 February 2020) and were in accordance with the Russian Federation Order Requirements N 267 M3 and the National Institutes of Health Guide for the Care and Use of Laboratory Animals.

4.2. Behavior

During cued fear conditioning mice were placed into the fear conditioning chamber (MED Associates Inc.) for a 3-min exploration of context A. Then seven conditioned sound stimuli followed by a foot shock (2 s, 0.75 mA for wild type mice or 1 mA for the transgenic mice) were presented with ITIs 40–60 s. Each CS consisted of 5 presentations of a tone (2 s, 9 kHz, 80 dB) with 2 s intervals. 24 h later cue memory was tested in context B. Mice were placed in context B for 3-min exploration, and then presented with the CS for 3 min (45 tone signals with 2 s intervals). Context A and B were cleaned before and after each session with 70% ethanol or 53% ethanol solution of peppermint, respectively. Context A was an IR light illuminated plastic box (30 cm × 23 cm × 21 cm) with a grid floor. To change the context for memory retrieval we placed the black plastic A-shaped insert into the FC chamber and covered the grid floor with a plastic sheet and wood sawdust on top (context B). Context B was illuminated with white and IR light. Freezing behavior was quantified using an automatic detection system (Video Freeze, MED Associates Inc., Fairfax, VT, USA).

For c-Fos immunostaining three groups of mice were used: Paired (n(FC) = 12, n(test) = 11), CS-only(n(FC) = 10, n(test) = 11) and Home cage group (n(FC) = 12, n(test) = 13). Mice from the Paired group were conditioned using the protocol described above. Mice from the CS-only group were submitted to the same protocol but without foot shock. Half of the mice were sacrificed 90 min after FC (n(FC)), while the other half was tested for memory retrieval and sacrificed 90 min after a test (n(test)). The Home cage mice were sacrificed in parallel with experimental animals without any behavioral manipulations.

Fos-EGFP mice were divided into the same groups: Paired (n = 7), CS-only (n = 6) and Home cage (n = 3). One month after cranial window implantation mice undergo FC training and memory retention test as described above.

Fos-Cre-GCaMP mice were FC trained 24 h after TM injection according to the described protocol.

4.3. Genotyping

Fos-Cre-GCaMP mice were genotyped at the age of 30–60 days. For genotyping we extracted DNA from the tail tissue in lysis buffer (0.01 M Tris-HCl, pH = 7.5; 0.01 M EDTA, pH = 8.0; 0.1 M NaCl; 1% SDS; 0.2 mg/mL proteinase K), performed PCR (in PCR buffer with Taq polymerase, DNTP, forward and reverse primers) and identified DNA sites using electrophoresis in agarose gel (2.5%). Primers oIMR4981, oIMR8038, 34,319, 34,962 were used for GCaMP genotyping, and 17,016–17,018 for Fos-Cre genotyping (The Jackson Laboratory).

4.4. Surgery

For cranial window surgery, transgenic mice were anesthetized with an intraperitoneal injection of zoletil (0.04 mg/g body weight) and xylazine (0.5 μg/g body weight). Dexamethasone (4 mg/kg) was administered subcutaneously 5 min before surgery to prevent tissue stress and cerebral edema. Viscotears moisturizing gel (Novartis Healthcare) was applied to prevent eye drying. Mice were fixed in a stereotaxic frame (Stoelting) and 37 °C body temperature was maintained by a heating plate (Physitemp). 3 mm craniotomy over the PtA (centered 1.0 mm lateral and 1.7 mm posterior to the Bregma) [3] was performed as described previously [39]. A 5-mm round glass coverslip (Menzel, Thermo Fisher) was attached to the skull using cyanoacrylate glass glue (Henkel). A Neurotar head post (Neurotar Ltd., Helsinki, Finland) was cemented to the skull with dental cement (Stoelting) and was later used for head fixation in the Mobile Home Cage system (MHC, Neurotar Ltd., Helsinki, Finland).

Two weeks after surgery Fos-EGFP and Fos-Cre-GCaMP mice were head-fixed in the MHC each day (5 to 40 min) for two weeks for habituation to imaging conditions. In this system, a head-fixed mouse can move around a lifted MHC and freely explore its environment [40].

4.5. Tamoxifen Injection

Fos-Cre-GCaMP mice received a single i.p. injection (150 mg/kg) of TM (Sigma) 24 h before FC training. TM was dissolved in the corn oil (Sigma) (10 mL/kg) and 96% ethanol (1.3 mL/kg) at 65 °C for 1–2 h. The dose of TM and timing of injection were in accordance with the previously described protocol [32].

4.6. Immunohistochemistry

Mice were sacrificed 90 min after FC training or test. Brains were removed and immediately frozen in liquid nitrogen vapor. 20-μm coronal sections were prepared on a cryostat (Leica) and fixed in 4% paraformaldehyde. For c-Fos immunostaining, primary rabbit polyclonal antibodies against the c-Fos protein (sc-52, Santa Cruz Biotechnology, Dallas, TX, USA, dilution 1:500) and horse secondary antibodies against rabbit conjugated with avidin-biotin complex (ImPRESS reagent kit anti-rabbit, Vector Laboratories) were used. Sections were stained in 0.06% diaminobenzidine solution (Sigma). Brain sections were taken at a distance of −1.7 mm from the bregma. Whole section images were acquired through the fluorescence microscope scanner (Olympus, VS110).

4.7. In Vivo Two-Photon Imaging

Two-photon imaging was performed on Fos-EGFP and Fos-Cre-GCaMP mice 30–60 days after cranial window implantation using an Olympus MPE1000 two-photon microscope equipped with a Mai Tai Ti:Sapphire femtosecond-pulse laser (Spectra-Physics) and a water-immersion objective lens, 20 × 1.05 NA (Olympus). 960 nm wavelength was used for excitation. Series of images were recorded with the Olympus Fluoview Software Version 3.1.

4.7.1. Fos-EGFP Mice

The volume series of images (0–350 μm under the pia) were recorded at 0.82 frame per second (fps) continuously, with 500 × 500 μm field of view and 512 × 512 pixels resolution. PtA two-photon imaging was performed three days before FC (basal level of c-Fos expression in the home cage), 90 min after FC, and 90 min after memory test. Home cage mice were imaged together with other groups but without any prior experience.

4.7.2. Fos-Cre-GCaMP Mice

To determine the dynamics of GCaMP accumulation after Cre recombination, we visualized the PtA at different time points starting 2 h after FC in 4 mice. Volume series of images (0–300 μm under the pia) were recorded at 0.2 fps continuously, with 500 × 500 μm field of view and 512 × 512 pixels resolution.

To analyze the activity of trapped cells PtA layer 2/3 neurons (approximately 100–200 μm deep from pia) were imaged three days after FC. During the calcium imaging, mice received seven series of the CS (one series consisted of 10 short tones (2 s, 9 kHz, 80 dB) with 2 s intervals). Time series of images were recorded at 1.78 fps continuously, with 253 × 253 μm field of view, 256 × 256 pixels resolution and 2× zoom.

For spontaneous calcium activity in dendrites, we recorded a time series of images at 1.92 fps, with 35 × 44 μm field of view, 180 × 144 pixels resolution and zoom 4× at the depth of 5–10 μm without any stimulation.

The image analysis was performed in Olympus Fluoview Software Version 3.1, FIJI, Imaris 7.4.2 and a custom Python plugin. Regions of interest (ROIs) corresponding to identifiable cell bodies or spines were selected manually in the Olympus Fluoview Software Version 3.1. ΔF/F was calculated by subtracting each value with the mean of the lower 50% of previous 10-s values and dividing it by the mean of the lower 50% of previous 10-s values.Calcium events detection was performed whenever the difference between a trace amplitude and its median value crossed the threshold of 4 median absolute deviations, calculated for each cell over the whole trace.

Author Contributions: O.I.I. analyzed mice behavior and ex vivo c-Fos expression; O.I.I. and K.A.T. breed transgenic mice; M.A.R. and A.M.G. performed In vivo investigation of c-Fos expression; A.M.G. and M.A.R. performed the calcium imaging registration and data analysis; A.M.G., O.I.I. and K.V.A. analyzed and interpreted the data; A.M.G., O.I.I., K.A.T. and K.V.A. wrote the manuscript. All authors reviewed the manuscript. All authors have read and agreed to the published version of the manuscript.

Funding: This research was funded by grant No 075-15-2020-801 from the Ministry of Science and Higher Education of the Russian Federation.

Institutional Review Board Statement: The study was conducted according to the guidelines of the Declaration of Helsinki, and approved by the Committee on Animal Care of the National Research Center "Kurchatov Institute" (protocol code NG-1/109PR, date of approval 13 February 2020) and were done in accordance with the *Russian* Federation Order Requirements N *267* M3 and the National Institutes of Health Guide for the Care and Use of Laboratory Animals.

Informed Consent Statement: Not applicable.

Data Availability Statement: Data is contained within the article.

Conflicts of Interest: The authors declare no conflict of interest. The funders had no role in the design of the study; in the collection, analyses, or interpretation of data; in the writing of the manuscript, or in the decision to publish the results.

References

1. Freedman, D.J.; Ibos, G. An Integrative Framework for Sensory, Motor, and Cognitive Functions of the Posterior Parietal Cortex. *Neuron* **2018**, *97*, 1219–1234. [CrossRef] [PubMed]
2. Berlucchi, G.; Vallar, G. The history of the neurophysiology and neurology of the parietal lobe. *Handb. Clin. Neurol.* **2018**, *151*, 3–30. [CrossRef] [PubMed]
3. Lyamzin, D.; Benucci, A. The mouse posterior parietal cortex: Anatomy and functions. *Neurosci. Res.* **2019**, *140*, 14–22. [CrossRef] [PubMed]
4. Hovde, K.; Gianatti, M.; Witter, M.P.; Whitlock, J.R. Architecture and organization of mouse posterior parietal cortex relative to extrastriate areas. *Eur. J. Neurosci.* **2018**, *49*, 1313–1329. [CrossRef]
5. Franklin, B.J.; Paxinos, G. *The Mouse Brain in Stereotaxic Coordinates*, 3rd ed.; Academic Press: New York, NY, USA, 2007.
6. Dong, H.W. *The Allen Reference Atlas: A Digital Color Brain Atlas of the C57Bl/6J Male Mouse*; John Wiley & Sons Inc.: Hoboken, NJ, USA, 2008.
7. Zingg, B.; Hintiryan, H.; Gou, L.; Song, M.Y.; Bay, M.; Bienkowski, M.S.; Foster, N.N.; Yamashita, S.; Bowman, I.; Toga, A.W.; et al. Neural Networks of the Mouse Neocortex. *Cell* **2014**, *156*, 1096–1111. [CrossRef]
8. Lippert, M.T.; Takagaki, K.; Kayser, C.; Ohl, F.W. Asymmetric Multisensory Interactions of Visual and Somatosensory Responses in a Region of the Rat Parietal Cortex. *PLoS ONE* **2013**, *8*, e63631. [CrossRef]
9. Mohajerani, M.H.; Chan, A.W.; Mohsenvand, M.; LeDue, J.; Liu, R.; McVea, D.A.; Boyd, J.D.; Wang, Y.T.; Reimers, M.; Murphy, T.H. Spontaneous cortical activity alternates between motifs defined by regional axonal projections. *Nat. Neurosci.* **2013**, *16*, 1426–1435. [CrossRef] [PubMed]
10. Kuroki, S.; Yoshida, T.; Tsutsui, H.; Iwama, M.; Ando, R.; Michikawa, T.; Miyawaki, A.; Ohshima, T.; Itohara, S. Excitatory Neuronal Hubs Configure Multisensory Integration of Slow Waves in Association Cortex. *Cell Rep.* **2018**, *22*, 2873–2885. [CrossRef]
11. Whitlock, J.R. Navigating actions through the rodent parietal cortex. *Front. Hum. Neurosci.* **2014**, *8*, 293. [CrossRef]
12. Goard, M.J.; Pho, G.N.; Woodson, J.; Sur, M. Distinct roles of visual, parietal, and frontal motor cortices in memory-guided sensorimotor decisions. *eLife* **2016**, *5*, e13764. [CrossRef] [PubMed]
13. Zhong, L.; Zhang, Y.; Duan, C.A.; Deng, J.; Pan, J.; Xu, N.-L. Causal contributions of parietal cortex to perceptual decision-making during stimulus categorization. *Nat. Neurosci.* **2019**, *22*, 963–973. [CrossRef]
14. Lim, D.H.; Mohajerani, M.H.; LeDue, J.; Boyd, J.; Chen, S.; Murphy, T.H. In vivo Large-Scale Cortical Mapping Using Channelrhodopsin-2 Stimulation in Transgenic Mice Reveals Asymmetric and Reciprocal Relationships between Cortical Areas. *Front. Neural Circuits* **2012**, *6*, 11. [CrossRef] [PubMed]
15. Frankland, P.W.; Bontempi, B.; Talton, L.E.; Kaczmarek, L.; Silva, A.J. The Involvement of the Anterior Cingulate Cortex in Remote Contextual Fear Memory. *Science* **2004**, *304*, 881–883. [CrossRef] [PubMed]
16. Vann, S.D.; Aggleton, J.P.; Maguire, E.A. What does the retrosplenial cortex do? *Nat. Rev. Neurosci.* **2009**, *10*, 792–802. [CrossRef]
17. Goshen, I.; Brodsky, M.; Prakash, R.; Wallace, J.; Gradinaru, V.; Ramakrishnan, C.; Deisseroth, K. Dynamics of Retrieval Strategies for Remote Memories. *Cell* **2011**, *147*, 678–689. [CrossRef]
18. Czajkowski, R.; Jayaprakash, B.; Wiltgen, B.; Rogerson, T.; Guzman-Karlsson, M.C.; Barth, A.L.; Trachtenberg, J.T.; Silva, A.J. Encoding and storage of spatial information in the retrosplenial cortex. *Proc. Natl. Acad. Sci. USA* **2014**, *111*, 8661–8666. [CrossRef] [PubMed]

19. Nakayama, D.; Baraki, Z.; Onoue, K.; Ikegaya, Y.; Matsuki, N.; Nomura, H. Frontal Association Cortex Is Engaged in Stimulus Integration during Associative Learning. *Curr. Biol.* **2015**, *25*, 117–123. [CrossRef]
20. Park, K.; Chung, C. Systemic Cellular Activation Mapping of an Extinction-Impaired Animal Model. *Front. Cell. Neurosci.* **2019**, *13*, 99. [CrossRef] [PubMed]
21. Silva, B.A.; Burns, A.M.; Gräff, J. A cFos activation map of remote fear memory attenuation. *Psychopharmacology* **2018**, *236*, 369–381. [CrossRef]
22. Quiñones-Laracuente, K.; Vega-Medina, A.; Quirk, G.J. Time-Dependent Recruitment of Prelimbic Prefrontal Circuits for Retrieval of Fear Memory. *Front. Behav. Neurosci.* **2021**, *15*, 665116. [CrossRef]
23. Save, E.; Moghaddam, M. Effects of lesions of the associative parietal cortex on the acquisition and use of spatial memory in egocentric and allocentric navigation tasks in the rat. *Behav. Neurosci.* **1996**, *110*, 74–85. [CrossRef]
24. Maviel, T.; Durkin, T.P.; Menzaghi, F.; Bontempi, B. Sites of Neocortical Reorganization Critical for Remote Spatial Memory. *Science* **2004**, *305*, 96–99. [CrossRef]
25. Josselyn, S.A.; Köhler, S.; Frankland, P.W. Finding the engram. *Nat. Rev. Neurosci.* **2015**, *16*, 521–534. [CrossRef]
26. Tonegawa, S.; Pignatelli, M.; Roy, D.S.; Ryan, T.J. Memory engram storage and retrieval. *Curr. Opin. Neurobiol.* **2015**, *35*, 101–109. [CrossRef]
27. Salery, M.; Godino, A.; Nestler, E.J. Drug-activated cells: From immediate early genes to neuronal ensembles in addiction. *Adv. Pharmacol.* **2021**, *90*, 173–216. [CrossRef]
28. Miyashita, T.; Kikuchi, E.; Horiuchi, J.; Saitoe, M. Long-Term Memory Engram Cells Are Established by c-Fos/CREB Transcriptional Cycling. *Cell Rep.* **2018**, *25*, 2716–2728. [CrossRef] [PubMed]
29. Barth, A.L.; Gerkin, R.C.; Dean, K.L. Alteration of Neuronal Firing Properties after In Vivo Experience in a FosGFP Transgenic Mouse. *J. Neurosci.* **2004**, *24*, 6466–6475. [CrossRef]
30. Milczarek, M.M.; Vann, S.D.; Sengpiel, F. Spatial Memory Engram in the Mouse Retrosplenial Cortex. *Curr. Biol.* **2018**, *28*, 1975–1980. [CrossRef] [PubMed]
31. Brebner, L.S.; Ziminski, J.J.; Margetts-Smith, G.; Sieburg, M.C.; Reeve, H.M.; Nowotny, T.; Hirrlinger, J.; Heintz, T.G.; Lagnado, L.; Kato, S.; et al. The Emergence of a Stable Neuronal Ensemble from a Wider Pool of Activated Neurons in the Dorsal Medial Prefrontal Cortex during Appetitive Learning in Mice. *J. Neurosci.* **2019**, *40*, 395–410. [CrossRef]
32. Guenthner, C.J.; Miyamichi, K.; Yang, H.; Heller, H.C.; Luo, L. Permanent Genetic Access to Transiently Active Neurons via TRAP: Targeted Recombination in Active Populations. *Neuron* **2013**, *78*, 773–784. [CrossRef]
33. Ivashkina, O.I.; Vorobyeva, N.S.; Gruzdeva, A.M.; Roshchina, M.A.; Toropova, K.A.; Anokhin, K.V. Cognitive tagging of neurons: CRE-mediated genetic labeling and characterization of cells involved in learning and memory. *Acta Nat.* **2018**, *10*, 37–47. [CrossRef]
34. Denny, C.A.; Kheirbek, M.A.; Alba, E.L.; Tanaka, K.F.; Brachman, R.A.; Laughman, K.B.; Tomm, N.K.; Turi, G.F.; Losonczy, A.; Hen, R. Hippocampal Memory Traces Are Differentially Modulated by Experience, Time, and Adult Neurogenesis. *Neuron* **2014**, *83*, 189–201. [CrossRef]
35. Guzowski, J.F.; McNaughton, B.L.; Barnes, C.A.; Worley, P.F. Environment-specific expression of the immediate-early gene Arc in hippocampal neuronal ensembles. *Nat. Neurosci.* **1999**, *2*, 1120–1124. [CrossRef]
36. Saidov, K.; Tiunova, A.; Subach, O.; Subach, F.; Anokhin, K. New Tools in Cognitive Neurobiology: Biotin-Digoxigenin Detection of Overlapping Active Neuronal Populations by Two-Color c-fos Compartment Analysis of Temporal Activity by Fluorescent in situ Hybridization (catFISH) and c-Fos Immunohistochemistry. *OBM Genet.* **2019**, *3*, 18. [CrossRef]
37. Frankland, P.W.; Bontempi, B. The organization of recent and remote memories. *Nat. Rev. Neurosci.* **2005**, *6*, 119–130. [CrossRef] [PubMed]
38. DeNardo, L.A.; Liu, C.D.; Allen, W.E.; Adams, E.L.; Friedmann, D.; Fu, L.; Guenthner, C.J.; Tessier-Lavigne, M.; Luo, L. Temporal evolution of cortical ensembles promoting remote memory retrieval. *Nat. Neurosci.* **2019**, *22*, 460–469. [CrossRef]
39. Holtmaat, A.; Bonhoeffer, T.; Chow, D.K.; Chuckowree, J.; De Paola, V.; Hofer, S.B.; Hübener, M.; Keck, T.; Knott, G.; Lee, W.-C.; et al. Long-term, high-resolution imaging in the mouse neocortex through a chronic cranial window. *Nat. Protoc.* **2009**, *4*, 1128–1144. [CrossRef]
40. Kislin, M.; Mugantseva, E.; Molotkov, D.; Kulesskaya, N.; Khirug, S.; Kirilkin, I.; Pryazhnikov, E.; Kolikova, J.; Toptunov, D.; Yuryev, M.; et al. Flat-floored Air-lifted Platform: A New Method for Combining Behavior with Microscopy or Electrophysiology on Awake Freely Moving Rodents. *J. Vis. Exp.* **2014**, *88*, e51869. [CrossRef]

International Journal of
Molecular Sciences

Article

Activation of Neuronal Nicotinic Receptors Inhibits Acetylcholine Release in the Neuromuscular Junction by Increasing Ca^{2+} Flux through Ca$_V$1 Channels

Nikita Zhilyakov [1,*], Arsenii Arkhipov [1], Artem Malomouzh [1] and Dmitry Samigullin [1,2,*]

[1] Kazan Institute of Biochemistry and Biophysics, FRC Kazan Scientific Center, Russian Academy of Sciences, P.O. Box 261, 420111 Kazan, Russia; senjaarh@rambler.ru (A.A.); artur57@gmail.com (A.M.)

[2] Department of Radiophotonics and Microwave Technologies, Federal State Budgetary Educational Institution of Higher Education "Kazan National Research Technical University Named after A.N. Tupolev–KAI", 420111 Kazan, Russia

* Correspondence: kiosak71@gmail.com (N.Z.); samid75@mail.ru (D.S.); Tel./Fax: +7-843-292-7347 (N.Z.); +7-843-231-0158 (D.S.)

Abstract: Cholinergic neurotransmission is a key signal pathway in the peripheral nervous system and in several branches of the central nervous system. Despite the fact that it has been studied extensively for a long period of time, some aspects of its regulation still have not yet been established. One is the relationship between the nicotine-induced autoregulation of acetylcholine (ACh) release with changes in the concentration of presynaptic calcium levels. The mouse neuromuscular junction of m. Levator Auris Longus was chosen as the model of the cholinergic synapse. ACh release was assessed by electrophysiological methods. Changes in calcium transients were recorded using a calcium-sensitive dye. Nicotine hydrogen tartrate salt application (10 µM) decreased the amount of evoked ACh release, while the calcium transient increased in the motor nerve terminal. Both of these effects of nicotine were abolished by the neuronal ACh receptor antagonist dihydro-beta-erythroidine and Ca$_V$1 blockers, verapamil, and nitrendipine. These data allow us to suggest that neuronal nicotinic ACh receptor activation decreases the number of ACh quanta released by boosting calcium influx through Ca$_V$1 channels.

Keywords: neuromuscular junction; neurotransmitter release; acetylcholine; nicotinic receptor; calcium channel; calcium transient

Citation: Zhilyakov, N.; Arkhipov, A.; Malomouzh, A.; Samigullin, D. Activation of Neuronal Nicotinic Receptors Inhibits Acetylcholine Release in the Neuromuscular Junction by Increasing Ca^{2+} Flux through Ca$_V$1 Channels. *Int. J. Mol. Sci.* **2021**, 22, 9031. https://doi.org/10.3390/ijms22169031

Academic Editors: Piotr D. Bregestovski and Carlo Matera

Received: 10 August 2021
Accepted: 19 August 2021
Published: 21 August 2021

Publisher's Note: MDPI stays neutral with regard to jurisdictional claims in published maps and institutional affiliations.

1. Introduction

Acetylcholine (ACh) is the main neurotransmitter in the peripheral nervous system of vertebrates and humans. In particular, it is responsible for the transmission of signals from the motor nerve to the skeletal muscles [1,2]. Since the neuromuscular junction is a key linker in the initiation of any motor act (from voluntary movement of the limbs, to breathing, to contraction of the vocal cords), an investigation of the regulation of neuromuscular transmission is of great importance for both fundamental neurobiology and applied medicine.

Since the midst of the 20th century, the data began to accumulate indicating that ACh, released in the synaptic cleft from the nerve endings, activates presynaptic cholinergic receptors, thus exerting a modulatory effect on the neurotransmission process by changing the amount and/or dynamics of subsequent portions of neurotransmitter release [2–7]. Initially pharmacologically, and later by other methods it has been shown that both ionotropic nicotinic and metabotropic muscarinic cholinergic receptors are present in the motor nerve terminal, and the activation of these receptors can lead to autoregulation of ACh release [4,5,8–11]. According to the data of morphological and functional analysis, presynaptic cholinergic receptors can be located both near the active zones and relatively far from the synaptic cleft [4,12,13].

When studying autoregulation mediated by muscarinic cholinergic receptors, it was found that the activation of the M1-subtype receptors led to facilitation of the release. In contrast, activation of the M2-subtype caused inhibition of the ACh quanta release [14,15]. Both the M1- and M2-mediated mechanisms depend on calcium influx [14,16–18].

Studies of the mechanisms of the autoregulation of ACh release mediated by nicotinic cholinergic receptors are complicated by the fact that the predominant population of these proteins is located in the postsynaptic membrane. Their activation is accompanied by the depolarization of sarcolemma and the subsequent generation of action potential, which ultimately leads to muscle contraction. The data indicate that the activation of presynaptic nicotinic cholinergic receptors leads to inhibition of the process of ACh release [19–21].

Also, experimental evidence was obtained indicating the possible involvement of voltage-gated calcium channels (VGCCs) of the L-type (Ca_V1) in the modulation of neurotransmission [20,22]. Meanwhile, the results of a number of studies demonstrate that neither the Ca_V1 type nor the N-type ($Ca_V2.2$) VGCCs participate in the evoked release of ACh in mammalian neuromuscular contacts [23–26].

Thus, the question of the role of calcium channels in the mechanisms of the regulation of ACh release, as mediated by nicotinic cholinergic receptors, remains open as of now.

In the present study, using a pharmacological approach, electrophysiological techniques and the method of the optical registration of changes in calcium levels in the motor nerve endings, we made the following observations. An agonist of nicotinic receptors (at a concentration not significantly affecting the state of the postsynaptic membrane) leads to a decrease in the amount of released ACh quanta. This effect is accompanied not by a decrease, but by an increase of calcium ion entry into the motor nerve terminal. Our data suggest that the nicotinic cholinergic receptors responsible for the mechanism of ACh release autoregulation are the receptors of neuronal type. Activation of these receptors leads to upregulation of the Ca_V1 type of VGCCs, resulting in the enhancement of Ca^{2+} entry into the nerve ending.

2. Results

2.1. Effects of Nicotine on the Electrophysiological Parameters of the Neuromuscular Junction

Using the intracellular microelectrode technique, we recorded the resting membrane potential (RMP) of the muscle fiber, amplitude of miniature endplate potentials (mEPPs), frequency of occurrence of mEPPs, and the amplitude of the evoked potentials of the end plate (EPPs).

Alterations in RMP and mEPP amplitude indicate the postsynaptic action of the pharmacological agent. Changes in the frequency of occurrence of mEPPs suggest the presynaptic action of the drug. The EPP amplitude, in turn, can vary due to changes at both the pre- and post-synaptic levels. Therefore, it was necessary to assess the effect of nicotine on every parameter mentioned above to determine the optimal effective concentration of nicotine to study the autoregulation.

The control RMP value of muscle fibers was -71.48 ± 0.77 mV ($n = 30$). Application of nicotine at concentrations of 0.1 μM, 1 and 5 μM did not affect the RMP significantly, providing values of -69.72 ± 0.90 mV ($n = 30$), -70.91 ± 0.78 mV ($n = 30$), and -70.85 ± 0.87 mV ($n = 30$), respectively (Figure 1a). A slight significant depolarization was observed when nicotine concentration was increased to 10 μM (-67.04 ± 0.79 mV; $n = 30$). At a concentration of 50 μM, a more pronounced depolarization was observed, and the mean RMP value decreased to -56.94 ± 1.29 mV ($n = 30$; Figure 1a).

Another sign of the postsynaptic action of nicotine was the change in the amplitude of mEPP. The mean value of the amplitude of the spontaneous signal in the control was 0.88 ± 0.05 mV ($n = 30$). Application of nicotine in concentration up to 10 μM did not affect mEPP amplitude significantly, with the values being 0.83 ± 0.04 mV ($n = 30$) for 0.1 μM, 0.84 ± 0.05 mV ($n = 30$) for 1 μM, 0.96 ± 0.06 mV ($n = 30$) for 5 μM, and 0.83 ± 0.05 mV ($n = 30$) for 10 μM (Figure 1b). Asignificant decrease in mEPP amplitude to 0.55 ± 0.04 mV ($n = 30$) was observed only with 50 μM nicotine (Figure 1b).

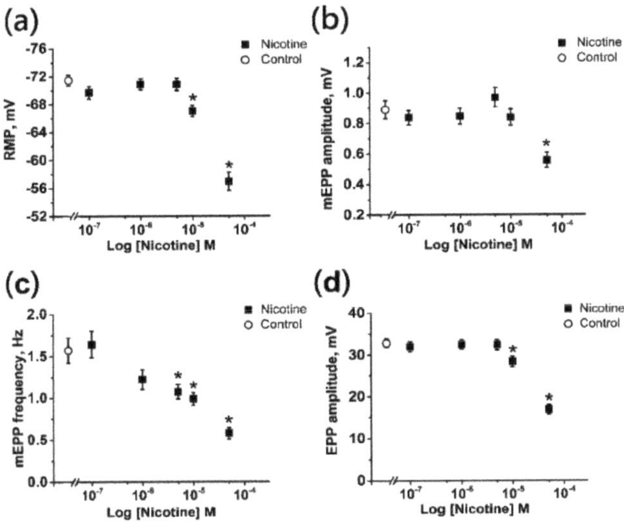

Figure 1. Effects of nicotine on the electrophysiological parameters registered at the mouse neuromuscular junction. Changes in absolute values are shown (**a**) resting membrane potential (RMP) of muscle fibers, (**b**) amplitudes of miniature endplate potentials (mEPP), (**c**) frequency of the mEPPs, and (**d**) the amplitudes of evoked endplate potentials (EPP) in the control and 15 min after nicotine application (the range from 0.1 to 50 μM). Results are expressed as mean ± SEM. Asterisks (*) indicate significant effects ($p < 0.05$, one-way ANOVA test with Dunnet's post-hoc comparison).

In contrast to the amplitude of mEPPs, the effect of nicotine on the frequency of occurrence of spontaneous signals was detected at significantly lower concentrations. That is, the average values of the frequency of mEPPs upon application of 0.1 and 1 μM nicotine were 1.64 ± 0.15 Hz ($n = 30$) and 1.22 ± 0.12 Hz ($n = 30$), respectively, and did not differ from the control value of 1.57 ± 0.14 Hz ($n = 30$; Figure 1c). After application of 5 μM nicotine, the frequency significantly decreased to 1.07 ± 0.08 Hz ($n = 30$), and inhibition was further enhanced to 0.98 ± 0.07 Hz ($n = 30$) for 10 μM and 0.57 ± 0.06 Hz ($n = 30$) for 50 μM (Figure 1c).

The amplitude of the EPP, which reflects the level of evoked ACh release and depends on changes in the sensitivity of the postsynaptic membrane in the area of the neuromuscular contact, was 32.80 ± 1.02 mV ($n = 30$) in the control. Application of nicotine at concentrations of 0.1 μM, 1 and 5 μM did not alter the average amplitudes of EPP, which were equal to 31.93 ± 1.21 mV ($n = 30$), 32.43 ± 1.11 mV ($n = 30$), and 32.33 ± 1.18 mV ($n = 30$), respectively (Figure 1d). However, nicotine produced a decrease in EPP amplitude, starting at the concentration of 10 μM (28.36 ± 1.27 mV; $n = 30$), while at 50 μM the amplitude decreased almost twofold to 16.89 ± 1.09 mV ($n = 30$; Figure 1d).

Thus, for further investigations of the ACh release autoregulation mechanisms, concentration of 10 μM nicotine was chosen. When using nicotine at this concentration, a decrease in the EPP amplitude was observable, while there were no changes in the mEPP amplitude (with only a slight depolarization of the sarcolemma).

2.2. Activation of Neuronal Nicotinic Receptors Leads to Downregulation of the EPP Quantal Content

Under control conditions, the quantal content (QC) was 46.8 ± 4.5. The bath application of nicotine (10 μM) decreased the QC of EPP significantly by $12.0 \pm 4.4\%$ ($n = 7$; Figure 2).

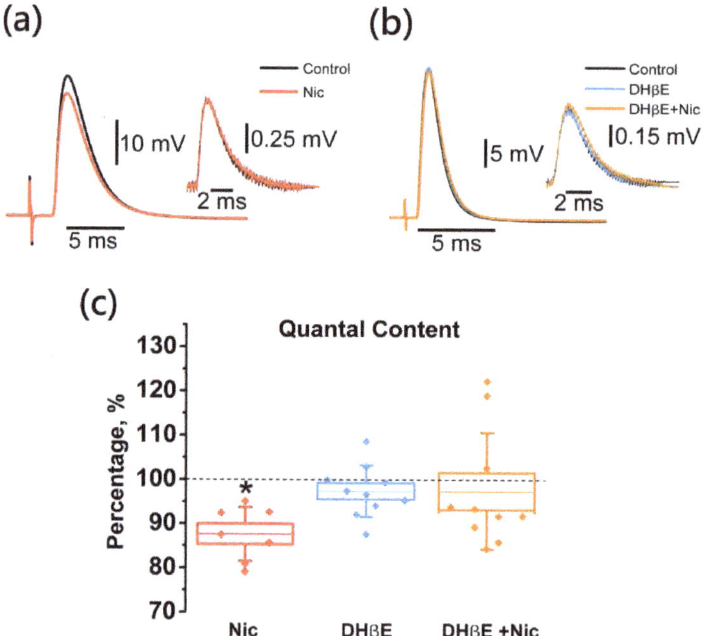

Figure 2. Nicotine inhibits the evoked release of ACh quanta (quantal content, QC) by activating nNAChRs. Panels on the top are representative traces of EPP and mEPP (50 signals averaged) in separate experiments with nicotine application (Nic, 10 μM; (**a**)), and nicotine application after pretreatment with the neuronal cholinergic receptor antagonist DHβE (1 μM; (**b**)). (**c**) Results are expressed as mean ± SEM and SD of QC, as percentages with nicotine ($n = 7$), DHβE ($n = 9$), and DHβE plus nicotine ($n = 9$) applications versus control. Asterisks (*) indicate significant effects ($p < 0.05$, one-way repeated ANOVA test with Tukey's post-hoc comparison).

The nicotine-induced decrease in the number of ACh quanta released in response to stimulation of the motor nerve suggests the involvement of presynaptic cholinergic receptors. Using the antagonist of neuronal nicotinic ACh receptors (nNAChRs) DHβE [27,28], we obtained the data supporting this suggestion. The application of DHβE alone at a concentration of 1 μM did not change the QC of EPP ($101.5 \pm 1.3\%$; $n = 9$, Figure 2), however, after pretreatment with DHβE, the inhibitory effect of nicotine on the quantal release of ACh was completely abolished ($105.5 \pm 6.6\%$; $n = 9$; Figure 2).

2.3. Activation of Neuronal Nicotinic Receptors Induces an Increase of the Calcium Level in the Motor Nerve Terminal

Since the process of the evoked release of a neurotransmitter is triggered by the entry of calcium ions into the nerve ending [29,30], it was suggested that the inhibitory effect of nicotine on ACh release could be related to a decrease in Ca^{2+} influx.

The amplitude changes of the optical signal ($\Delta F/F_0$) from the calcium dye loaded into the nerve terminal in response to a single stimulus (with the same characteristics as during EPP registration) averaged about 30% (Figure 3). Nicotine application did not lead to a decrease, as expected, but instead caused a significant increase in the amplitude of the calcium transient by $13.7 \pm 4.3\%$ ($n = 8$; Figure 4). Thus, in the presence of a nicotinic receptor agonist, the presynaptic calcium level increases more strongly in response to nerve stimulation than in its absence. Is this increase indeed triggered by nNAChRs, whose activation leads to a decrease in the subsequent ACh release? The answer to this question was obtained in the experiments with an antagonist of nicotinic receptors, DHβE.

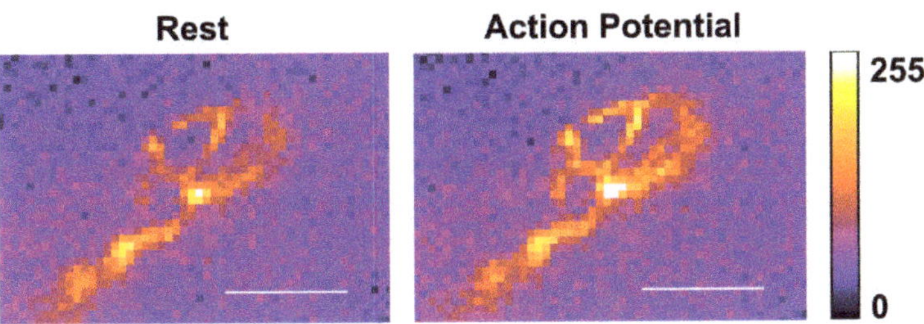

Figure 3. Pseudo-color calcium images of a motor nerve terminal loaded with Oregon Green 488 BAPTA-1 Hexapotassium Salt. The axon is imaged before and during single electrical stimulus (0.2 ms duration). Scale bar, 20 µm.

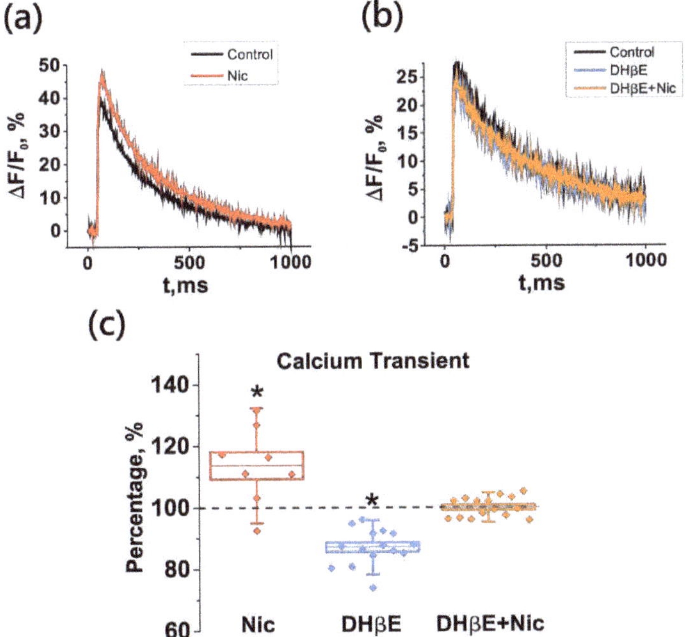

Figure 4. Nicotine increases the calcium transient in the motor nerve ending by the activation of neuronal ACh receptors. Blockade of the receptors leads to a decrease in the amplitude of the calcium signal. Panels on the top are representative traces of calcium transient from separate experiments with nicotine application (Nic, 10 µM; (**a**)), and nicotine application after pretreatment with an nNAChR antagonist DHβE (1 µM; (**b**)). (**c**) Mean ± SEM and SD of the amplitude of the calcium signal, expressed as a percentage of control when applying nicotine ($n = 8$), DHβE ($n = 15$) and DHβE plus nicotine ($n = 15$). Asterisks (*) indicate significant effects ($p < 0.05$, one-way repeated ANOVA test with Tukey's post-hoc comparison).

Application of the antagonist alone led to a decrease in the calcium transient significantly by 12.9 ± 1.5% ($n = 15$; Figure 4). However, after pretreatment with DHβE, the calcium signal-enhancing effect of nicotine was completely abolished (100.0 ± 0.8%; $n = 15$; Figure 4).

2.4. Neuronal Nicotinic Receptors Alter Calcium Level in Presynaptic Terminal by Gating L-Type (Ca_v1) Calcium Channels

To identify the source of the increase in the calcium signal upon the activation of presynaptic nNAChRs, cadmium chloride, a nonselective blocker of calcium-permeable channels, was used at a concentration of 10 µM. After the application of cadmium chloride, a decrease in the amplitude of the calcium transient was observed by $54.5 \pm 2.7\%$ ($n = 9$). In the presence of cadmium, the effect of nicotine on the alterations in calcium levels was completely abolished ($101.4 \pm 3.4\%$; $n = 9$; Figure 5). Therefore, the observed increase in the presynaptic calcium level upon the activation of nNAChRs is mediated by proteins (channels) which are permeable for Ca^{2+}. Further experiments were carried out to establish which type of VGCCs are involved in nicotine-induced increases in calcium transients.

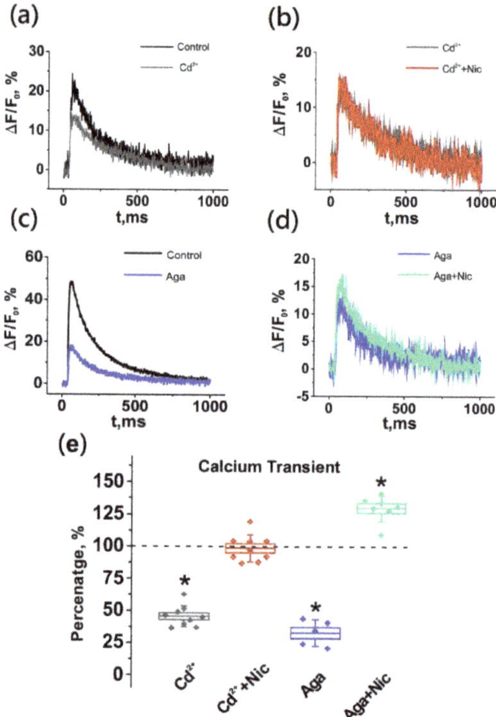

Figure 5. The calcium transient-enhancing effect of nicotine is abolished after a nonspecific calcium channel blockade, but not after inhibition of $Ca_v2.1$-type calcium channels. Panels on the top are representative traces of calcium transient from individual experiments: (**a**) the effect of the nonspecific calcium channel blocker $CdCl_2$ (Cd^{2+}, 10 µM), (**b**) no effect of nicotine (Nic, 10 µM) after pretreatment with $CdCl_2$, (**c**) the effect of the $Ca_v2.1$ VGCC blocker ω-agatoxin IVA (Aga, 40 nM), and (**d**) the effect of nicotine on the calcium transient after pre-incubation with ω-agatoxin IVA. (**e**) Mean $\pm$ SEM and SD of calcium signal amplitudes obtained in the above series and expressed as a percentage of control or value after $CdCl_2$ ($n = 9$), $CdCl_2$ + Nic ($n = 9$), Aga ($n = 5$;) and Aga + Nic ($n = 7$) application. Asterisks (*) indicate significant effects ($p < 0.05$, one-way repeated ANOVA test).

The application of the specific P/Q-type ($Ca_v2.1$) VGCCs blocker ω-agatoxin IVA at a concentration of 40 nM that blocks only a certain proportion of channels [31] led to a significant decrease in the calcium transient by $67.0 \pm 4.4\%$ ($n = 5$; Figure 5). In the case of a partial blockade of the main type of VGCCs $Ca_v2.1$, nicotine application (10 µM) led to an increase in the amplitude of the calcium transient significantly by $29.9 \pm 3.8\%$ ($n = 10$,

Figure 5). Therefore, the effect of the activation of nNAChRs on the intracellular calcium level is not mediated by $Ca_V2.1$ channels.

Ca_V1 calcium channel blockers such as verapamil (50 μM) and nitrendipine (25 μM), produced significant calcium transient decreases of $25.0 \pm 4.4\%$ ($n = 9$) and $18.8 \pm 1.1\%$ ($n = 17$), respectively (Figure 6). The application of nicotine after pretreatment by these blockers did not cause any changes in the calcium transient: the amplitudes were $101.8 \pm 1.5\%$ ($n = 9$) and $100.7 \pm 1.2\%$ ($n = 17$), respectively (Figure 6).

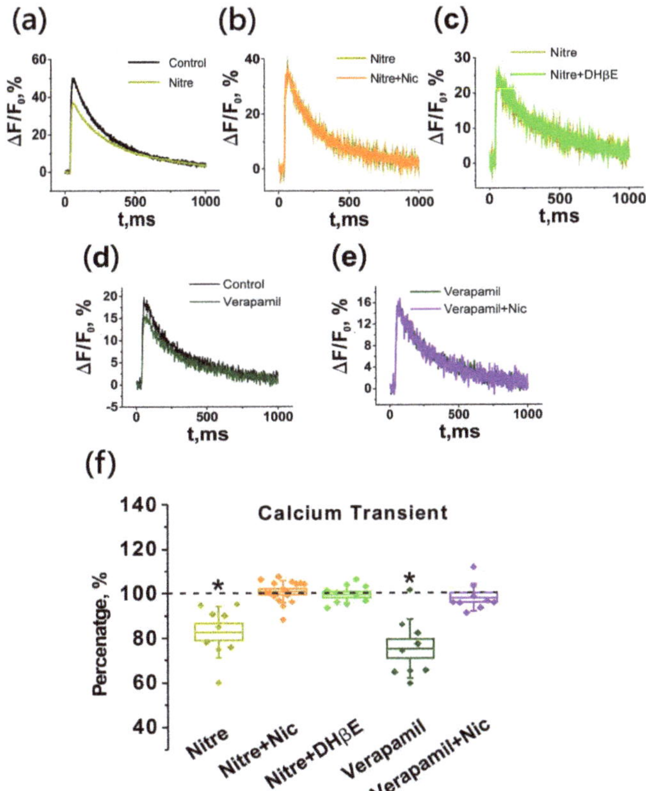

Figure 6. Lack of effect of nicotine (an increase in the amplitude of the calcium transient) and DHβE (a decrease in the amplitude of the calcium transient) after blockade of the Ca_V1 channels. Panels on the top are representative traces of calcium transient from individual experiments: (**a**) effect of Ca_V1 calcium channel blocker nitrendipine (Nitre, 25 μM), (**b**) lack of nicotine (Nic, 10 μM) effect after pre-application of nitrendipine, (**c**) lack of DHβE (1 μM) effect after nitrendipine pre-treatment, (**d**) effect of Ca_V1 type VGCCs blocker verapamil (50 μM), (**e**) no effect of nicotine after pre-application of verapamil. (**f**) Mean $\pm$ SEM and SD of calcium signal amplitudes obtained in the above series and expressed as a percentage of control or value after Nitre ($n = 17$), Nitre plus Nic ($n = 17$), Nitre plus DHβE ($n = 7$), verapamil ($n = 9$) and verapamil plus Nic ($n = 9$) application. Asterisks (*) indicate significant effects ($p < 0.05$, one-way repeated ANOVA test).

These data allow us to conclude that the observed increase in the calcium level in the nerve ending upon activation of nNAChRs by an exogenous agonist is due to mediation by Ca_V1 type VGCCs. Therefore, the phenomenon of the endogenous activation of presynaptic cholinergic receptors discovered by us should also be mediated by calcium channels of this type. Indeed, the calcium transient-reducing effect of DHβE, when Ca_V1 channels were blocked by nitrendipine, was completely abolished ($100.0 \pm 0.8\%$; $n = 7$; Figure 6).

Thus, the results obtained allow us to conclude that activation of nNAChRs leads to an additional increase in the entry of Ca^{2+} into the nerve ending through VGCCs of the Ca_v1 type. Therefore, if this is the mechanism underlying the decrease in the quantal content upon activation of this type of cholinergic receptor, then it should be expected that the blockade of Ca_v1 type of calcium channels will eliminate the nicotine-induced decrease in the amount of released ACh quanta. Examining this assumption became the scope of the next step of the study.

2.5. Nicotine-Induced Decrease in Acetylcholine Release Is Mediated by L-Type (Ca_v1) Calcium Channels and Not Coupled to Apamin-Sensitive K_{Ca} Channels

To assess the possible role of Ca_v1 type of calcium channels in the nicotine-induced mechanism of ACh release autoinhibition, verapamil and nitrendipine were used at the same concentrations as in experiments with calcium transients.

Verapamil and nitrendipine application resulted in a significant decrease in the QC by $14.2 \pm 3.2\%$ ($n = 6$) and $11.2 \pm 2.9\%$ ($n = 6$), respectively (Figure 7). Nicotine application after pre-treatment with Ca_v1 channel blockers did not cause any changes in the evoked ACh release, and the QCs were $101.6 \pm 1.9\%$ ($n = 6$) and $103.8 \pm 2.4\%$ ($n = 6$), respectively (Figure 7).

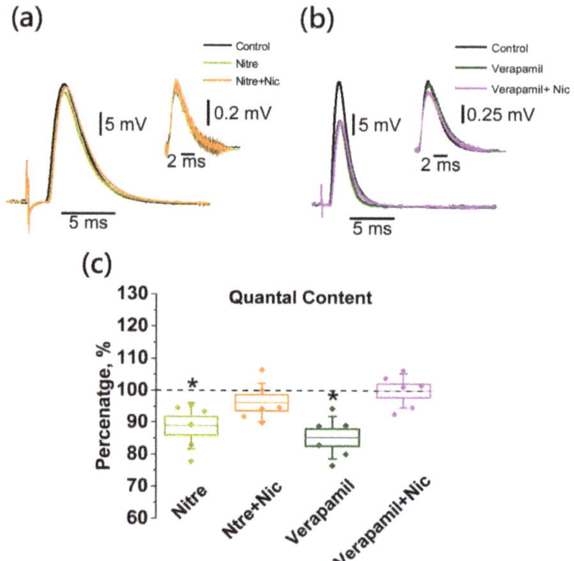

Figure 7. Nicotine-induced decrease in the ACh release (quantal content, QC) involves L-type Ca_v1 channels. Panels on the top are representative traces of EPP and mEPP (50 signals averaged) from individual experiments: (**a**) and (**b**) lack of nicotine (Nic, 10 µM) effect after pre-application with the Ca_v1 VGCC blockers nitrendipine (Nitre, 25 µM) and verapamil (50 µM); (**c**) mean $\pm$ SEM and SD of QC, expressed as a percentage of control or value after Nitre ($n = 6$), Nitre plus Nic ($n = 6$), Verapamil ($n = 6$) and Verapamil plus Nic ($n = 6$) application. Asterisks (*) indicate significant effects ($p < 0.05$, one-way repeated ANOVA test with Tukey's post-hoc comparison).

One of the possible calcium-activated targets that may be responsible for the nicotine-induced depression of ACh release is the apamin-sensitive K_{Ca} channel [21]. To examine the involvement of this channel into the nicotine-induced decrease of ACh release from the motor nerve terminal, experiments with K_{Ca} blocker apamin were conducted.

The application of this blocker to small-conductance Ca^{2+}-activated K^+ channels alone at a concentration of 100 nM did not change the QC of EPP ($95.9 \pm 1.8\%$; $n = 6$, Figure 8).

In the presence of apamin the effect of nicotine on the QC of EPP remained unchanged ($10.5 \pm 1.4\%$; $n = 6$, Figure 8). Therefore, the effect of the activation of nNAChRs on the quantal release of ACh is mediated by Ca_V1 channels and not coupled to apamin-sensitive K_{Ca} channels.

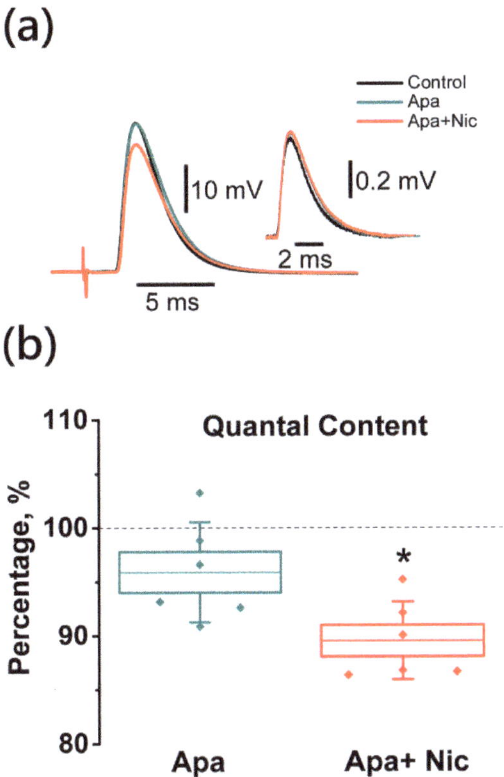

Figure 8. Nicotine-induced decrease in ACh release (quantal content, QC) is observed in the presence of apamin (Apa), the blocker of small conductance Ca^{2+}-activated K^+ channels. Panel on the top (**a**) is representative traces of EPP and mEPP (50 signals averaged) from individual experiments with apamin application (100 nM) and with nicotine (Nic, 10 µM) application after pretreatment with apamin; (**b**) Mean $\pm$ SEM and SD of QC, expressed as a percentage of control or value after apamin ($n = 6$), apamin + Nic ($n = 6$). Asterisks (*) indicate significant effects ($p < 0.05$, one-way repeated ANOVA test with Tukey's post-hoc comparison).

3. Discussion

3.1. Effects of Nicotine on ACh Release

The results of our study demonstrate that in the mouse neuromuscular preparation of m.LAL, nicotine at concentrations up to 1 µM has no effect on either the processes of ACh release from the nerve terminal or on the processes of its interaction with the postsynaptic membrane. At the concentration of 5 µM, the presynaptic effect of nicotine appears to become detectable (inhibition of the spontaneous release of ACh due to the activation of presynaptic cholinergic receptors). An increase in concentration to 10 µM enhances the presynaptic effect of the alkaloid (suppression of not only spontaneous, but also of the evoked ACh release) and leads to a weak postsynaptic effect (decrease in RMP due to the activation of postsynaptic cholinergic receptors). With an increase in nicotine concentration to 50 μM, dramatic changes become evident in all recorded parameters of neurotransmission.

Thus, nicotine at a concentration of 10 μM exerts both postsynaptic and presynaptic inhibitory effects on the neuromuscular synapse, causing a decrease in the number of ACh quanta released in response to action potential. A similar decrease in the QC during the activation of cholinergic receptors has been noted earlier [17,20,21,32], however, these results were obtained on other preparations, in conditions of initially reduced QC, or in cut fiber preparations.

3.2. Presynaptic Cholinergic Receptors and the Role of Calcium Influx in the Mechanism of ACh Release Autoinhibition

Since the inhibitory effect of nicotine on the QC was completely abolished by the application of the antagonist of neuronal cholinergic receptors DHβE, it was concluded that these receptors are involved in the cholinergic mechanism of regulation of ACh release. DHβE binds to the β2 subunits of neuronal receptors and is a selective antagonist for non-α7 nAChRs [33]. In the heteromeric receptors of ganglionic neurons the primary α subunit is α3, whereas in the rodent central nervous system the primary α subunit is α4 [34]. The α4β2 nNAChR is the most abundant subtype expressed in the brain, and studies have demonstrated that this receptor subtype is located presynaptically [35]. Therefore, we can assume that, in the neuromuscular synapse, neuronal cholinergic receptors have the α4β2 subunit composition. This assumption is supported by the data that at 1 μM DHβE, that we used in this study, the mouse α4β2 receptors are almost completely blocked, while the α3β4 subunit receptors remain essentially active [34].

In the next step of the mechanism of autoinhibition triggered by nNAChRs, it was necessary to answer the following key question: how is the activation of the presynaptic nicotinic receptors coupled to changes in the intracellular calcium level? Previous data [20,32,36] were indicating such a coupling, but there was no direct evidence found for this prior to our study. Using standard electrophysiological methods, combined with the fluorescent method for registration of calcium transients, which reflect changes in the calcium level within the presynaptic terminal upon action potential arrival, enabled us to obtain data on changes in the ACh release. Our results demonstrate that the activation of nNAChRs (sensitive to DHβE), leading to a decrease in ACh release, is accompanied by an increase in the level of calcium in the nerve terminal. Another important observation made was the significant effect of DHβE on the amplitude of the calcium transient when applied alone. This may indicate that there exists a background tonic activation of nNAChRs which results in a tonic increase in calcium entry into the nerve terminal. We suggest that the release of endogenous ACh during motor nerve stimulation and the spontaneous quantal release under physiological conditions results in the modulation of presynaptic Ca^{2+} entry and provides a physiologically important feedback [17]. In the absence of impulse activity, the largest amount of ACh is released from the nerve terminal through non-quantal release [37]. This ACh is able to tonic the activation of muscarinic cholinergic receptors at the neuromuscular synapse [38]. Therefore, it can be assumed that DHβE-sensitive nicotinic cholinergic receptors can also be activated by a non-quantal ACh.

The complete absence of the effect of nicotine on the calcium signal after pre-treatment with cadmium (10 μM), which is a nonselective blocker of all types of calcium Ca_V channels [39] allows two conclusions to be put forward: (i) the increase in the calcium level in the nerve terminal is mediated by transmembrane proteins permeable for Ca^{2+} from the environment; (ii) the observed entry of Ca^{2+} is mediated by channels other than those of nNAChRs. The last suggestion is very important, because it has been shown earlier that nNAChRs are more permeable to Ca^{2+} as compared to permeability of their muscle-type counterparts [40,41]. At the same time, it was shown that cadmium up to a concentration of 200 μM does not block currents through nNAChRs [42], but significantly blocks currents through VGCCs [43,44].

After the inactivation of VGCCs of $Ca_V2.1$ type, which are key to triggering the process of evoked ACh release [31,45,46] the effect of nicotine on calcium entry into the terminal was preserved, while after the blockade of Ca_v1 type channels, it was completely abolished. It should be noted that the possibility of the involvement of these channels in the evoked release of ACh quanta remained under debate until recently, as in [20,47–49] versus [23–26]. We have obtained clear evidence of the involvement of the Ca_v1 type of calcium channels in the regulation of the bulk calcium level in the terminal, and of the process of neurotransmission in the mammalian neuromuscular junction.

3.3. Critical Issues in Establishing the Coupling between nNAChRs and L-Type (Ca_v1) Calcium Channels while Using a Pharmacological Approach

The phenylalkylamine (verapamil), dihydropyridine (nitrendipine) and benzothiazepine classes of Ca_v1 type calcium channel blockers are capable of blocking nNAChRs [42,50]. The absence of the effects of nicotine (both on the calcium transient and on the QC) after the application of verapamil and nitrendipine may well be related to a simple direct blockade of nNAChRs (sensitive to DHβE and permeable to Ca^{2+}). Indeed, DHβE, verapamil, and nitrendipine all lead to a decrease in the transients. At the same time, all three pharmacological agents abolish the effect of nicotine.

If even the direct blockade of nNAChRs by verapamil and nitrendipine does take place, a number of additional facts still point to the involvement of the Ca_v1 calcium channels in the mechanism of modulation of calcium entry and the process of ACh release in the nerve terminal. So, after the application of all three agents, the calcium entry decreases, however, the effect of verapamil and nitrendipine is almost two times more pronounced than that of DHβE. Furthermore, the QC does not change after the addition of DHβE, whereas in the presence of verapamil and nitrendipine it is decreased by more than 10%. In addition, the activation of cholinergic receptors (by nicotine) and presumed blockade of cholinergic receptors (by verapamil and nitrendipine) have not an opposite, but a unidirectional effect: a decrease in the QC. And, last but not least, the absence of the effect of nicotine on the calcium transient after the application of cadmium, which does not affect the functioning of nNAChRs (including $\alpha4\beta2$ nNAChRs) or even potentiate them [42,51], indicates that in our case a pharmacological effect on two different targets took place.

3.4. How the Activation of nNAChRs Modulates L-Type (Ca_v1) Channel Functioning

nNAChRs have high Ca^{2+} permeability [41], which does not have a direct effect on the calcium transient amplitude. However, calcium entry through these receptors can lead to two potential outcomes and one can observe an increase in the amplitude of the calcium transient as a result of two mechanisms: (i) the CDF process (calcium dependent facilitation) of the Ca_v1 type channel is triggered [52] (Figure 9a); (ii) the CDI (calcium dependent inactivation) process of Ca_v1 type calcium channels, mediated by the interaction between CaM and Ca^{2+} channel, is disrupted by increasing CaMKII activity [53] (Figure 9b). However, according to [20], a decrease in ACh release caused by activation of nNAChRs is not associated with CaM.

The existence of a functional interaction between nNAChRs and the channels of the Ca_v1 type was established in a primary culture of neurons in the mouse cerebral cortex [54]. It should be noted that this work revealed the interaction of calcium channels with $\alpha4\beta2$ nNAChRs. Potentially, a similar interaction takes place in the muscle-nerve junction. The authors believe that activation of presynaptic receptors leads to depolarization sufficient for opening of Ca_v1 calcium channels and entry of calcium into the neuron [54] (Figure 9c).

The question that arises is how enhanced Ca^{2+} entry via Ca_v1 channels can result in the downregulation of the ACh release. Firstly, the results obtained on the neuromuscular preparations suggest that the calcium channels of Ca_v1 type are located far from the active zones [55], an therefore that Ca^{2+} entering through them cannot directly interact with the exocytosis machine. Secondly, it was shown repeatedly that the elevation of the intra-terminal Ca^{2+} level could activate calcium-sensitive proteins participating in the downregulation of neurotransmitter release. Among them, the most probable candidates are calmodulin, calcineurin, and Ca^{2+}-activated K^+ (K_{Ca}) channels [21,56,57]. However, according to our data, apamin-sensitive K_{Ca} channels are not involved into the nicotine-induced effects on the evoked ACh release.

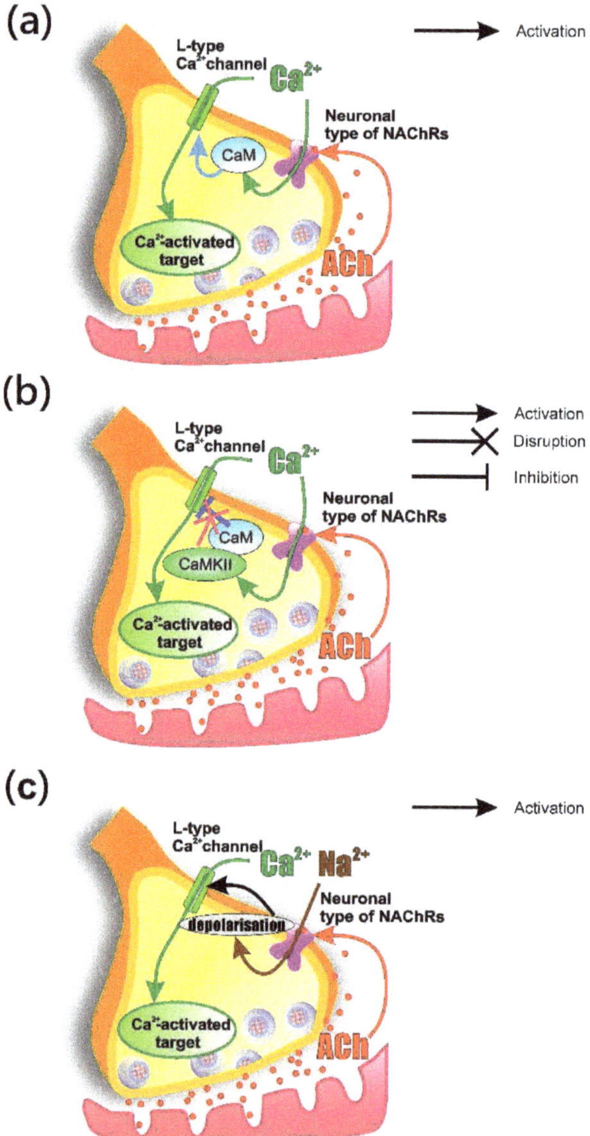

Figure 9. Schematic drawing showing possible mechanisms of the coupling between nAChRs and L-type (Ca_V1) calcium channels in the motor nerve terminal. (**a**) Calcium dependent facilitation of the Ca_V1 type channel, mediated by CaM and triggered by calcium [52], which enters through nNAChRs; (**b**) calcium dependent inactivation process of Ca_V1 type calcium channels, mediated by the interaction between CaM and the Ca^{2+} channel, is disrupted by an increase in CaMKII activity [53]; (**c**) opening of Ca_V1 calcium channels and entry of calcium into terminal caused by depolarization due to activation of nAChRs [54].

4. Materials and Methods

4.1. Animals

Mice BALB/C (20–23 g, 2–3 months old) of either sex were used in this study. All animal care and experimental protocols met the requirements of the European Communities Council Directive 86/609/EEC and was approved by the Ethical Committee of Kazan Medical University (No. 10; 20 December 2016). Animals were housed in groups of 10 animals divided by gender inside plastic cages with plenty of food (standard mice chow) and water ad libitum. The temperature (22 °C) of the room was kept constant and a 12 h light/dark cycle was imposed. Animal studies are reported in compliance with the ARRIVE guidelines [58] and with the recommendations made by the British Journal of Pharmacology [59].

4.2. Tissue Preparations and Solutions

Animals were euthanized by cervical dislocation in accordance with the approved project protocol. The *Levator auris longus* muscle (m. LAL) was quickly removed [60], then put into a Sylgard® chamber with bubbled (95% O_2 and 5% CO_2) Ringer solution (pH 7.4) containing (in mM): NaCl 137, KCl 5, $CaCl_2$ 2, $MgCl_2$ 1, NaH_2PO_4 1, $NaHCO_3$ 11.9, and glucose 11. Bath solution temperature was controlled by a Peltier semiconductor device. Experiments were performed at 20.0 ± 0.3 °C. Muscle contractions were prevented using μ-conotoxin GIIIB in a 2 μM concentration [61].

4.3. Electrophysiology

We used a standard intracellular recording technique [14,62]. Microelectrodes were prepared from borosilicate glass (World Precision Instruments, Sarasota, FL, USA) using a P97 micropipette puller (Sutter Instrument, Novato, CA, USA). Recording electrodes were 20–30 MΩ and filled with 3 M KCl. To record evoked and spontaneous (miniature) endplate potentials (EPPs and mEPPs, respectively) an Axoclamp 900 A amplifier and a DigiData 1440 A digitizer (Axon Instruments, San Jose, CA, USA) were used. Membrane potential was at −60 to −80 mV and recorded by a miniDigi 1B (Axon Instruments, San Jose, CA, USA). Experiments with membrane potential deviations over 7 mV were declined. The nerve was stimulated with rectangular suprathreshold stimuli (0.2 ms duration, 0.5 Hz frequency) via a suction electrode connected to a model 2100 isolated pulse stimulator (A-M Systems, Sequim, WA, USA). All electronic devices were driven by pClamp v10.4 software (Molecular Devices, San Jose, CA, USA). Bandwidth was from 1 Hz to 10 kHz. After collecting 35 EPPs, mEPPs during 2 min were recorded. Quantal content was estimated as ratio of averaged EPPs to mEPPs amplitude.

4.4. Calcium Transient Recording

Nerve motor endings were loaded with Oregon Green 488 BAPTA-1 Hexapotassium Salt 1 mM high-affinity calcium-sensitive dye (Molecular Probes, Eugene, OR, USA) through the nerve stump, as described previously [44]. The fluorescence signal was recorded using an imaging setup based on an Olympus BX-51 microscope with a ×40 water-immersion objective (Olympus, Tokyo, Japan). Calcium transient registration was performed via a high-sensitivity RedShirtImaging NeuroCCD-smq camera (RedShirtImaging, Decatur, GA, USA), at 500 fps (exposure time 2 ms) and 80 × 80 pixels, which was sufficient for calcium transient registration with good temporal resolution. As the source of light, a Polychrome V (Till Photonics, Munich, Germany) was used with a set light wavelength of 488 nm. The following filter set was used to isolate the fluorescent signal: 505DCXT dichroic mirror, E520LP emission (Chroma, Bellows Falls, VT, USA). We used Turbo-SM software (RedShirtImaging, Decatur, GA, USA) for data recording. In each experiment, 8 fluorescence responses were recorded, then averaged. This was an optimal amount to obtain data of sufficient quality and to reduce excitotoxicity and photobleaching of the fluorophore.

To analyze the recorded images, ImageJ software (NIH, Bethesda, MD, USA) was used. We picked regions of interest in the motor nerve ending image and background manually. Subsequent data processing was performed in Excel (Microsoft, Redmond, WA, USA). Background values were averaged and subtracted from signal values. Data were represented as a ratio: $(\Delta F/F_0 - 1) \times 100\ \%$, where ΔF is the fluorescence intensity during stimulation, and F_0 is the fluorescence intensity at rest [63].

4.5. Materials

Nicotine hydrogen tartrate salt (nicotine), nitrendipine, verapamil, ω-agatoxin IVA, cadmium chloride, and dimethylsulfoxide (DMSO) were obtained from Sigma Aldrich (St. Louis, MO, USA). Dihydro-β-erythroidine hydrobromide (DHβE), apamin (TOCRIS, Bristol, UK), and μ-conotoxin GIIIB were obtained from Peptide Institute Inc. (Osaka, Japan). Drugs were dissolved in distilled water with the exceptions of nitrendipine and verapamil, which were dissolved in DMSO. Further dilutions for all drugs were done in Ringer solution. In experiments with drugs which were dissolved in DMSO, the same concentration of DMSO was added to the control solution as was present in the solution with the agent. Finally, the DMSO concentration in the solution did not exceed 0.01%.

4.6. Data and Statistical Analysis

Data collection and statistical analysis comply with the recommendations of the British Journal of Pharmacology on experimental design and analysis in pharmacology [64]. The number of experiments in each experimental group was selected on the basis of observing a statistically significant effect while using the minimum number of animals (3R principles) and on experience from previous studies. Animals were randomly assigned to the different experimental groups, with each group having the same number of animals by design. Blinding of the operator was not feasible, but data analysis was performed semi-blinded by an independent analyst.

Statistical analysis was performed using Statistica 6.1 Base (Tulsa, OK, USA). A Shapiro–Wilk normality test was used to analyze the data distribution. Null-hypothesis testing was performed by ANOVA. One-way ANOVA followed by Dunnett's or Tukey's test for multiple comparison post hoc was used. For related groups, a one-way repeated ANOVA test was performed. Data are presented as mean $\pm$ SEM. Values of $p < 0.05$ were considered significant.

5. Conclusions

In the present study, we found that activation of presynaptic nNAChRs leads to a decrease in the quantal ACh release from the nerve ending. This negative feedback mechanism is mediated by the modulating of the function of VGCCs of the Ca_v1 type, which leads to an increase in the entry of Ca^{2+} into the nerve terminal (Figure 10). Understanding of the peculiarities of action of ACh (nicotine) on the nNAChR-containing nerve endings (not only in cholinergic synapses [51,65]) has a broad scientific and clinical significance, since cholinergic nicotinic signaling (in addition to neuromuscular transmission and synaptic transmission in ganglia) is involved in the setting of a variety of processes, including anxiety, depression, arousal, memory, and attention [66,67].

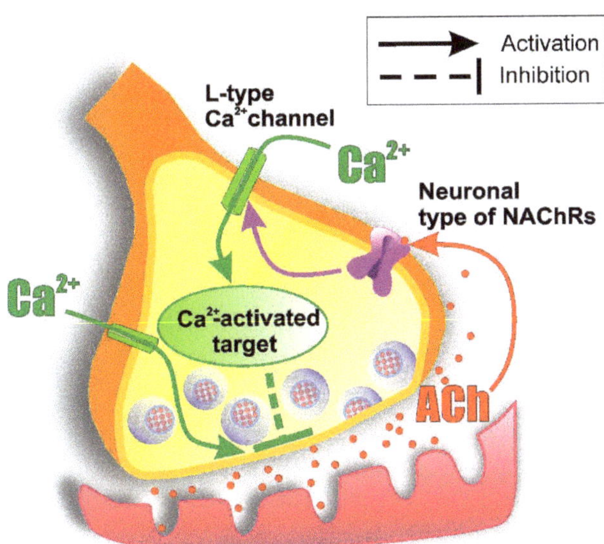

Figure 10. Working model of the mechanism of autoregulation of ACh release in the peripheral cholinergic synapse via nNAChRs. Activation of nNAChRs is accompanied by an increase in the entry of calcium ions into the motor nerve terminal through the L-type (Ca_v1) calcium channels. The latter are involved in both the process of evoked ACh release and its modulation.

Author Contributions: N.Z. and D.S. conceived and designed research; N.Z. and A.A. performed experiments; N.Z., A.A, A.M. and D.S. did the data analysis and interpretation. N.Z., D.S. and A.M. drafted the manuscript. All authors have read and agreed to the published version of the manuscript.

Funding: This research was funded by RFBR #19-04-00490 and by the government assignment for FRC Kazan Scientific Center of RAS AAAA-A18-118022790083-9.

Institutional Review Board Statement: The study was conducted according to the guidelines of the Declaration of Helsinki, and approved by the Local Ethics Committee of Kazan State Medical University (No. 10; 20 December 2016).

Informed Consent Statement: Not applicable.

Data Availability Statement: The data that support the findings of this study are available from the corresponding author upon reasonable request.

Acknowledgments: The authors are thankful to Isabel Bermudez-Diaz, Ellya Bukharaeva and Alexey Petrov for consulting us on some issues. The authors would like to thank Victor I. Ilyin, for his critical reading of this manuscript.

Conflicts of Interest: The authors declare no conflict of interest.

Abbreviations

ACh	acetylcholine
VGCCs	voltage-gated calcium channels
RMP	resting membrane potential
mEPPs	miniature endplate potentials
EPPs	end plate potentials
QC	quantal content
nNAChRs	neuronal nicotinic ACh receptors
DHβE	dihydro-β-erythroidine hydrobromide

References

1. Del Castillo, J.; Katz, B. Interaction at End-Plate Receptors between Different Choline Derivatives. *Proc. R. Soc. Lond. B Biol. Sci.* **1957**, *146*, 369–381. [CrossRef] [PubMed]
2. Malomouzh, A.I.; Nikolsky, E.E. Modern Concepts of Cholinergic Neurotransmission at the Motor Synapse. *Biochem. Suppl. Ser. A Membr. Cell Biol.* **2018**, *12*, 209–222. [CrossRef]
3. Starke, K.; Göthert, M.; Kilbinger, H. Modulation of Neurotransmitter Release by Presynaptic Autoreceptors. *Physiol. Rev.* **1989**, *69*, 864–989. [CrossRef] [PubMed]
4. Bowman, W.C.; Prior, C.; Marshall, I.G. Presynaptic Receptors in the Neuromuscular Junction. *Ann. N. Y. Acad. Sci.* **1990**, *604*, 69–81. [CrossRef]
5. Miller, R.J. Presynaptic Receptors. *Annu. Rev. Pharmacol. Toxicol.* **1998**, *38*, 201–227. [CrossRef]
6. Nikolsky, E.E.; Vyskocil, F.; Bukharaeva, E.A.; Samigullin, D.; Magazanik, L.G. Cholinergic Regulation of the Evoked Quantal Release at Frog Neuromuscular Junction. *J. Physiol.* **2004**, *560*, 77–88. [CrossRef]
7. Prior, C.; Tian, L.; Dempster, J.; Marshall, I.G. Prejunctional Actions of Muscle Relaxants: Synaptic Vesicles and Transmitter Mobilization as Sites of Action. *Gen. Pharmacol.* **1995**, *26*, 659–666. [CrossRef]
8. Santafé, M.M.; Salon, I.; Garcia, N.; Lanuza, M.A.; Uchitel, O.D.; Tomàs, J. Muscarinic Autoreceptors Related with Calcium Channels in the Strong and Weak Inputs at Polyinnervated Developing Rat Neuromuscular Junctions. *Neuroscience* **2004**, *123*, 61–73. [CrossRef]
9. Tsentsevitsky, A.N.; Zakyrjanova, G.F.; Petrov, A.M.; Kovyazina, I.V. Breakdown of Phospholipids and the Elevated Nitric Oxide Are Involved in M3 Muscarinic Regulation of Acetylcholine Secretion in the Frog Motor Synapse. *Biochem. Biophys. Res. Commun.* **2020**, *524*, 589–594. [CrossRef] [PubMed]
10. Tsentsevitsky, A.N.; Kovyazina, I.V.; Nurullin, L.F.; Nikolsky, E.E. Muscarinic Cholinoreceptors (M1-, M2-, M3- and M4-Type) Modulate the Acetylcholine Secretion in the Frog Neuromuscular Junction. *Neurosci. Lett.* **2017**, *649*, 62–69. [CrossRef]
11. Cilleros-Mañé, V.; Just-Borràs, L.; Polishchuk, A.; Durán, M.; Tomàs, M.; Garcia, N.; Tomàs, J.M.; Lanuza, M.A. M 1 and M 2 MAChRs Activate PDK1 and Regulate PKC BI and ε and the Exocytotic Apparatus at the NMJ. *FASEB J.* **2021**, *35*, e21724. [CrossRef] [PubMed]
12. Tsuneki, H.; Kimura, I.; Dezaki, K.; Kimura, M.; Sala, C.; Fumagalli, G. Immunohistochemical Localization of Neuronal Nicotinic Receptor Subtypes at the Pre- and Postjunctional Sites in Mouse Diaphragm Muscle. *Neurosci. Lett.* **1995**, *196*, 13–16. [CrossRef]
13. Malomouzh, A.I.; Arkhipova, S.S.; Nikolsky, E.E.; Vyskočil, F. Immunocytochemical Demonstration of M1 Muscarinic Acetylcholine Receptors at the Presynaptic and Postsynaptic Membranes of Rat Diaphragm Endplates. *Physiol. Res.* **2011**, *60*, 185–188. [CrossRef] [PubMed]
14. Santafé, M.M.; Salon, I.; Garcia, N.; Lanuza, M.A.; Uchitel, O.D.; Tomàs, J. Modulation of ACh Release by Presynaptic Muscarinic Autoreceptors in the Neuromuscular Junction of the Newborn and Adult Rat. *Eur. J. Neurosci.* **2003**, *17*, 119–127. [CrossRef] [PubMed]
15. Oliveira, L.; Timóteo, M.A.; Correia-de-Sá, P. Modulation by Adenosine of Both Muscarinic M1-Facilitation and M2-Inhibition of [3H]-Acetylcholine Release from the Rat Motor Nerve Terminals. *Eur. J. Neurosci.* **2002**, *15*, 1728–1736. [CrossRef]
16. Zhilyakov, N.V.; Khaziev, E.F.; Latfullin, A.R.; Malomouzh, A.I.; Bukharaeva, E.A.; Nikolsky, E.E.; Samigullin, D.V. Changes in Calcium Levels in Motor Nerve Endings in Mice on Activation of Metabotropic Cholinoreceptors and GABA Receptors. *Neurosci. Behav. Physiol.* **2019**, *49*, 1092–1095. [CrossRef]
17. Khaziev, E.; Samigullin, D.; Zhilyakov, N.; Fatikhov, N.; Bukharaeva, E.; Verkhratsky, A.; Nikolsky, E. Acetylcholine-Induced Inhibition of Presynaptic Calcium Signals and Transmitter Release in the Frog Neuromuscular Junction. *Front. Physiol.* **2016**, *7*, 1–10. [CrossRef] [PubMed]
18. Slutsky, I.; Wess, J.; Gomeza, J.; Dudel, J.; Parnas, I.; Parnas, H. Use of Knockout Mice Reveals Involvement of M2-Muscarinic Receptors in Control of the Kinetics of Acetylcholine Release. *J. Neurophysiol.* **2003**, *89*, 1954–1967. [CrossRef] [PubMed]
19. Van der Kloot, W. Nicotinic Agonists Antagonize Quantal Size Increases and Evoked Release at Frog Neuromuscular Junction. *J. Physiol.* **1993**, *468*, 567–589. [CrossRef] [PubMed]
20. Prior, C.; Singh, S. Factors Influencing the Low-Frequency Associated Nicotinic ACh Autoreceptor-Mediated Depression of ACh Release from Rat Motor Nerve Terminals. *Br. J. Pharmacol.* **2000**, *129*, 1067–1074. [CrossRef]
21. Balezina, O.P.; Fedorin, V.V.; Gaidukov, A.E. Effect of Nicotine on Neuromuscular Transmission in Mouse Motor Synapses. *Bull. Exp. Biol. Med.* **2006**, *142*, 17–21. [CrossRef] [PubMed]
22. Khaziev, E.F.; Samigullin, D.V.; Tsentsevitsky, A.N.; Bukharaeva, E.A.; Nikolsky, E.E. ATP Reduces the Entry of Calcium Ions into the Nerve Ending by Blocking L-Type Calcium Channels. *Acta Naturae* **2018**, *10*, 93–96. [CrossRef]
23. Atchison, W.D. Dihydropyridine-Sensitive and -Insensitive Components of Acetylcholine Release from Rat Motor Nerve Terminals. *J. Pharmacol. Exp. Ther.* **1989**, *251*, 672–678. [PubMed]
24. Protti, D.A.; Szczupak, L.; Scornik, F.S.; Uchitel, O.D. Effect of ω-Conotoxin GVIA on Neurotransmitter Release at the Mouse Neuromuscular Junction. *Brain Res.* **1991**, *557*, 336–339. [CrossRef]
25. Penner, R.; Dreyer, F. Two Different Presynaptic Calcium Currents in Mouse Motor Nerve Terminals. *Pflügers Arch. Eur. J. Physiol.* **1986**, *406*, 190–197. [CrossRef] [PubMed]

26. Bowersox, S.S.; Miljanich, G.P.; Sugiura, Y.; Li, C.; Nadasdi, L.; Hoffman, B.B.; Ramachandran, J.; Ko, C.P. Differential Blockade of Voltage-Sensitive Calcium Channels at the Mouse Neuromuscular Junction by Novel ω-Conopeptides and ω-Agatoxin-IVA. *J. Pharmacol. Exp. Ther.* **1995**, *273*, 248–256. [PubMed]
27. Moroni, M.; Zwart, R.; Sher, E.; Cassels, B.K.; Bermudez, I. A4β2 Nicotinic Receptors with High and Low Acetylcholine Sensitivity: Pharmacology, Stoichiometry, and Sensitivity to Long-Term Exposure to Nicotine. *Mol. Pharmacol.* **2006**, *70*, 755–768. [CrossRef]
28. Wonnacott, S. Nicotinic ACh Receptors. Available online: https://www.tocris.com/literature/scientific-reviews/nicotinic-ach-receptors (accessed on 27 August 2021).
29. Katz, B. *The Release of Neural Transmitter Substances*; Liverpool University Press: Liverpool, UK, 1969; pp. 5–39.
30. Crawford, A.C. The Dependence of Evoked Transmitter Release on External Calcium Ions at Very Low Mean Quantal Contents. *J. Physiol.* **1974**, *240*, 255–278. [CrossRef] [PubMed]
31. Protti, D.A.; Uchitel, O.D. Transmitter Release and Presynaptic Ca2+ Currents Blocked by the Spider Toxin ω-Aga-IVA. *Neuroreport* **1993**, *5*, 333–336. [CrossRef]
32. Tian, L.; Prior, C.; Dempster, J.; Marshall, I.G. Nicotinic Antagonist-produced Frequency-dependent Changes in Acetylcholine Release from Rat Motor Nerve Terminals. *J. Physiol.* **1994**, *476*, 517–529. [CrossRef] [PubMed]
33. Chavez-Noriega, L.E.; Gillespie, A.; Stauderman, K.A.; Crona, J.H.; Claeps, B.O.; Elliott, K.J.; Reid, R.T.; Rao, T.S.; Veliçelebi, G.; Harpold, M.M.; et al. Characterization of the Recombinant Human Neuronal Nicotinic Acetylcholine Receptors A3β2 and A4β2 Stably Expressed in HEK293 Cells. *Neuropharmacology* **2000**, *39*, 2543–2560. [CrossRef]
34. Papke, R.L.; Wecker, L.; Stitzel, J.A. Activation and Inhibition of Mouse Muscle and Neuronal Nicotinic Acetylcholine Receptors Expressed in Xenopus Oocytes. *J. Pharmacol. Exp. Ther.* **2010**, *333*, 501–518. [CrossRef] [PubMed]
35. Karadsheh, M.S.; Shah, M.S.; Tang, X.; Macdonald, R.L.; Stitzel, J.A. Functional Characterization of Mouse A4β2 Nicotinic Acetylcholine Receptors Stably Expressed in HEK293T Cells. *J. Neurochem.* **2004**, *91*, 1138–1150. [CrossRef] [PubMed]
36. Wang, X.; Michael McIntosh, J.; Rich, M.M. Muscle Nicotinic Acetylcholine Receptors May Mediate Trans-Synaptic Signaling at the Mouse Neuromuscular Junction. *J. Neurosci.* **2018**, *38*, 1725–1736. [CrossRef]
37. Vyskocil, F.; Malomouzh, A.; Nikolsky, E. Non-Quantal Acetylcholine Release at the Neuromuscular Junction. *Physiol. Res.* **2009**, *58*, 763–784. [CrossRef]
38. Malomouzh, A.; Mukhtarov, M.; Nikolsky, E.; Vyskočil, F. Muscarinic M1 Acetylcholine Receptors Regulate the Non-Quantal Release of Acetylcholine in the Rat Neuromuscular Junction via NO-Dependent Mechanism. *J. Neurochem.* **2007**, *102*, 2110–2117. [CrossRef]
39. Hess, P.; Lansman, J.B.; Tsien, R.W. Different Modes of Ca Channel Gating Behaviour Favoured by Dihydropyridine Ca Agonists and Antagonists. *Nature* **1984**, *311*, 538–544. [CrossRef]
40. Radford Deckera, E.; Dani, J.A. Calcium Permeability of the Nicotinic Acetylcholine Receptor: The Single-Channel Calcium Influx Is Significant. *J. Neurosci.* **1990**, *10*, 3413–3420. [CrossRef]
41. Gotti, C.; Clementi, F. Neuronal Nicotinic Receptors: From Structure to Pathology. *Prog. Neurobiol.* **2004**, *74*, 363–396. [CrossRef] [PubMed]
42. Wheeler, D.G.; Barrett, C.F.; Tsien, R.W. L-Type Calcium Channel Ligands Block Nicotine-Induced Signaling to CREB by Inhibiting Nicotinic Receptors. *Neuropharmacology* **2006**, *51*, 27–36. [CrossRef]
43. Lansman, J.B.; Hess, P.; Tsien, R.W. Blockade of Current through Single Calcium Channels by Cd^{2+}, Mg^{2+}, and Ca^{2+}s: Voltage and Concentration Dependence of Calcium Entry into the Pore. *J. Gen. Physiol.* **1986**, *88*, 321–347. [CrossRef]
44. Samigullin, D.V.; Khaziev, E.F.; Zhilyakov, N.V.; Sudakov, I.A.; Bukharaeva, E.A.; Nikolsky, E.E. Calcium Transient Registration in Response to Single Stimulation and during Train of Pulses in Mouse Neuromuscular Junction. *Bionanoscience* **2017**, *7*, 162–166. [CrossRef]
45. Nachshen, D.A.; Blaustein, M.P. The Effects of Some Organic "Calcium Antagonists" on Calcium Influx in Presynaptic Nerve Terminals. *Mol. Pharmacol.* **1979**, *16*, 576–586. [PubMed]
46. Katz, E.; Ferro, P.A.; Weisz, G.; Uchitel, O.D. Calcium Channels Involved in Synaptic Transmission at the Mature and Regenerating Mouse Neuromuscular Junction. *J. Physiol.* **1996**, *497*, 687–697. [CrossRef] [PubMed]
47. Pagani, R.; Song, M.; Mcenery, M.; Qin, N.; Tsien, R.W.; Toro, L.; Stefani, E.; Uchitel, O.D. Differential Expression of A1 and β Subunits of Voltage Dependent Ca^{2+} Channel at the Neuromuscular Junction of Normal and P/Q Ca^{2+} Channel Knockout Mouse. *Neuroscience* **2004**, *123*, 75–85. [CrossRef] [PubMed]
48. Perissinotti, P.P.; Tropper, B.G.; Uchitel, O.D. L-Type Calcium Channels Are Involved in Fast Endocytosis at the Mouse Neuromuscular Junction. *Eur. J. Neurosci.* **2008**, *27*, 1333–1344. [CrossRef]
49. Urbano, F.J.; Rosato-Siri, M.D.; Uchitel, O.D. Calcium Channels Involved in Neurotransmitter Release at Adult, Neonatal and P/Q-Type Deficient Neuromuscular Junctions. *Mol. Membr. Biol.* **2002**, *19*, 293–300. [CrossRef]
50. Houlihan, L.M.; Slater, E.Y.; Beadle, D.J.; Lukas, R.J.; Bermudez, I. Effects of Diltiazem on Human Nicotinic Acetylcholine and GABA(A) Receptors. *Neuropharmacology* **2000**, *39*, 2533–2542. [CrossRef]
51. Garduño, J.; Galindo-Charles, L.; Jiménez-Rodríguez, J.; Galarraga, E.; Tapia, D.; Mihailescu, S.; Hernandez-Lopez, S. Presynaptic A4β2 Nicotinic Acetylcholine Receptors Increase Glutamate Release and Serotonin Neuron Excitability in the Dorsal Raphe Nucleus. *J. Neurosci.* **2012**, *32*, 15148–15157. [CrossRef] [PubMed]

52. Kim, E.Y.; Rumpf, C.H.; Fujiwara, Y.; Cooley, E.S.; Van Petegem, F.; Minor, D.L. Structures of CaV2 Ca^{2+}/CaM-IQ Domain Complexes Reveal Binding Modes That Underlie Calcium-Dependent Inactivation and Facilitation. *Structure* **2008**, *16*, 1455–1467. [CrossRef]
53. Abiria, S.A.; Colbran, R.J. CaMKII Associates with Ca$_V$ 1.2 L-Type Calcium Channels via Selected β Subunits to Enhance Regulatory Phosphorylation. *J. Neurochem.* **2010**, *112*, 150–161. [CrossRef] [PubMed]
54. Katsura, M.; Mohri, Y.; Shuto, K.; Hai-Du, Y.; Amano, T.; Tsujimura, A.; Sasa, M.; Ohkuma, S. Up-Regulation of L-Type Voltage-Dependent Calcium Channels after Long Term Exposure to Nicotine in Cerebral Cortical Neurons. *J. Biol. Chem.* **2002**, *277*, 7979–7988. [CrossRef] [PubMed]
55. Polo-Parada, L.; Bose, C.M.; Landmesser, L.T. Alterations in Transmission, Vesicle Dynamics, and Transmitter Release Machinery at NCAM-Deficient Neuromuscular Junctions. *Neuron* **2001**, *32*, 815–828. [CrossRef]
56. De Lorenzo, S.; Veggetti, M.; Muchnik, S.; Losavio, A. Presynaptic Inhibition of Spontaneous Acetylcholine Release Mediated by P2Y Receptors at the Mouse Neuromuscular Junction. *Neuroscience* **2006**, *142*, 71–85. [CrossRef]
57. Tarasova, E.; Gaydukov, A.E.; Balezina, O.P. Calcineurin and Its Role in Synaptic Transmission. *Biochemistry* **2018**, *83*, 674–689. [CrossRef]
58. Percie du Sert, N.; Hurst, V.; Ahluwalia, A.; Alam, S.; Avey, M.T.; Baker, M.; Browne, W.J.; Clark, A.; Cuthill, I.C.; Dirnagl, U.; et al. The Arrive Guidelines 2.0: Updated Guidelines for Reporting Animal Research. *PLoS Biol.* **2020**, *18*, 1769–1777. [CrossRef]
59. Lilley, E.; Stanford, S.C.; Kendall, D.E.; Alexander, S.P.H.; Cirino, G.; Docherty, J.R.; George, C.H.; Insel, P.A.; Izzo, A.A.; Ji, Y.; et al. ARRIVE 2.0 and the British Journal of Pharmacology: Updated Guidance for 2020. *Br. J. Pharmacol.* **2020**, *177*, 3611–3616. [CrossRef] [PubMed]
60. Angaut-Petit, D.; Molgo, J.; Connold, A.L.; Faille, L. The Levator Auris Longus Muscle of the Mouse: A Convenient Preparation for Studies of Short- and Long-Term Presynaptic Effects of Drugs or Toxins. *Neurosci. Lett.* **1987**, *82*, 83–88. [CrossRef]
61. Hill, J.M.; Alewood, P.F.; Craik, D.J. Three-Dimensional Solution Structure of μ-Conotoxin GIIIB, a Specific Blocker of Skeletal Muscle Sodium Channels. *Biochemistry* **1996**, *35*, 8824–8835. [CrossRef] [PubMed]
62. Thesleff, S. A Study of the Interaction between Neuromuscular Blocking Agents and Acetylcholine at the Mammalian Motor End-Plate. *Acta Anaesthesiol. Scand.* **1958**, *2*, 69–79. [CrossRef]
63. Samigullin, D.V.; Khaziev, E.F.; Zhilyakov, N.V.; Bukharaeva, E.A.; Nikolsky, E.E. Loading a Calcium Dye into Frog Nerve Endings through the Nerve Stump: Calcium Transient Registration in the Frog Neuromuscular Junction. *J. Vis. Exp.* **2017**, *125*, e55122. [CrossRef] [PubMed]
64. Curtis, M.J.; Alexander, S.; Cirino, G.; Docherty, J.R.; George, C.H.; Giembycz, M.A.; Hoyer, D.; Insel, P.A.; Izzo, A.A.; Ji, Y.; et al. Experimental Design and Analysis and Their Reporting II: Updated and Simplified Guidance for Authors and Peer Reviewers. *Br. J. Pharmacol.* **2018**, *175*, 987–993. [CrossRef] [PubMed]
65. Seth, P.; Cheeta, S.; Tucci, S.; File, S.E. Nicotinic–Serotonergic Interactions in Brain and Behaviour. *Pharmacol. Biochem. Behav.* **2002**, *71*, 795–805. [CrossRef]
66. Hogg, R.C.; Raggenbass, M.; Bertrand, D. Nicotinic acetylcholine receptors: From structure to brain function. In *Reviews of Physiology, Biochemistry and Pharmacology*; Springer: Berlin/Heidelberg, Germany, 2003; Volume 147, pp. 1–46.
67. Picciotto, M.R. Nicotine as a Modulator of Behavior: Beyond the Inverted U. *Trends Pharmacol. Sci.* **2003**, *24*, 493–499. [CrossRef]

International Journal of
Molecular Sciences

Article

Simultaneous Monitoring of pH and Chloride (Cl^-) in Brain Slices of Transgenic Mice

Daria Ponomareva [1,2,3,†], Elena Petukhova [2,3,†] and Piotr Bregestovski [1,2,3,*]

1   Institut de Neurosciences des Systèmes, Aix-Marseille University, INSERM, INS, 13005 Marseille, France; Ponomareva_DN@mail.ru
2   Institute of Neurosciences, Kazan State Medical University, 420111 Kazan, Russia; petukhovaeo@mail.ru
3   Department of Normal Physiology, Kazan State Medical University, 420111 Kazan, Russia
*   Correspondence: pbreges@gmail.com
†   These authors contributed equally to this work.

Abstract: Optosensorics is the direction of research possessing the possibility of non-invasive monitoring of the concentration of intracellular ions or activity of intracellular components using specific biosensors. In recent years, genetically encoded proteins have been used as effective optosensory means. These probes possess fluorophore groups capable of changing fluorescence when interacting with certain ions or molecules. For monitoring of intracellular concentrations of chloride ($[Cl^-]_i$) and hydrogen ($[H^+]_i$) the construct, called ClopHensor, which consists of a H^+- and Cl^--sensitive variant of the enhanced green fluorescent protein (E^2GFP) fused with a monomeric red fluorescent protein (mDsRed) has been proposed. We recently developed a line of transgenic mice expressing ClopHensor in neurons and obtained the map of its expression in different areas of the brain. The purpose of this study was to examine the effectiveness of transgenic mice expressing ClopHensor for estimation of $[H^+]_i$ and $[Cl^-]_i$ concentrations in neurons of brain slices. We performed simultaneous monitoring of $[H^+]_i$ and $[Cl^-]_i$ under different experimental conditions including changing of external concentrations of ions (Ca^{2+}, Cl^-, K^+, Na^+) and synaptic stimulation of Shaffer's collaterals of hippocampal slices. The results obtained illuminate different pathways of regulation of Cl^- and pH equilibrium in neurons and demonstrate that transgenic mice expressing ClopHensor represent a reliable tool for non-invasive simultaneous monitoring of intracellular Cl^- and pH.

Keywords: genetically encoded biosensors; optopharmacology; transgenic mice; intracellular pH; intracellular chloride; brain slices; pH and Cl^- transporters

Citation: Ponomareva, D.; Petukhova, E.; Bregestovski, P. Simultaneous Monitoring of pH and Chloride (Cl^-) in Brain Slices of Transgenic Mice. *Int. J. Mol. Sci.* **2021**, *22*, 13601. https://doi.org/10.3390/ijms222413601

Academic Editor: Carlo Matera

Received: 28 November 2021
Accepted: 14 December 2021
Published: 18 December 2021

Publisher's Note: MDPI stays neutral with regard to jurisdictional claims in published maps and institutional affiliations.

1. Introduction

Neuronal activity is accompanied by dynamical changes in the intra- and extracellular concentration of ions. A number of studies demonstrated that activation or inhibition of neurons can cause a rapid shift of hydrogen (H^+) and chloride (Cl^-) [1–3]. The physiological intracellular pH range in different eukaryotic cells is 6.5–8.0 [4], which corresponds to a very low free H^+ concentration, from 10 nM to 300 nM. In the mammalian brain the intracellular pH is 7.0–7.4 [5,6] reflecting the importance of maintaining the intracellular H^+ concentration ($[H^+]_i$) in a narrow range.

Intracellular Cl^- concentration ($[Cl^-]_i$) in different cell types varies from 3 mM to 60 mM, being around 5–10 mM in the majority of mammalian neurons [7]. Deviations from this physiological range can alter the excitability of cells, modulate the function of a variety of proteins including ion channels [8–10]. Abnormal changes in concentrations of these ions are associated with the development of pathological processes and some disorders including neurodegeneration, epilepsy and brain ischemia [11–13].

To maintain H^+ and Cl^- in physiological ranges, various transporters, cotransporters, and other ion regulating proteins are expressed in cells of biological organisms. The regulation of intracellular $[H^+]_i$ is primarily driven by Na^+/H^+ exchange, Na^+-driven Cl^-/HCO_3^- exchange, Na^+/HCO_3^- cotransport, and Cl^-/HCO_3^- exchange [1,14–16].

The level of $[Cl^-]_i$ is maintained predominantly by a K^+/Cl^- cotransporter (KCC2), which pumps out Cl^- from the cells, a $Na^+/K^+/Cl^-$ cotransporter NKCC1, which loads Cl^- into the cell, and a Cl^-/HCO_3^- exchanger [3]. Activation of the Cl^-/HCO_3^- exchanger leads to local simultaneous pH and Cl^- changes. These interrelated relationships highlight the necessity of simultaneous monitoring of changes in Cl^- and H^+ concentrations.

To measure pH and Cl^- in cells of biological organisms, several methods have been proposed. The most used are ion-selective microelectrodes [17–19], fluorescent dyes [20–22] and genetically encoded probes [7,23–26]. Especially promising are genetically encoded fluorescent sensors, which allow non-invasive monitoring of intracellular ion concentrations in specific cell types.

For simultaneous registration of $[H^+]_i$ and $[Cl^-]_i$, the construct ClopHensor has been developed, which consists of a H^+- and Cl^--sensitive variant of the enhanced green fluorescent protein (E^2GFP) fused with a monomeric red fluorescent protein (mDsRed) [27]. The sensor was studied at the heterologous expression in different cells. It has been shown that the construct possesses a pKa = 6.8 for H^+ and K_d for Cl^- in the range 40–50 mM at physiological pH ~ 7.3 [24,27,28]. Improved variants of ClopHensor have been developed and tested in cultured brain slices using biolistical transfection [29]. Another biosensor optimized for the simultaneous $[Cl^-]_i$ and pH_i imaging, called LSSmClopHensor, has been developed [30]. It was expressed in the cortex of rats using in utero electroporation and used for analysis in brain slices and in vivo [30,31].

Recently, we presented a line of transgenic mice expressing ClopHensor in neurons due to the neuronal-specific promoter Thy1 [32]. The pattern of ClopHensor expression across the brain has been revealed using the CLARITY method in combination with confocal and light-sheet microscopy. It showed a robust expression of ClopHensor in the hippocampal formation, thalamus, ponds, medulla, cerebellum and other areas [32].

The present study is devoted to the analysis of H^+- and Cl^--transients at synaptic stimulation of neurons in hippocampal slices of Thy1: ClopHensor mice. We analyzed changes in both ion concentrations at different approaches including inhibition of GABA-ergic synaptic transmission, extracellular Ca^{2+}-free and low Cl^- or Na^+ conditions. Our observations demonstrate the efficiency of transgenic mice expressing ChlopHensor for reliable non-invasive monitoring of intracellular Cl^- and pH in normal and pathological conditions.

2. Results

2.1. Simultaneous Monitoring of pH and Cl^- Using ClopHensor

Experiments were performed on brain slices of transgenic mice expressing ClopHensor using the mouse neuronal promoter Thy1. ClopHensor consists of a modified enhanced green fluorescent protein E^2GFP connected with a monomeric DsRed (mDsRed) via a 20-amino-acid flexible linker (Figure 1A). E^2GFP carries a specific anion-binding site engineered by the single substitution T203Y and elevation of Cl^- causes static quenching of E2GFP's fluorescence [33]. Like all green fluorescent proteins, E^2GFP is also sensitive to pH: its emission decreases when it is acidified. However, E^2GFP's emission intensity does not depend on pH when excited at its isosbestic point of 458 nm, allowing one to perform ratiometric analysis. In addition, at wavelengths above 540 nm, signals of both fluorescent proteins, E^2GFP and mDsRed, are pH and Cl^--independent (Figure 1B). This allows the simultaneous ratiometric analysis of changes in $[H^+]_i$ and $[Cl^-]_i$ at monitoring upon excitation at three wavelengths: 488 nm: pH- and Cl^--dependent E^2GFP signal, 458 nm: pH-independent E^2GFP signal, and above 543 nm: Cl^-- and pH-independent mDsRed signal [27].

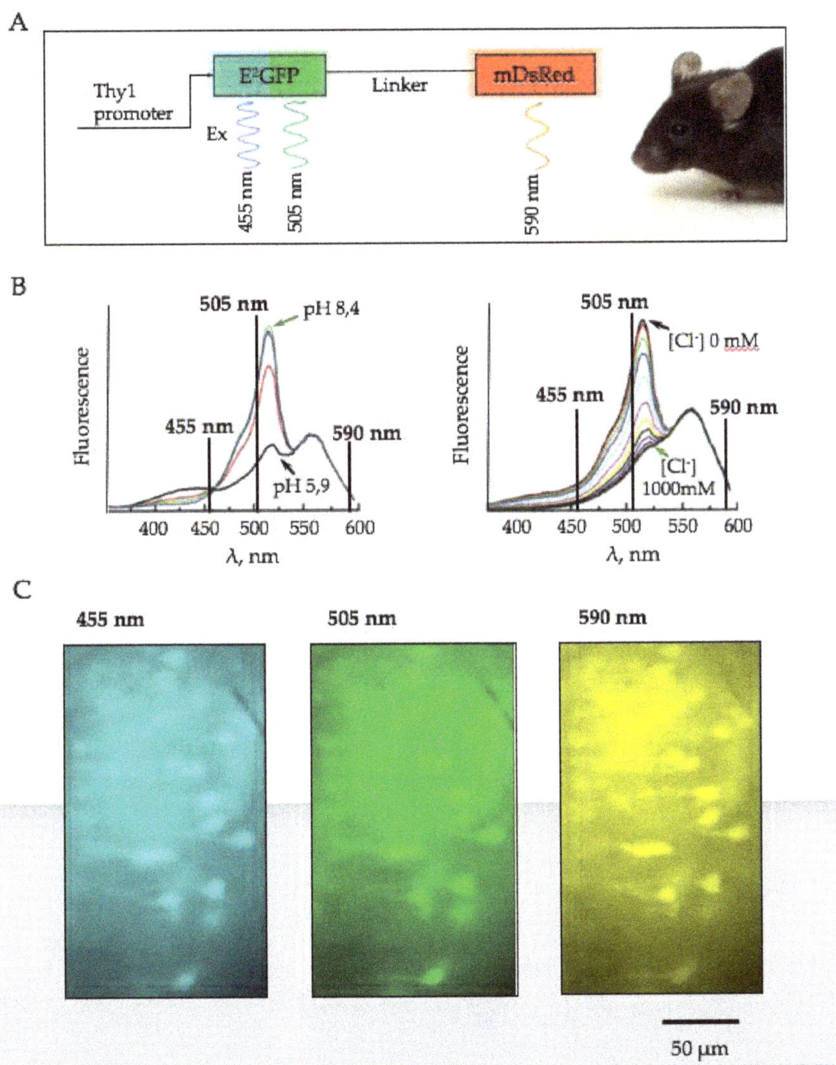

Figure 1. Monitoring pH and Cl⁻ in brain slices of mice expressing ClopHensor. (**A**) Schematic representation of ClopHensor. (**B**) Excitation spectra of ClopHensor collected at different pH values (5.9, 6.9, 7.4, and 8.4) in the absence of Cl⁻ (left) and at increasing Cl⁻ concentration (0–1 M) and constant pH = 6.9 (right) (modified from Arosio et al., 2010 [27]). (**C**) Micrographs of pyramidal cells of CA1 hippocampal region, expressing ClopHensor, under illumination with excitation light with wavelengths of 455 nm, 505 nm and 590 nm (age P7).

In the present study, we recorded fluorescence at slightly different excitation wavelengths (455 nm, 505 nm and 590 nm) and estimated changes in concentrations of ions using the following ratios:

$$R_{pH} = \Delta F_{505nm} / \Delta F_{455nm} \tag{1}$$

$$R_{Cl} = \Delta F_{590nm} / \Delta F_{455nm} \tag{2}$$

Examples of fluorescence at different excitation wavelengths are shown in Figure 1C. Details of calibration are presented in Section 4.

2.2. Effect of Bicuculline on Synaptically Induced Cl^- and pH-Transients in Hippocampal Neurons

To examine the properties of intracellular Cl^- and H^+ transients during synaptic activation, we performed experiments on hippocampal slices of transgenic mice expressing ClopHensor. The first task was to establish how the changes in intracellular concentrations of Cl^- and H^+, evoked by electrical stimulation, are amenable to pharmacological modulation of synaptic transmission. Fluorescent emission signals were recorded from the hippocampal CA1 area upon sustained tetanic stimulation of Schaffer's collaterals (100 Hz for 20 s). Stimulus intensities ranged from 20 µA to 320 µA. The experiments were carried out on adult transgenic mice aged 2–8 months.

High-frequency stimulation of Schaffer's collaterals caused strong changes of fluorescence signals excited at 455 nm and 505 nm without effect on excitation at 590 nm (not shown). Ratiometric analysis revealed that these changes correspond to an elevation of both $[Cl^-]_i$ and $[H^+]_i$. In this set of experiments, in hippocampal CA1 neurons, the base level of $[Cl^-]_i$ was 8.0 ± 1.1 mM ($n = 5$) and the mean intracellular pH was 7.30 ± 0.02 ($n = 11$). High frequency synaptic stimulation increased the $[Cl^-]_i$ to 0.8 ± 0.3 mM ($n = 5$) and acidification of neurons to 0.024 ± 0.006 units pH ($n = 11$).

In order to test the assumption that during synaptic stimulation, the accumulation of Cl^- in the hippocampal neurons is, at least partially, due to Cl^- influx via activated GABAA receptors, we applied bicuculline, an antagonist of these receptors [34]. The traces shown in Figure 2A illustrate that upon the synaptic stimulation, bicuculline strongly decreased the amplitude of synaptically induced changes of $[Cl^-]_i$. On average, the mean amplitude of Cl^- influx decreased to $41.2 \pm 12.2\%$ (Figure 2B, $n = 7$, $p < 0.01$). In contrast, upon application of bicuculline the amplitude of synaptically induced pH transients increased by $90.7 \pm 17.4\%$ (Figure 2C, $n = 8$, $p < 0.01$). Picrotoxin, a blocker of Cl^--selective GABA receptor channels, caused similar changes in the amplitudes of Cl^- and pH transients (data not shown).

The augmentation of pH transients can be, at least partially, explained by the fact that bicuculline blocks inhibitory transmission, which leads to an increase in the general excitability [35]. To check the involvement of this mechanism in the potentiation of synaptically induced H^+ responses upon inhibition of GABA receptors, we analyzed the effect of bicuculline on the amplitude of evoked local field potentials (eLFP). After stimulation of Schaffer's collaterals, a population of CA1 neurons are activated simultaneously and fire an action potential in synchrony, giving rise to a single eLFP. As illustrated in Figure 2D, the addition of bicuculline (40 µM) led to an elevation of eLFPs. The mean potentiation was $24 \pm 3.8\%$ (Figure 2E). Consequently, in the presence of bicuculline, the effectiveness of the stimulation increased, causing stronger depolarization and augmentation of pH transients. A competitive AMPA/kainate receptor antagonist (CNQX, 40 µm) completely abolished stimulation-induced changes of ratiometric emission signals, confirming that they have emerged from synaptic activation (not shown).

Altogether, these data demonstrate that: (i) bicuculline modulates in opposite ways the synaptically evoked Cl^-- and pH-transients in neurons expressing ClopHensor; (ii) bicuculline only partially inhibits synaptically induced elevation of $[Cl^-]_i$; (iii) inhibition of GABA receptor by bicuculline causes potentiation of eLFPs.

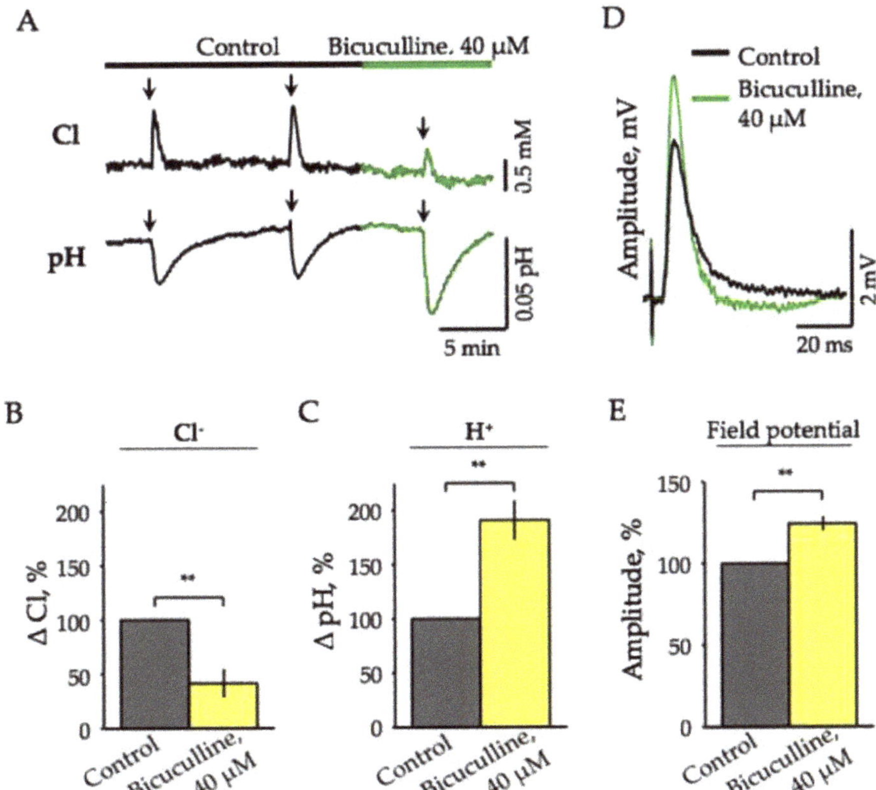

Figure 2. Effect of bicuculline on Cl⁻- and pH-transients induced by synaptic stimulation of hippocampal CA1 neurons in brain slices of transgenic ClopHensor mice. (**A**) Traces of Cl⁻- (top) and pH-specific emission signals (bottom) in control (black) and the presence of 40 µM bicuculline (green). Arrows indicate stimulation (100 µA, 100 Hz, 20 s). (**B**) Mean inhibition by bicuculline (40 µM) of Cl⁻ transients induced by synaptic stimulation of CA1 hippocampal neurons (mean percentage ± SEM, *n* = 7, 2–8 months). ** Significant difference with *p* < 0.01 (Paired Sample Wilcoxon Signed Rank Test). (**C**) Mean potentiation by bicuculline (40 µM) of stimulation-induced pH-transients. Summary data from 8 slices (mean percentage ± SEM, *n* = 8, 2–8 months). ** Significant difference with *p* < 0.01 (Paired Sample Wilcoxon Signed Rank Test). Age of mice 2–8 months. (**D**) Superimposed traces of evoked local field potentials (eLFP) induced by single stimulation of Schaffer's collaterals in control (black) and the presence of 40 µM bicuculline (green). Presented averaged traces of 10 individual eLFPs (stimulation: 20 µA, single pulse width 200 µs; age 7 months). (**E**) Summary of the eLFP amplitude potentiation by 40 µM bicuculline. 100% is the amplitude of eLFPs in control condition. Values are mean ± SEM (*n* = 7). ** Significant difference with *p* < 0.01 (Paired Sample Wilcoxon Signed Rank Test). Age of mice 2–8 months.

2.3. Effect of Bicuculline on Intracellular pH Changes after Tetanic Stimulation, Assessed Using BCECF-AM

To ensure the specificity of the pH changes when using ClopHsensor, we performed a series of experiments with the pH-sensitive dye BCECF-AM [36]. This compound is well tolerated by cells and offers ratiometric estimation of pH values.

We determined the pH changes by calculating the ratio of two emission signals obtained after illumination by light at 455 nm and 505 nm. The intracellular pH changes after tetanic stimulation and its modulation by bicuculline were analyzed. Experiments were conducted on wild type juvenile mice (P10–P12). Registration conditions were similar to those performed on slices from transgenic ClopHensor mice. Examples of

CA1 hippocampal neurons loaded with BCECF and excited at different wavelengths are presented in Figure 3A. Evoked pH-specific ratiometric emission signals were obtained by giving the same pattern of electrical stimulation (100 Hz for 20s); the stimulus intensities ranged from 200 µA to 250 µA.

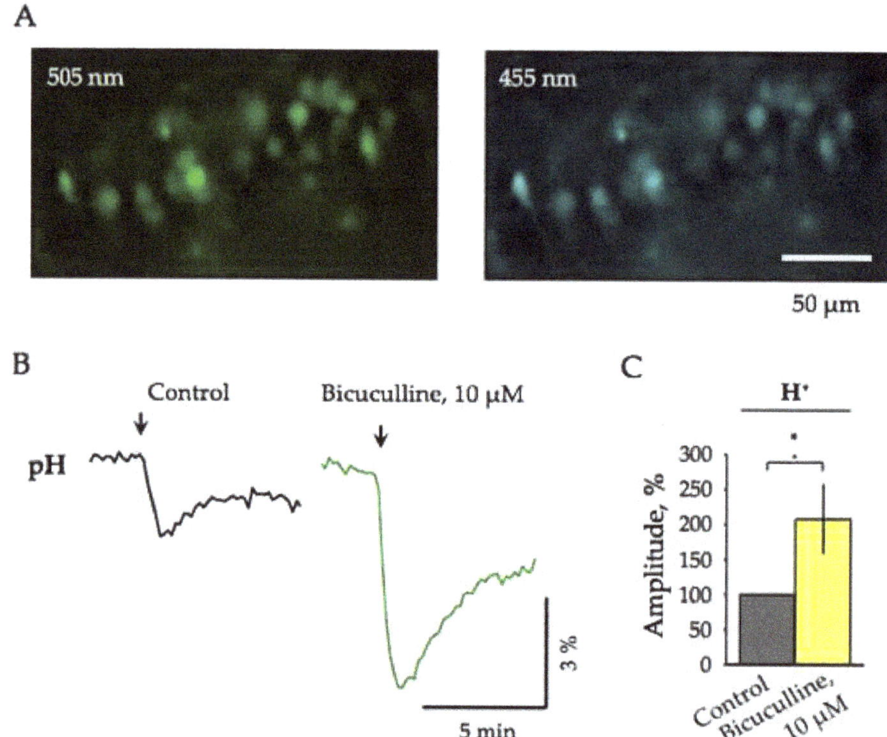

Figure 3. The effect of bicuculline on the evoked pH signals recorded using BCECF-AM. (**A**) Micro-graphs of pyramidal cells of the CA1 hippocampal region, loaded by BCECF-AM, under illumination with excitation wavelengths of 505 nm and 455 nm (P12). (**B**) Representative traces of pH-specific ratiometric emission transients evoked by high-frequency electrical stimulation (200 µA, 100 Hz for 20 s) in control (left trace, black) and after addition of 10 µM bicuculline (right trace, green). The arrows indicate the moments of stimulation (P11). (**C**) Summary of bicuculline action on the amplitude of pH changes induced by synaptic stimulation. Normalization to the control values (mean $\pm$ SEM, P10–12, $n = 5$). * Significant difference with $p < 0.05$ (Paired Sample Wilcoxon Signed Rank Test).

Similar to the observations with ClopHensor, on neurons loaded with BCECF-AM, the addition of bicuculline resulted in an increase of the pH transients evoked by tetanic stimulation (Figure 3B). Summary data from 5 slices showed that upon addition of bicu-culline, the mean amplitude of the pH signals increased to 208.6 $\pm$ 48.8% (Figure 3C, $n = 5$, $p < 0.05$).

These results demonstrate that: (i) sustained high-frequency stimulation induces acid-ification of cells and bicuculline potentiates the amplitude of pH transients; (ii) pH-specific emission signals are similar at recording either with BCECF-AM or using ClopHensor.

2.4. Analysis of the Decay Kinetics of Cl⁻ and pH Transients

We then performed a comparative analysis of the kinetics of synaptically induced Cl⁻ and pH transients. It was evaluated by determining the decay time constants, i.e., the time during which the peak amplitude of Cl⁻ and H⁺ components decreases by e-times (τ_{decay}).

The τ_{decay} of pH-fluorescent signals recorded from hippocampal slices of mice expressing ClopHensor was 157 ± 20.3 s, $n = 16$ (Figure 4A,B). This was close to values obtained on slices from wild type mice loaded with BCECF ($\tau_{\text{decay}} = 188.6 \pm 16.5$ s, $n = 14$; Figure 4A,B). The concentration of intracellular Cl^- ions was restored considerably faster than the pH after high-frequency stimulation of synaptic inputs (Figure 4A, red trace). The mean τ decay of Cl^- transients was 35 ± 4.2 s ($n = 9$, Figure 4B), i.e., about 4.5 times shorter than that of pH signals.

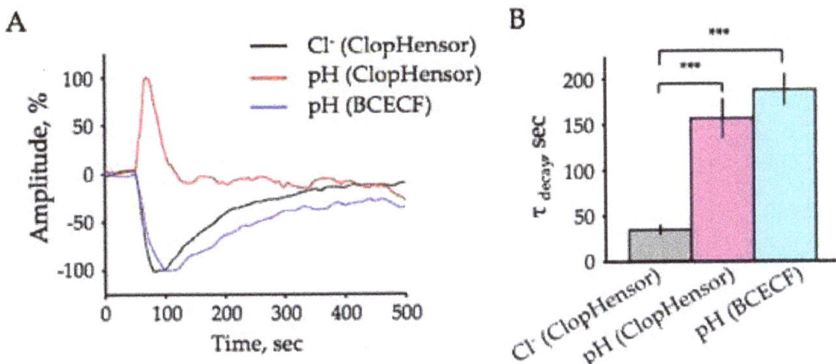

Figure 4. Comparison of the decay kinetics of stimulation-induced Cl^- and pH transients in hippocampal CA1 neurons. (**A**) Representative traces of normalized Cl^--(red trace) and pH-(black trace) transients registered using the transgenic ClopHensor mouse (stimulation 100 Hz for 20 s; age 6 months), and pH-specific signals using BCECF-AM (blue trace; stimulation: 100 Hz for 20 s; age P11). (**B**) The mean time constants of decay of the Cl^- and pH stimulation-induced responses were recorded with ClopHensor (age 2–7 months) and BCECF (age P10-12). Data from 9–16 slices. Values are mean $\pm$ SEM. *** Significant difference with $p < 0.001$ (Mann Whitney U test).

These results suggest that the mechanisms involved in the withdrawing of ions from neurons operate much faster for Cl^- than for H^+ and different mechanisms are involved in the maintenance of normal physiological transmembrane concentrations of these ions.

Next, we analyzed how changes of physiologically important ions, involved in regulation of Cl^- and H^+ transporters, modulate the amplitude of Cl^- and pH transients induced by depolarization.

2.5. Changes of $[H^+]_i$- and $[Cl^-]_i$-Induced by High $[K^+]_o$-Depolarization

Several studies used $[K^+]_o$-induced depolarization as a tool for stimulation of pH changes in different cell types [37–40]. We used this approach to analyze the properties of pH_i and $[Cl^-]_i$ transients in neurons of hippocampal slices under different experimental conditions.

In all experiments, the high $[K^+]_o$ solution contained 20 mM, i.e., 17.5 mM of potassium gluconate was added to normal ASCF containing 2.5 mM K^+. Depolarization induced by application of high $[K^+]_o$ ASCF caused a remarkable elevation of $[Cl^-]_i$ in all slices, with an average of 11.4 ± 1.1 mM ($n = 11$). Concerning pH, two types of responses were observed. In the majority of slices, $[K^+]_o$-induced depolarization caused acidification of neurons by, on average, 0.15 ± 0.02 pH units ($n = 23$). In 8 cases, biphasic pH_i-shifts composed of an early short-lasting alkalinisation that turned into a longer-lasting acidification were observed. We have not analyzed yet the reasons for this difference in the pH responses.

2.6. $[K^+]_o$-Induced Changes of pH and Cl^- in a Ca^{2+}-Free ACSF

Observations of cells in culture conditions demonstrated that elevation of external K^+ causes strong depolarization of neurons, accompanied, in addition to acidification, by a rise in intracellular Ca^{2+} [38,39]. In these studies, acidic $[K^+]_o$-induced pH responses were either completely prevented by the use of a Ca^{2+}-free medium [38], or even inversed and became alkalizing [39], suggesting a key role of Ca^{2+} ions in depolarization-induced pH changes.

We performed an analysis of the removal of external Ca^{2+} on depolarization-induced transients of $[Cl^-]_i$ and pH_i in brain slices of ClopHensor mice. Under control conditions, when high K^+ was applied to slices for 3–6 min, an elevation in $[Cl^-]_i$ and decrease in pH were observed (Figure 5A). In this set of experiments, $[K^+]_o$-induced intracellular acidification occurred, on average by 0.12 ± 0.11 pH ($n = 4$, Figure 5C), and an increase in $[Cl^-]_i$ occurred by 10.12 ± 1.21 mM ($n = 4$, Figure 5B). After washing, $[Cl^-]_i$ recovered toward the baseline, while the base level of pH_i sometimes followed to more alkaline values (Figure 5A).

Applying a Ca^{2+}-free ACSF containing 1 mM EGTA did not change the base level of $[Cl^-]_i$, while a decrease in pH, i.e., acidification of the cell cytoplasm was observed (Figure 5A). On average, the base level of pH_i decreased by 0.09 ± 0.02 (Figure 5D, $n = 8$, $p < 0.01$). $[K^+]_o$-induced depolarization caused an elevation of $[Cl^-]_i$ by 12.75 ± 0.83 mM ($n = 4$, Figure 5B). Surprisingly, unlike reports on preparations in culture [38,39], in our conditions, the pH responses to high K^+ were acidifying, only with reduced amplitudes compared to the control. On average, the $[K^+]_o$-induced pH decrease was by 0.10 ± 0.08 pH ($n = 4$, Figure 5C).

2.7. The Effect of Extracellular Cl^- on Synaptically-Induced Changes of $[Cl^-]_i$ and pH_i

Our next task was to determine the contribution of Cl^- ions to the $[Cl^-]_i$ and pH_i transients. For this, we analyzed the effects of decreasing Cl^- concentration in ACSF from 134.7 mM to 7.2 mM ("low Cl^-" conditions). The changes in base levels of pH_i and $[Cl^-]_i$, as well as the amplitudes of Cl^- and H^+ transients evoked by a high-frequency stimulation, were determined. On average, the mean amplitude of Cl^- influx on high-frequency stimulation was 0.8 ± 0.3 mM (Figure 6A,B). At the same time, stimulation caused acidification by 0.012 ± 0.005 pH (Figure 6A,C). After adding the "low-Cl^-" ACSF, a diminishing base level of $[Cl^-]_i$, synchronously with alkalinisation of neurons, was observed. New levels of $[Cl^-]_i$ and pH_i were completely stabilized in about 15 min (Figure 6A). On average, $[Cl^-]_i$ diminished by 7.1 ± 1.1 mM ($p < 0.05$, $n = 5$) and $_i$ increased by 0.19 ± 0.01 units ($p < 0.05$, $n = 5$; Figure 6D,E).

During the first minutes after establishing new steady-state levels in "low-Cl^-" external solution, high-frequency stimulation of Schaffer's collaterals caused the acidification directed pH responses, in which the amplitude was much higher than in control ACSF (Figure 6A). On average, ΔpH_i increased by 5 times in comparison with that recorded in the normal ACSF and became 0.063 ± 0.013 units pH ($p < 0.05$, $n = 5$, Figure 6C). The Cl^- transients on stimulation were nearly completely suppressed (Figure 6A,B). With an increase in the duration of incubation of slices in the "low-Cl^-" ACSF, the amplitude of the evoked pH responses decreased until they completely disappeared (data not shown). Our preliminary observations suggest that the reason for this elimination of pH responses is a continuous strong decrease in the amplitude of evoked local field potentials upon transition to "low-Cl^-" ACSF conditions.

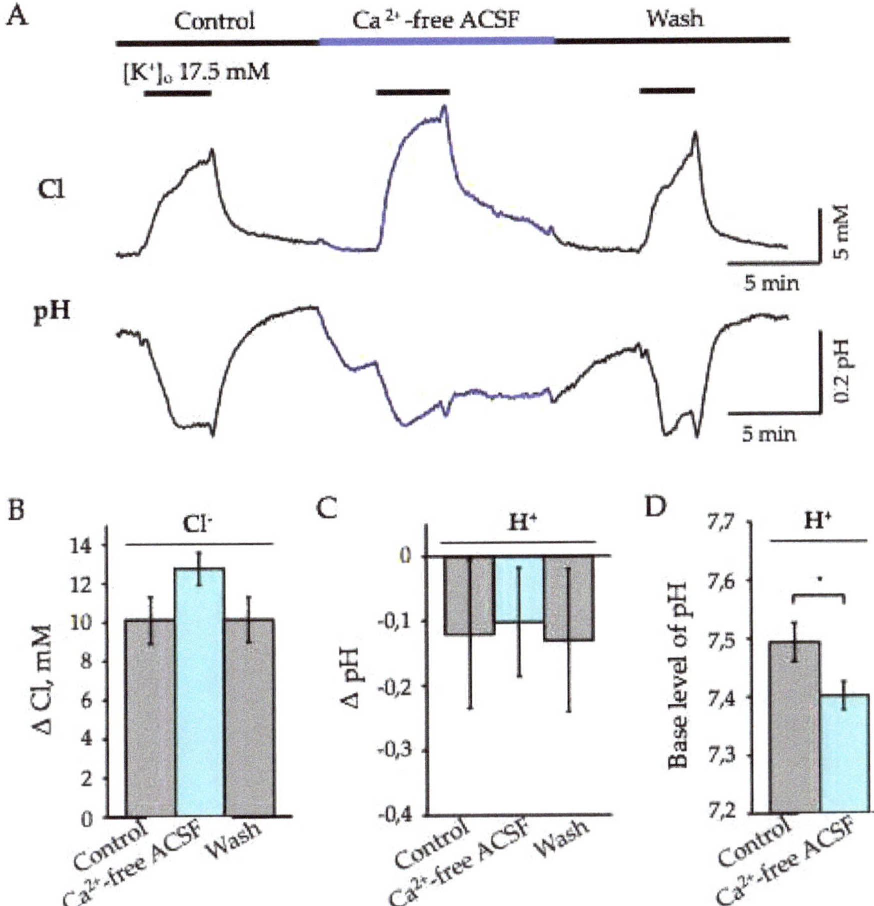

Figure 5. Effect of external Ca^{2+} removal on depolarization-induced changes in $[Cl^-]_i$ and pH_i in neurons of hippocampal slices. (**A**) Typical changes in $[Cl^-]_i$ (top) and pH_i (bottom) evoked by the application of 17.5 mM K^+ in the presence of 2.3 mM external Ca^{2+} (Control) and in Ca^{2+}-free ASCF. The periods of K^+ application and changes in external Ca^{2+} are indicated by the bars above the traces. Note the decrease of pH upon elimination of external Ca^{2+}. (**B,C**) The mean $[K^+]_o$-induced changes of $[Cl^-_i$ (**B**) and pH_i (**C**) in control and in Ca^{2+}-free ASCF. Summary of data from 4 slices (mean percentage $\pm$ SEM, age 4 months). (**D**) Mean values of base level of pH_i in the control condition (2.3 mM of external Ca^{2+}) and in Ca^{2+}-free ASCF. * Significant difference with $p < 0.05$.

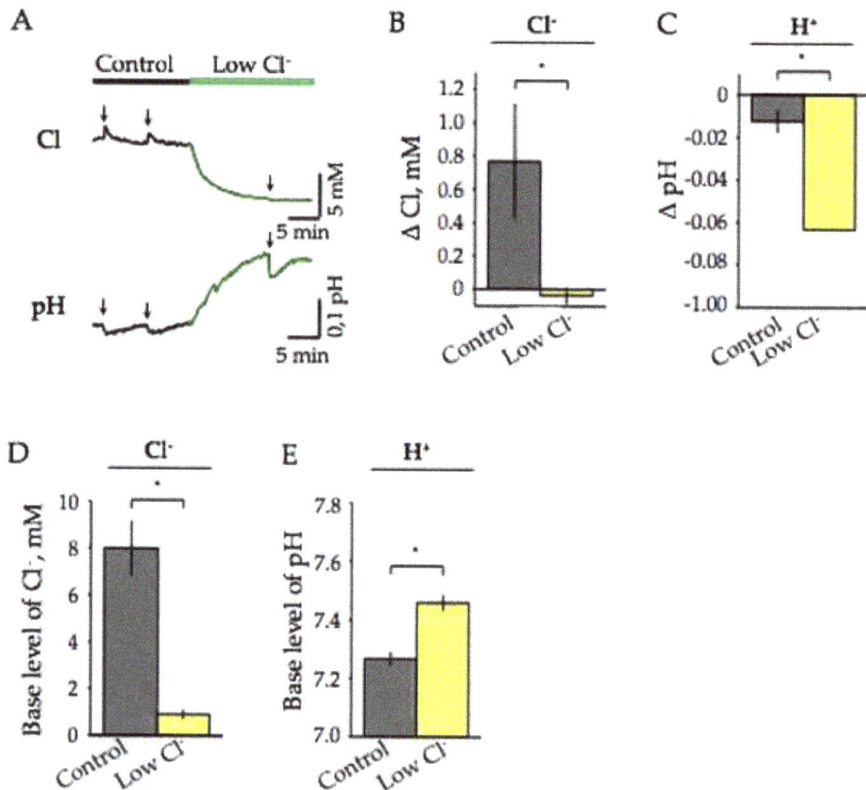

Figure 6. Effect of reduction of extracellular Cl^- and high-frequency stimulation on $[Cl^-]_i$ and pH_i in the hippocampal CA1 neurons of ClopHensor mice. (**A**) Typical traces of continuous recording of $[Cl^-]_i$ (top) and pH_i (bottom) illustrating the effects of the transition to "low Cl^-" ASCF and responses on high-frequency stimulation (age 6 months). The "low Cl^-" ACSF application is highlighted by yellow. Arrows indicate moments of stimulation. (**B,C**) Mean changes of synaptically induced $[Cl^-]_i$ (**B**) and pH_i (**C**) in conditions of high and low external Cl^- (mean $\pm$ SEM, $n = 5$, age 6 months). (**D,E**) Summary of the effect of reduction of external Cl^- concentration on base levels of $[Cl^-]_i$ (**D**) and pH_i (**E**) (mean percentage $\pm$ SEM, $n = 5$, 6 months). * Significant difference with $p < 0.05$ (Paired Sample Wilcoxon Signed Rank Test).

2.8. Changes of $[Cl^-]_i$ and pH_i during $[K^+]_o$-Induced Depolarization in the Low-Cl^- ACSF

We then tested how a decrease in extracellular Cl^- affects the amplitude of $[Cl^-]_i$ and pH_i changes caused by the $[K^+]_o$-induced depolarization. Similarly to previously described observations (Section 2.5), the addition of 17.5 mM K^+ in normal ACSF induced reversible transients of acidification and synchronous elevation of $[Cl^-]_i$ in neurons (Figure 7A). In this set of experiments, the mean K^+-induced decrease in pH_i was 0.15 ± 0.02 pH and the elevation of $[Cl^-]_i$ was 4.36 ± 0.77 mM ($n = 5$).

As previously described, upon perfusion of slices with "low Cl^-" ACSF, the base level of pH increased and $[Cl^-]_i$ synchronously decreased (Figure 7A). In "low Cl^-" ACSF conditions, depolarization caused by the addition of 17.5 mM K^+, resulted in a significant increase in the amplitude of pH acidification responses, while $[Cl^-]_i$ responses were nearly completely suppressed or even oppositely directed, i.e., showed a small efflux of intracellular Cl^- (Figure 7A). On average, the mean amplitude of "high K^+"-induced pH_i responses was 0.33 ± 0.06 ($p < 0.05$, $n = 5$), i.e., about 2 times higher than in the control ACSF, while the mean change of $[Cl^-]_i$ was -0.14 ± 0.03 ($p < 0.05$, $n = 5$), i.e., about 30 times smaller in comparison with control conditions. After washing with normal ACSF, the base levels of pH$_i$ and [Cl$^-$]$_i$ returned to initial values; in addition, the amplitudes of K$^+$-induced transients were fully recovered (Figure 7B,C).

These results are in accord with the above-described effect of lowering external Cl$^-$ on synaptically induced changes of pH$_i$ and [Cl$^-$]$_i$.

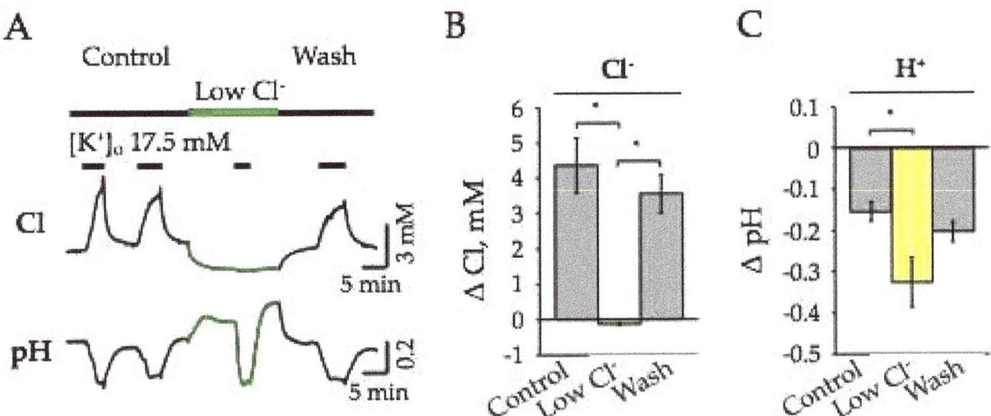

Figure 7. Effect of extracellular Cl$^-$ on changes of [Cl$^-$]$_i$ and pH$_i$ caused by K$^+$-induced depolarization. (**A**) Typical traces of continuous recording of [Cl$^-$]$_i$ (top) and pH$_i$ (bottom) illustrating the effects of the transition to "low Cl$^-$" ASCF and responses to the addition of 17.5 mM K$^+$ to hippocampal slices (age 7 months). The "low Cl$^-$" ACSF application is highlighted by the green bar. (**B,C**) Summary of the effect of K+-induced depolarization on changes of [Cl$^-$]$_i$ (**B**) and pH$_i$ (**C**). Data from 5 slices (mean ± SEM, *n* = 5, 7 months). * Significant difference with *p* < 0.05 (Paired Sample Wilcoxon Signed Rank Test).

2.9. Changes in [Cl$^-$]$_i$ and [H$^+$]$_i$ during [K$^+$]$_o$-Induced Depolarization in the Low-Na$^+$ ACSF

Extracellular Na$^+$ is one of the important participants and regulators of transporters of intracellular proton concentration [1]. For instance, the electrogenic sodium bicarbonate cotransporter NBCe1 (SLC4A4) contributes to intracellular as well as extracellular acid/base homeostasis in the brain and its dysfunctions are associated with pathophysiological states [16,41,42]. In addition, the Na$^+$-dependent 2HCO$_3$$^-$/Cl$^-$-exchanger (NDCBE) represents an important pH- Cl$^-$-coupled physiological controller of ions [15].

Lastly, we analyzed how a decrease in external Na$^+$ affects changes in pH$_i$ and [Cl$^-$]$_i$ caused by high K$^+$ depolarization. Similar to the above-presented results, in ASCF containing normal Na$^+$ concentration, K$^+$-induced depolarization caused intracellular elevation of both H$^+$ and Cl$^-$ ions (Figure 8A). On average, the increase in [Cl$^-$]$_i$ was 12.06 ± 1.19 mM (Figure 8B) and acidification was 0.18 ± 0.04 pH (Figure 8C).

Changing the solution to a low-Na$^+$ ASCF (replacement of NaCl with choline Cl) led to a decrease in the base level of intracellular Cl$^-$ and H$^+$ (Figure 8A). The mean changes were for [Cl$^-$]$_i$ by 1.98 ± 0.15 mM (*n* = 5, *p* < 0.05, 6 months) and for H+ by 0.08 ± 0.01 pH units (*n* = 5, *p* < 0.05, 6 months). In low-Na$^+$ ASCF, responses to K$^+$-induced depolarization were potentiated by more than 2 times for both ions, resulting in the mean changes of 25.77 ± 2.74 mM and 0.40 ± 0.05 pH units (*n* = 5, Figure 8B,C).

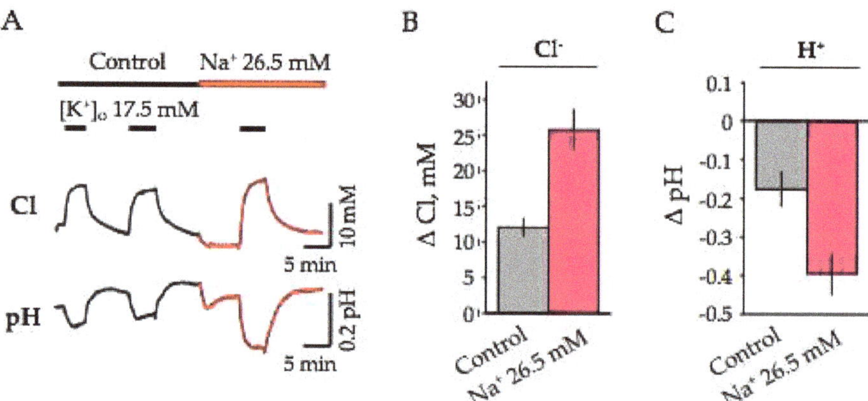

Figure 8. Effect of extracellular Na$^+$ on [Cl$^-$]$_i$ and pH$_i$ responses induces by the high K$^+$ depolarization. (**A**) Typical traces of continuous recording of [Cl$^-$]$_i$ (top) and pH$_i$ (bottom) illustrating the effects of the transition to "low Na$^+$" ASCF and responses to the addition of 17.5 mM K$^+$ to hippocampal slices (age 6 months). (**B,C**) Summary of the high K$^+$-induced changes in concentrations of Cl$^-$ (**B**) and H$^+$ (**C**) in control (grey columns) and in ACSF containing low Na$^+$ (26.5 mM) (red columns). Summary from 5 slices (mean $\pm$ SEM, 6 months).

3. Discussion

Our study provides evidence that the optosensoric reporter, ClopHensor, expressed in transgenic mice, represents an efficient tool for non-invasive monitoring of intracellular Cl$^-$ and H$^+$ ions. Applying different conditions for modulation of neuronal activity in brain slices from transgenic mice, we demonstrated the ability of ClopHensor to simultaneously monitor and analyze changes in [Cl$^-$]$_i$ and [H$^+$]$_i$ during artificial changes of its equilibrium.

Maintaining physiologically relevant concentrations of H$^+$ and Cl$^-$ has a pivotal role in controlling neuronal excitability in the adult brain and during development, and is likely to be crucial in pathophysiological conditions. Both ions play an important role in many cellular processes, including neurotransmission, regulation of membrane potential, cell volume and water–salt balance, modulation of the functions of different proteins, including voltage-gated and receptor-operated channels [1,12,43–46]. Because of high metabolic activity, accompanied by depolarization, neurons may be susceptible to acidification-induced injury, as can happen in excessive synaptic activation or during conditions of anoxia or ischemia.

The steady-state physiologically relevant dynamic range of pH$_i$ and [Cl$^-$]$_i$ is determined by the balance of both passive transport, through ion channels and by active mechanisms via exchangers or cotransporters [1,46,47].

The intracellular Cl$^-$ in neurons of the mammalian brain is primary regulated by two cation–Cl$^-$ cotransporters, the Na$^+$/K$^+$/2Cl$^-$ cotransporter 1 (NKCC-1) and the K$^+$/Cl$^-$ co-transporter 2 (KCC-2) [46–49]. NKCC1 pumps Cl$^-$ into neurons, while KCC2 uses the energy of the K$^+$ gradient to extrude Cl$^-$ from neurons and maintain relatively low [Cl$^-$]$_i$ (Figure 9). Dysfunction or changes in the expression of these cotransporters is critically linked to the etiology of several neurologic disorders including epilepsy, acute trauma, ischemia, autoimmune disorders, and neuropathic pain [48,50,51].

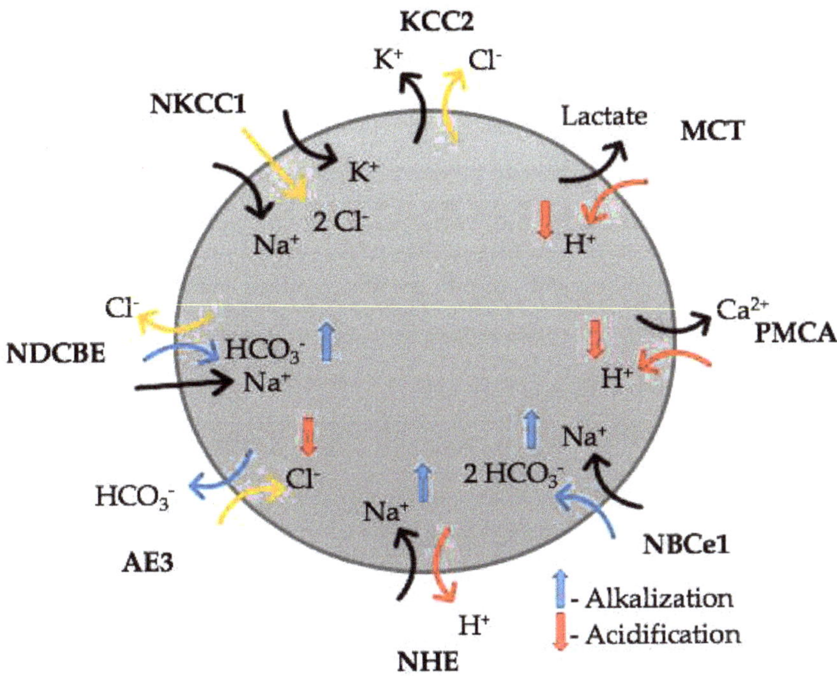

Figure 9. Scheme of key transmembrane transport elements controlling intracellular Cl^- and pH in brain cells.

Similarly, a modest shift of intra- and extracellular H^+ ion concentration can have significant effects on brain functions, such as neuronal excitability, synaptic transmission and metabolism [1,52]. This is the result of the sensitivity of a large number of processes to protons, such as ion channel gating and conductance, synaptic transmission, cell-to-cell communication via gap junctions, and enzymatic activity in brain energy metabolism [52–55]. The major pH regulating transporters identified in the mammalian brain so far comprise the Na^+/H^+ exchanger (NHE1), electrogenic Na^+/HCO_3^- cotransporter 1 (NBCe1), Na^+-dependent Cl^-/HCO_3^- exchange (NDCBE), and Na^+-independent anion exchanger (AE3) (Figure 9; [1,15,56]. NBce-1, NDCBE and AE3 transporters perform translocation of HCO_3^-, thus controlling the physiological range of pH_i. The electrogenic Na^+/HCO_3^- cotransporter NBCe1 is one of the major regulators of $[H^+]_i$ and is expressed in most brain cell types, with the most prominent expression being astrocytes [1]. It has been shown that NBCe1 is not only an acid extruder/base loader, as suggested in early studies [57,58], but also an acid loader/base extruder [16]. Anion Cl^-/HCO_3^- exchange (AE3) is known as an important acid-loading performer in brain cells [16,59]. The finding that hippocampal neurons of knockdown AE3 mice ($Ae3^{-/-}$) lack detectable Cl^-/HCO_3^- exchange activity [60] suggests that AE3 plays a critical role in the maintenance of the Cl^- equilibrium potential and/or pH_i in neurons.

The role of monocarboxylate transporters (MCTs) in the regulation of the functional integrity of synaptic transmission has been demonstrated. In excitatory synapses, MCT constitutively supports synaptic transmission, even under conditions when there is a sufficient amount of glucose and intracellular ATP [61]. Monocarboxylates cause a decrease in the pH of neurons, which is associated with changes in bioelectric activity [62]. MCT's involvement in Cl modulation requires future analysis.

These features of transporters emphasize the need for simultaneous monitoring of $[Cl^-]_i$ and $[H^+]_i$ when analyzing the mechanisms of ion homeostasis and neuronal activity, particularly using neuropathological models. A genetically encoded sensor, named ClopHensor, was proposed for simultaneous measurement of Cl^- and H^+ ion concentrations [27]. This construct has shown its effectiveness at the heterologous expression in cell lines, demonstrating good sensitivity and high stability to bleaching during long fluorescence measurements [28]. We recently presented a line of transgenic mice expressing ClopHensor in neurons and obtained a detailed map of its distribution in the mouse brain [32]. Expression of this probe in transgenic mice was found to be highly specific and reproducible in different animals, suggesting that this experimental model represents a promising tool for analysis of dynamic changes in $[Cl^-]_i$ and pH_i.

In the present study, we analyzed the effectiveness of ClopHensor expressed in transgenic mice for monitoring $[Cl^-]_i$ and $[H^+]_i$ in neurons of hippocampal slices upon changing the ionic equilibrium by pharmacological modulation of neuronal activity, changes in extracellular concentrations of Ca^{2+}, Cl^- or Na^+, and by depolarization caused by high-frequency synaptic stimulation or the use of high K^+.

3.1. Kinetics of Synaptically Induced Transients of $[Cl^-]_i$ and $[H^+]_i$

High frequency induced stimulation of Schaffer's collaterals resulted in the elevation of $[Cl^-]_i$ and $[H^+]_i$ and slow recovery. Importantly, the decay kinetics of Cl^- transients was nearly 5-fold faster than pH. The experiments on wild type mice using the chemical pH sensor, BCECF, confirmed remarkably slow kinetics of pH_i recovery. These differences in the kinetics of recovery of Cl^- and H^+ responses may reflect the involvement of different cellular participants in control of these ions' homeostasis.

In the study using the ClopHensorN indicator heterologously expressed in cultured hippocampal slices, about 1.5 times faster decay of intracellular Cl^- concentrations were indicated [29]. A profound analysis is necessary to reveal relationships and efficacy of different transporters and other mechanisms involved in the recovery of $[Cl^-]_i$ and $[H^+]_i$ from synaptically induced disequilibrium.

3.2. Effect of GABA Receptor Inhibition on Synaptically Induced Transients of $[Cl^-]_i$ and $[H^+]_i$

Results of our study demonstrate that in hippocampal slices of transgenic mice, high frequency stimulation of Shaffer's collaterals causes an intracellular increase of both ions, Cl^- and H^+. Synaptic stimulation causes the release of neurotransmitters, primary glutamate and GABA, which activates corresponding receptors, leading to the opening of cation and anion-selective ion channels and resulting depolarization of neurons.

As predicted, inhibition of GABA-mediated chloride conductance by bicuculline resulted in a decrease of the synaptically induced influx of Cl^-. On the other side, this decrease was accompanied by potentiation of acidific $[H^+]_i$ transients. This could be a consequence of two main things: (i) increased synaptic stimulation due to blockade of inhibitory GABA receptors; (ii) changes in the activity of transporters.

In support of the first possibility, the electrophysiological analysis showed that bicuculline causes an elevation in the amplitude of evoked local field potentials. On the other side, experiments that lowered the extracellular Cl showed that in spite of the strong reduction of evoked local field potentials, the amplitude of synaptically induced pH responses was potentiated nearly 10 times in comparison with control.

We suggest that this may be a consequence of the weakening of the acidifying activity of the $Cl^-HCO_3^-$ exchange. This assumption was tested in experiments with high K^+-induced depolarization and changes of external concentrations of ion participating in the functioning of transporters.

3.3. Analysis of High K^+-Induced Depolarization on Changes of $[Cl^-]_i$ and $[H^+]_i$ in External Ca^{2+}-Free Conditions

Depolarization induced by application of high K^+ or by other means has been used in a number of studies for analysis of pH_i changes in cultured and freshly dissociated neurons, and brain slices [37–40,63]. In most studies, depolarization induced by application of high external K^+ was accompanied by elevation of intracellular Ca^{2+} and decrease of pH, i.e., acidification [37,39]. Early observations performed on cultured cells showed that under external Ca^{2+}-free conditions, the acidification responses were either completely prevented [38] or even inversed, i.e., exhibited an increase in pH_i [39]. Moreover, a blocker of voltage-gated Ca^{2+} channels inhibited K^+-induced responses, suggesting that an increase in $[Ca^{2+}]_i$ is a key factor for acidosis to occur [38].

We tested this suggestion in experiments on brain slices of ClopHensor-expressing mice and obtained distinct results. In our experiments, transition to external Ca^{2+}-free conditions alone caused a decrease in base pH, i.e., acidification. Surprisingly, K^+-induced depolarization produced additional acidification. The responses were only partially attenuated in comparison with control, suggesting an only partial contribution of Ca^{2+}-dependent processes in pH_i responses induced by membrane depolarization.

The reasons for these contrast observations have to be clarified. Among them may be the use of HEPES solution in cell culture experiments and different energy states of neurons.

3.4. Effect of Decreasing Extracellular Cl^- and Na^+ on K^+-Induced Changes in $[Cl^-]_i$ and $[H^+]_i$

3.4.1. Decreasing Extracellular Cl^-

In our experiments, decreasing Cl^- in ASCF resulted in a slow-developing (10–15 min) strong decrease in the base level of cytoplasmic Cl^- and elevation of pH, i.e., extruding of H+ from neurons. However, in "low-Cl^-" conditions, high K^+-induced depolarization caused remarkable potentiation of acidification pH responses, while Cl^- responses were nearly completely suppressed. These observations are in accordance with results obtained at high-frequency synaptic stimulation and indicate that different mechanisms are involved in the control of $[Cl^-]_i$ and pH_i at changes of extracellular ion concentrations and during depolarization of neurons. A decrease in extracellular Cl^- weakens the ability of the Cl^-/HCO_3^- transporter to pump Cl^- into neurons in exchange for HCO_3^-, which leads to slow alkalinization. At high K^+ application, depolarization induces elevation of intracellular Ca^{2+}, resulting in stimulation of PMCA transporter activity and consequent acidification.

3.4.2. Decreasing Extracellular Na^+

Since the NHE, NBCe1 and NDCBE transporters operate on a Na^+ gradient, a decrease in the extracellular Na^+ concentration should lead to a decline in the operation of these pumps and, as a result, to a weakening of protons extruding from the cytoplasm, and a decrease in intracellular HCO_3^- We assume that for these reasons, in our experiments, there was a decrease in the pH_i of neurons during the transition to low Na^+ and a potentiation of acidic responses to depolarization caused by increased K^+. In the future, these processes will be analyzed in detail.

In conclusion, the baseline level and fluctuations of intracellular Cl^- and pH play a crucial role for intercellular and intracellular signaling, as well as for cellular and synaptic plasticity. Our study demonstrates that transgenic mice expressing ClopHensor provide ample opportunities for studying the homeostasis of chloride and hydrogen not only under normal conditions, but also in pathology.

In particular, Cl^- gradients are disrupted in epilepsy, especially in early childhood, which significantly complicates treatment. Hydrogen gradients in the central nervous system are disrupted by neuroinflammatory processes of various origins and ischemia. Research in these areas is relevant today, and ClopHensor mice can be a reliable tool for in-depth analysis of the mechanisms controlling the physiological ranges of concentrations of these ions.

4. Materials and Methods

4.1. Animals

Experiments were conducted on laboratory ICR CD-1 outbreed mice of both sexes at postnatal days 10–12 and adult transgenic mice, aged 2–6 months, strain C57BL/6N expressing ClopHensor (Diuba et al., 2020). Use of animals was carried out in accordance with the Guide for the Care and Use of Laboratory Animals (NIH Publication No. 85–23, revised 1996) and European Convention for the Protection of Vertebrate Animals used for Experimental and other Scientific Purposes (Council of Europe No. 123; 1985). All animal protocols and experimental procedures were approved by the Local Ethics Committee of Kazan State Medical University. Mice had free access to food and water and were kept under natural day length fluctuations. Animals were not involved in any previous procedures.

4.2. Solutions and Drugs

Brain slices were prepared in ice-cold high potassium solution containing (in mM): K-gluconate 120, HEPES acid 10, Na-gluconate 15, EGTA 0.2, NaCl 4 (pH 7.2, 290–300 mOsm). After cutting, slices were incubated in a high magnesium artificial cerebrospinal fluid (ACSF) containing (in mM): NaCl 125, KCl 2.5, $CaCl_2$ 0.8, $MgCl_2$ 8, $NaHPO_4$ 1.25, glucose 14, $NaHCO_3$ 24 (pH 7.3–7.4, 290–300 mOsm). Storage of slices and performing of experiments were conducted in ACSF, containing (in mM): NaCl 125, KCl 2.5, $CaCl_2$ 2.3, $MgCl_2$, 1.3, $NaHPO_4$ 1.25, glucose 14, $NaHCO_3$ 24 (pH 7.3–7.4, 290–300 mOsm). The ACSF was continuously oxygenated with 95% O_2 and 5% CO_2 to maintain the physiological pH. The following drugs were used: APV (40 μM, Hello Bio, Bristol, UK, Cat# HB0225), CNQX (10–40 μM, Sigma Aldrich, St. Quentin Fallavier, France, CAS: 115066-14-3), (-)-Bicuculline methochloride (10 μM or 40 μM, Tocris, Science Park Abingdon, UK, Cat#0131), BCECF/AM (10 μM, Sigma Aldrich, St. Quentin Fallavier, France, CAS: 117464-70-7). All drugs were diluted on the day of the experiment from the 1000× to 4000× stocks kept at −20 °C.

4.3. Preparation of Brain Slices

Sagittal 350 μm thick sections of the cerebral hemispheres containing the hippocampus were obtained with the use of a vibratome of Model NVSLM1, World Precision Instruments. Mice were euthanized by decapitation. The brain was quickly removed and placed in a Petri dish filled with ice-cold high-K^+ solution. The cerebellum and olfactory bulbs were cut off using a scalpel; the cerebral hemispheres were separated by cutting along the longitudinal fissure and mounted onto the vibratome specimen disc using superglue, orienting them downward with the sagittal cut surface and the cortex facing the razor blade. Sagittal 350 μm thick sections were prepared in ice-cold high K^+ solution. After being cut, slices were incubated for 15 min at room temperature in a resting high magnesium oxygenated ACSF. Then, slices were placed in a chamber filled with oxygenated ACSF. Before use, slices were allowed to recover for at least 1 h at room temperature.

Experiments were conducted during the period of 1–6 h after slicing. For the recordings, the brain slices were placed in a chamber perfused with an oxygenated ACSF. Recordings were carried out at 30–31 °C with the speed of perfusion of 25 mL/min.

4.4. Fluorescence Imaging in Brain Slices

4.4.1. Monitoring of $[Cl^-]_i$ and pH on Brain Slices ClopHensor Mice

Fluorescence images were obtained using an upright microscope Olympus BX51WI equipped with the iXon Life 897 EMCCD camera (Andor, Oxford Instruments, Abingdon, UK), a 4-Wavelength LED Source (LED4D001, Thorlabs, Newton, NJ, USA) accompanied with a Four-Channel LED Driver (DC4100, Thorlabs), a quad-band filter set (Cat# 9403, Chroma, Foothill Ranch, CA, USA), and a water-immersion objective (60× magnification, 1 numerical aperture; LumPlanFL N, Olympus, Tokyo, Japan) (Figure 10).

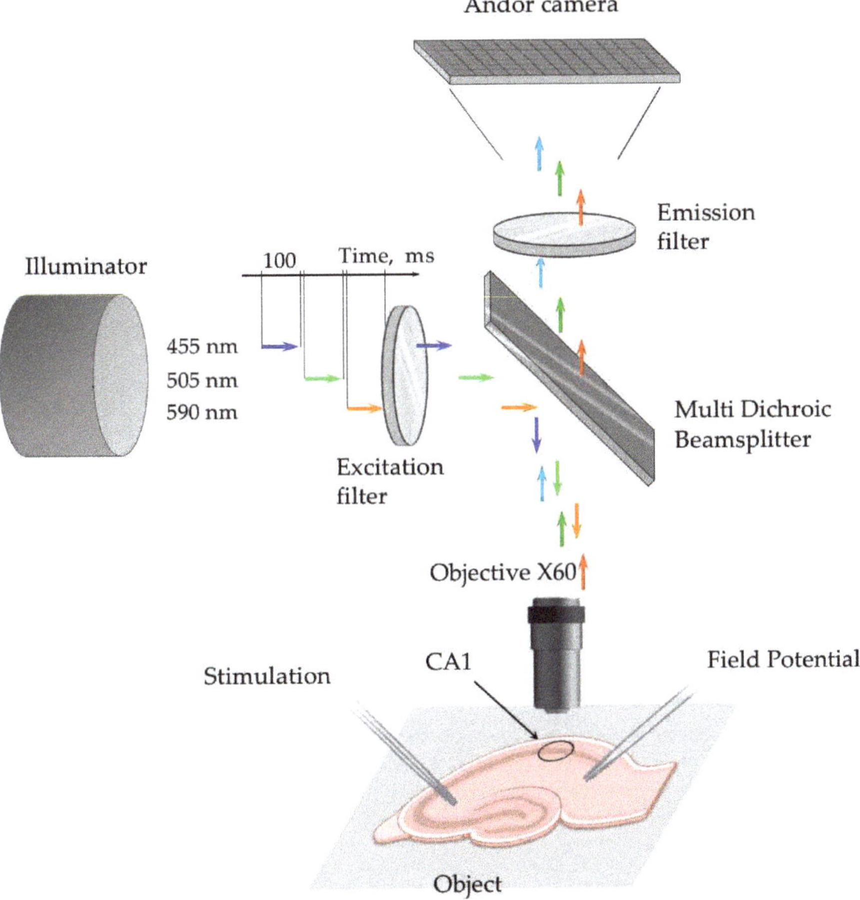

Figure 10. Setup for simultaneous monitoring of pH$_i$ and [Cl$^-$]$_i$ using ClopHensor.

The 4-Wavelength LED Source was supplied by light-emitting diodes (LEDs) at 365 nm, 455 nm, 505 nm and 590 nm, the last three of which were used in our study. The capacity of lighting and the LED switching order were adjusted via the DC4100 driver. Cyan channel fluorescence (455 nm excitation) was detected from 460 nm to 485 nm, green channel (505 nm excitation) from 527 nm to 551 nm, and red channel (590 nm excitation) from 600 nm to 680 nm.

All peripheral hardware control, image acquisition and the average fluorescence intensity measurement were achieved using DriverLed software (KSMU, Kazan). Regions of interest (ROI) were set around pyramidal cell bodies of the CA1 hippocampal region. The average fluorescence intensity of the ROI was tracked online at the time of the real-time fluorescence imaging. Numerical data were output by the DriverLed software on the computer in the form of text documents, which were subsequently parsed using Excel 2016 (Microsoft).

Excitation of E2GFP was provided by light-emitting diodes (LEDs) at 455 nm and 505 nm, and a LED at 590 nm provided excitation of DsRed. The duration of excitation at each wavelength was usually 70–110 ms. Examples of the images taken are shown in Figure 1C. Fluorescent emission was recorded continuously from the hippocampal CA1 pyramidal cells with a sampling interval of 10 s. Evoked emission signals were obtained by stimulation of Schaffer's collaterals by glass bipolar electrode, filled with ACSF, placed in the stratum radiatum of the CA2 hippocampal area (100 Hz for 20 s, 20–320 µA, single pulse width 200 µs).

4.4.2. Monitoring of pH Using BCECF

As an additional control of proper pH monitoring of intracellular pH in hippocampal neurons, we used the pH-sensitive dye, 2′,7′-Bis (2-carboxyethyl)- 5 (6)-carboxyfluorescein-acetoxymethyl ester (BCECF-AM) (Figure 11A). The fluorescent dye, BCECF, was introduced for measuring cytoplasmic pH by Roger Tsien and co-workers [36]. Presently used BCECF-AM is a mixture of three types of cell-permeable non-fluorescent molecules, which are converted to fluorescent non-membrane-penetrating form by intracellular esterases [21]. This dye provides dual-excitation ratiometric monitoring of intracellular pH.

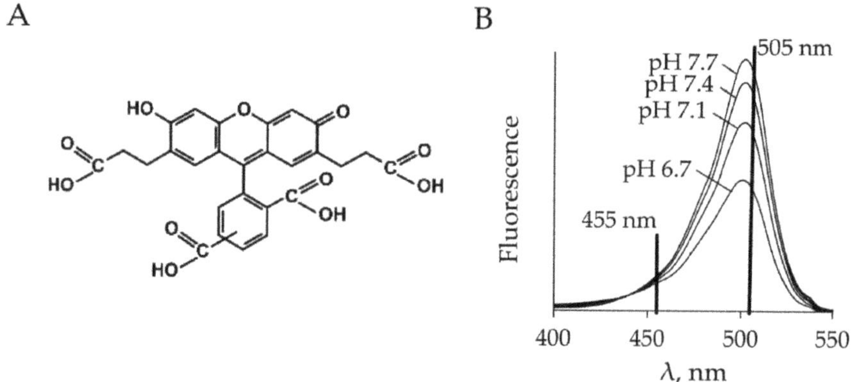

Figure 11. (**A**) The chemical structure of one of the molecules of 2′,7′-Bis (2-carboxyethyl)- 5 (6)-carboxyfluorescein-acetoxymethyl ester (BCECF-AM). (**B**) Excitation spectra of BCECF collected at different pH values. The BCECF pH measurements were made by determining the pH-dependent ratio of emission intensity when it was excited at 505 nm versus the emission intensity when excited at 455 nm.

We determined the pH changes by calculating the ratio of two emission signals obtained after illumination by light at 455 nm and 505 nm (Figure 11B). The intracellular pH changes after tetanic stimulation and its modulation by bicuculline were analyzed. Experiments were conducted on wild type juvenile mice (P10-P12). Registration conditions were similar to those performed on slices from transgenic ClopHensor mice. Evoked pH-specific ratiometric emission signals were obtained by giving the same pattern of electrical stimulation (100 Hz for 20s); the stimulus intensities ranged from 200 µA to 250 µA.

Hippocampal slices were stained with 10 µM BCECF-AM for 40 min in the oxygenated ACSF. Then the dye was washed out and slices were superfused with normal ACSF for at least 40 min to allow esterases to cleave AM and stabilise pHi. For fluorescent monitoring, slices were transferred to the optical recording chamber, which was mounted on the stage of an upright microscope (Olympus BX51WI) (Figure 10).

The setup consisted of a microscope (Olympus BX51WI equipped with the iXon Life 897 EMCCD camera (Andor camera)), a 4-Wavelength Thorlabs LED Source (Light source), and a quad-band filter set (excitation filter). Excitation light with wavelengths of 505 nm, 455 nm and 590 nm, each for 70–100 ms, was sequentially applied to slices and the emission signals from CA1 pyramidal cells were registered by an EMCCD camera. The stimulation electrode was placed in the stratum radiatum of the CA2 hippocampal area. The evoked local field potentials were recorded using a glass micropipette electrode, placed in the stratum radiatum of the CA1 hippocampal area and an HEKA EPC 10 Patch Clamp Amplifier (not shown).

Slices were alternately illuminated at 455 nm and 505 nm with duration of 100 ms. Light from both wavelengths was passed through appropriate excitation and emission filters. Fluorescence image pairs were captured every 10 s by an intensified camera. Evoked emission signals were obtained in the same way as for ClopHensor: by sustained high-frequency electrical stimulation of Schaffer's collaterals (100 Hz for 20 s, 20–320 µA, single pulse width 200 µs).

4.5. Electrophysiological Recording

Evoked local field potentials (eLFP) were recorded in the stratum radiatum of the CA1 hippocampal region using glass micropipette electrodes filled with ACSF (resistance 1–2 MΩ) and an HEKA EPC 10 Patch Clamp Amplifier (HEKA Elektronik, Lambrecht, Germany). The DS3 Constant Current Isolated Stimulator (Digitimer, Welwyn Garden, UK) and a bipolar stimulating electrode placed on the Schaffer's collaterals at the hippocampal area CA2 were used for the induction of eLFPs. Current pulses 20–300 µA in amplitude and 200 µs in duration were applied to obtain reliable eLFPs. Single eLFPs were recorded continuously every 20 s. PatchMaster software (HEKA Electronik, Lambrecht, Germany) was used to record eLFPs, control the HEKA EPC 10 Patch Clamp Amplifier and the DS3 Constant Current Isolated Stimulator.

4.6. Intracellular pH and Cl^- Calibration of ClopHensor on Hippocampal Slices

To perform pH and Cl^- calibration, we used a double ionophore technique [64]. The brain slices were exposed to the antibiotic nigericin, which acts as H^+/K^+ antiporter and the Cl^-/OH^- antiporter tributyltin, which forms pores in the cell membrane and allows external Cl^- to equilibrate with intracellular Cl^- [65]. For the action of the compounds, before fluorescent monitoring, the hippocampal slices were kept for 1–28 h at +4 °C in the following solution: 150 mM K-Gluconate, 20 mM HEPES, and 10 mM D-glucose) with addition of nigericin (20 mkM, pH = 7.22–7.25) and tributyltin (20 mkM).

For pH calibration, the solutions with different pH values (from 6.23 pH to 8.04 pH) were prepared by adding KOH for alkalization. For calibration of Cl^-, the solutions were prepared at pH = 7.25 and contained a different concentration of Cl^- (0, 3, 10, 30, 100 mM). To keep the osmolality of the solutions, KCl was correspondently substituted by the K-gluconate (150, 147, 140, 120, 50 mM). All of these solutions contained 20 mM HEPES, and 10 mM D-glucose.

Fluorescent signals were recorded from the CA1 zone of the hippocampus with the following lighting parameters in the Drive LED program: Exposition 70 ms, LED on 100 ms, binning 2, Voltage (590 nm) 70 mV, Voltage (505 nm) 10 mV, Voltage (455 nm) 15 mV. The slices were incubated in each solution until the fluorescence stabilized, usually for 20–30 min.

We obtained a linear dependence for pH (from 6.7 pH to 7.6 pH) on the fluorescence ratio (F_{505nm}/F_{455nm}):

$$pH = F_{505nm}/F_{455nm} \times K_1 + K_2, \tag{3}$$

where $K_1 = 0.44$, $K_2 = 6.06$.

Coefficients were obtained by fitting data for 6 slices.

To estimate the $[Cl^-]_i$ from calibration data obtained on 4 slices, the following equation was used:

$$[Cl^-]_i = K_0 + \frac{K_1 - K_0}{1 + \left(\frac{K_3}{F_{590\ nm}/F_{455\ nm}}\right)^{K_2}} \tag{4}$$

where $K_0 = -0.33$, $K_1 = 33.62$, $K_2 = 10.1$, $K_3 = 0.87$.

All coefficients were obtained by fitting data using the Igor Pro 6.02 software.

4.7. Data Analysis and Statistics

Amplitudes of evoked local field potentials were measured using the Online Analysis function of PatchMaster software (HEKA Electronik, Lambrecht, Germany). Superimposed average traces of evoked local field potentials were generated in PatchMaster software and processed for further presentation in Igor Pro 6.02 software (WaveMetrics, Tigard, OR, USA).

The average emission intensity of the region of interest (ROI) was measured online at the time of the real-time fluorescence imaging or, if necessary, recalculated using DriverLed software. Excel 2016 (Microsoft) software was used to compute and plot Cl^- and pH-dependent ratios of emission intensities and to measure the amplitudes evoked by tetanic stimulation pH- and Cl^--specific ratiometric fluorescence emission signals.

Origin 15 software was used to perform a statistical analysis of the data, to plot the graphs and compute the decay times of the evoked pH- and Cl^--specific fluorescence emission signals.

Data are represented as means $\pm$ SEM. Statistical significance was determined using Paired Sample Wilcoxon Signed Rank Test and Mann Whitney U test. Differences were considered significant at $p < 0.05$.

Author Contributions: For Conceptualization, P.B.; methodology and investigation E.P., D.P. and P.B.; formal analysis, E.P., D.P.; writing—original draft preparation, E.P., D.P. and P.B. supervision, P.B.; project administration, P.B.; funding acquisition, P.B. All authors have read and agreed to the published version of the manuscript.

Funding: This study was supported by the Russian Science Foundation (Grant: 18-15-00313).

Institutional Review Board Statement: The study was conducted according to the Guide for the Care and Use of Laboratory Animals (NIH Publication No. 85–23, revised 1996) and European Convention for the Protection of Vertebrate Ani-mals used for Experimental and other Scientific Purposes (Council of Europe No. 123; 1985). All animal protocols and experimental procedures were approved by the Local Ethics Committee of Kazan State Medical University.

Informed Consent Statement: Not applicable.

Data Availability Statement: Not applicable.

Acknowledgments: We cordially thank Zakharov, A. for developing the program for the analysis of fluorescent signals. We are also grateful to Yu, N. Davidyuk and A. Yusupova for PCR analysis of transgenic mice, and Kiani, N. for useful text comments.

Conflicts of Interest: The authors declare no conflict of interest.

References

1. Chesler, M. Regulation and modulation of pH in the brain. *Physiol. Rev.* **2003**, *83*, 1183–1221. [CrossRef]
2. Ruusuvuori, E.; Kaila, K. Carbonic anhydrases and brain pH in the control of neuronal excitability. *Subcell. Biochem.* **2014**, *75*, 271–1290. [CrossRef]
3. Doyon, N.; Vinay, L.; Prescott, S.A.; De Koninck, Y. Chloride Regulation: A Dynamic Equilibrium Crucial for Synaptic Inhibition. *Neuron* **2016**, *89*, 1157–1172. [CrossRef]
4. Roos, A.; Boron, W.F. Intracellular pH. *Physiol. Rev.* **1981**, *61*, 296–434. [CrossRef]
5. Madshus, I.H. Regulation of intracellular pH in eukaryotic cells. *Biochem. J.* **1988**, *250*, 1–8. [CrossRef]
6. Putnam, R.W. Intracellular pH regulation of neurons in chemosensitive and nonchemosensitive areas of brain slices. *Respir. Physiol.* **2001**, *129*, 37–56. [CrossRef]
7. Bregestovski, P.; Waseem, T.; Mukhtarov, M. Genetically encoded optical sensors for monitoring of intracellular chloride and chloride-selective channel activity. *Front. Mol. Neurosci.* **2009**, *2*, 15. [CrossRef] [PubMed]
8. Tong, C.K.; Chen, K.; Chesler, M. Kinetics of activity-evoked pH transients and extracellular pH buffering in rat hippocampal slices. *J. Neurophysiol.* **2006**, *95*, 3686–3697. [CrossRef] [PubMed]
9. Du, J.; Reznikov, L.R.; Price, M.P.; Zha, X.M.; Lu, Y.; Moninger, T.O.; Wemmie, J.A.; Welsha, M.J. Protons are a neurotransmitter that regulates synaptic plasticity in the lateral amygdala. *Proc. Natl. Acad. Sci. USA* **2014**, *111*, 8961–8966. [CrossRef]
10. Pardo, A.C.; Díaz, R.G.; González Arbeláez, L.F.; Pérez, N.G.; Swenson, E.R.; Mosca, S.M.; Alvarez, B.V. Benzolamide perpetuates acidic conditions during reperfusion and reduces myocardial ischemia-reperfusion injury. *J. Appl. Physiol.* **2018**, *125*, 340–352. [CrossRef] [PubMed]
11. Chiacchiaretta, M.; Latifi, S.; Bramini, M.; Fadda, M.; Fassio, A.; Benfenati, F.; Cesca, F. Neuronal hyperactivity causes $Na+/H+$ exchanger-induced extracellular acidification at active synapses. *J. Cell Sci.* **2017**, *130*, 1435–1449. [CrossRef] [PubMed]
12. Jentsch, T.J.; Pusch, M. CLC Chloride Channels and Transporters: Structure, Function, Physiology, and Disease. *Physiol. Rev.* **2018**, *98*, 1493–1590. [CrossRef]

13. Hwang, J.Y.; Zukin, R.S. REST, a master transcriptional regulator in neurodegenerative disease. *Curr. Opin. Neurobiol.* **2018**, *48*, 193–200. [CrossRef] [PubMed]
14. Kaila, K. Ionic basis of GABAA receptor channel function in the nervous system. *Prog. Neurobiol.* **1994**, *42*, 489–537. [CrossRef]
15. Alper, S.L. Molecular physiology and genetics of Na+-independent SLC4 anion exchangers. *J. Exp. Biol.* **2009**, *212*, 1672–1683. [CrossRef] [PubMed]
16. Theparambil, S.M.; Naoshin, Z.; Thyssen, A.; Deitmer, J.W. Reversed electrogenic sodium bicarbonate cotransporter 1 is the major acid loader during recovery from cytosolic alkalosis in mouse cortical astrocytes. *J. Physiol.* **2015**, *593*, 3533–3547. [CrossRef]
17. Thomas, R.C. Experimental displacement of intracellular pH and the mechanism of its subsequent recovery. *J. Physiol.* **1984**, *354*, 3–22. [CrossRef]
18. Lee, S.K.; Boron, W.F.; Parker, M.D. Monitoring ion activities in and around cells using ion-selective liquid-membrane microelectrodes. *Sensors* **2013**, *13*, 984–1003. [CrossRef]
19. DiFranco, M.; Quinonez, M.; Dziedzic, R.M.; Spokoyny, A.M.; Cannon, S.C. A highly-selective chloride microelectrode based on a mercuracarborand anion carrier. *Sci. Rep.* **2019**, *9*, 18860. [CrossRef]
20. Geddes, C.D.; Apperson, K.; Karolin, J.; Birch, D.J.S. Chloride-sensitive fluorescent indicators. *Anal. Biochem.* **2001**, *293*, 60–66. [CrossRef]
21. Han, J.; Burgess, K. Fluorescent indicators for intracellular pH. *Chem. Rev.* **2010**, *110*, 2709–2728. [CrossRef]
22. Le Guern, F.; Mussard, V.; Gaucher, A.; Rottman, M.; Prim, D. Fluorescein Derivatives as Fluorescent Probes for pH Monitoring along Recent Biological Applications. *Int. J. Mol. Sci.* **2020**, *21*, 9217. [CrossRef] [PubMed]
23. Tantama, M.; Hung, Y.P.; Yellen, G. Imaging intracellular pH in live cells with a genetically encoded red fluorescent protein sensor. *J. Am. Chem. Soc.* **2011**, *133*, 10034–10037. [CrossRef] [PubMed]
24. Bregestovski, P.; Arosio, D.; Bregestovski, P.; Arosio, D.; Jung, G. Green fluorescent protein-based chloride ion sensors for in vivo imaging. *Springer* **2012**, *12*, 99–124. [CrossRef]
25. Burgstaller, S.; Bischof, H.; Gensch, T.; Stryeck, S.; Gottschalk, B.; Ramadani-Muja, J.; Eroglu, E.; Rost, R.; Balfanz, S.; Baumann, A.; et al. pH-Lemon, a Fluorescent Protein-Based pH Reporter for Acidic Compartments. *ACS Sens.* **2019**, *4*, 883–891. [CrossRef]
26. Gao, L.; Lin, X.; Zheng, A.; Shuang, E.; Wang, J.; Chen, X. Real-time monitoring of intracellular pH in live cells with fluorescent ionic liquid. *Anal. Chim. Acta* **2020**, *1111*, 132–138. [CrossRef]
27. Arosio, D.; Ricci, F.; Marchetti, L.; Gualdani, R.; Albertazzi, L.; Beltram, F. Simultaneous intracellular chloride and pH measurements using a GFP-based sensor. *Nat. Methods* **2010**, *7*, 516–518. [CrossRef]
28. Mukhtarov, M.; Liguori, L.; Waseem, T.; Rocca, F.; Buldakova, S.; Arosio, D.; Bregestovski, P. Calibration and functional analysis of three genetically encoded Cl(-)/pH sensors. *Front. Mol. Neurosci.* **2013**, *6*, 9. [CrossRef]
29. Raimondo, J.V.; Joyce, B.; Kay, L.; Schlagheck, T.; Newey, S.E.; Srinivas, S.; Akerman, C.J. A genetically-encoded chloride and pH sensor for dissociating ion dynamics in the nervous system. *Front. Cell. Neurosci.* **2013**, *7*, 202. [CrossRef]
30. Paredes, J.M.; Idilli, A.I.; Mariotti, L.; Losi, G.; Arslanbaeva, L.R.; Sato, S.S.; Artoni, P.; Szczurkowska, J.; Cancedda, L.; Ratto, G.M.; et al. Synchronous Bioimaging of Intracellular pH and Chloride Based on LSS Fluorescent Protein. *ACS Chem. Biol.* **2016**, *11*, 1652–1660. [CrossRef]
31. Sato, S.S.; Artoni, P.; Landi, S.; Cozzolino, O.; Parra, R.; Pracucci, E.; Trovato, F.; Szczurkowska, J.; Luin, S.; Arosio, D.; et al. Simultaneous two-photon imaging of intracellular chloride concentration and pH in mouse pyramidal neurons in vivo. *Proc. Natl. Acad. Sci. USA* **2017**, *114*, E8770–E8779. [CrossRef] [PubMed]
32. Diuba, A.V.; Samigullin, D.V.; Kaszas, A.; Zonfrillo, F.; Malkov, A.; Petukhova, E.; Casini, A.; Arosio, D.; Esclapez, M.; Gross, C.T.; et al. CLARITY analysis of the Cl/pH sensor expression in the brain of transgenic mice. *Neuroscience* **2020**, *439*, 181–194. [CrossRef] [PubMed]
33. Arosio, D.; Garau, G.; Ricci, F.; Marchetti, L.; Bizzarri, R.; Nifosì, R.; Beltram, F. Spectroscopic and structural study of proton and halide ion cooperative binding to gfp. *Biophys. J.* **2007**, *93*, 232–244. [CrossRef]
34. Curtis, D.R.; Duggan, A.W.; Felix, D.; Johnston, G.A.R.; McLennan, H. Antagonism between bicuculline and GABA in the cat brain. *Brain Res.* **1971**, *33*, 57–73. [CrossRef]
35. Steidl, E.; Gleyzes, M.; Maddalena, F.; Debanne, D.; Buisson, B. Neuroservice proconvulsive (NS-PC) set: A new platform of electrophysiology-based assays to determine the proconvulsive potential of lead compounds. *J. Pharmacol. Toxicol. Methods* **2019**, *99*, 106587. [CrossRef]
36. Rink, T.I.; Tsien, R.Y.; Pozzan, T. Cytoplasmic pH and free Mg2+ in lymphocytes. *J. Cell Biol.* **1982**, *95*, 189–196. [CrossRef]
37. Zhan, R.Z.; Fujiwara, N.; Tanaka, E.; Shimoji, K. Intracellular acidification induced by membrane depolarization in rat hippocampal slices: Roles of intracellular Ca2+ and glycolysis. *Brain Res.* **1998**, *780*, 86–94. [CrossRef]
38. Wu, M.L.; Chen, J.H.; Chen, W.H.; Chen, Y.U.J.; Chu, K.C. Novel role of the Ca(2+)-ATPase in NMDA-induced intracellular acidification. *Am. J. Physiol.* **1999**, *277*, C717–C727. [CrossRef]
39. Cheng, Y.M.; Kelly, T.; Church, J. Potential contribution of a voltage-activated proton conductance to acid extrusion from rat hippocampal neurons. *Neuroscience* **2008**, *151*, 1084–1098. [CrossRef]
40. Svichar, N.; Esquenazi, S.; Chen, H.Y.; Chesler, M. Preemptive regulation of intracellular pH in hippocampal neurons by a dual mechanism of depolarization-induced alkalinization. *J. Neurosci.* **2011**, *31*, 6997–7004. [CrossRef]

41. Suzuki, M.; Van Paesschen, W.; Stalmans, I.; Horita, S.; Yamada, H.; Bergmans, B.A.; Legius, E.; Riant, F.; De Jonghe, P.; Li, Y.; et al. Defective membrane expression of the Na(+)-HCO(3)(-) cotransporter NBCe1 is associated with familial migraine. *Proc. Natl. Acad. Sci. USA* **2010**, *107*, 15963–15968. [CrossRef]

42. Russell, M.B.; Ducros, A. Sporadic and familial hemiplegic migraine: Pathophysiological mechanisms, clinical characteristics, diagnosis, and management. *Lancet. Neurol.* **2011**, *10*, 457–470. [CrossRef]

43. Ahmed, Z.; Connor, J.A. Intracellular pH changes induced by calcium influx during electrical activity in molluscan neurons. *J. Gen. Physiol.* **1980**, *75*, 403–426. [CrossRef]

44. Drapeau, P.; Nachshen, D.A. Effects of lowering extracellular and cytosolic pH on calcium fluxes, cytosolic calcium levels, and transmitter release in presynaptic nerve terminals isolated from rat brain. *J. Gen. Physiol.* **1988**, *91*, 305–315. [CrossRef]

45. Takahashi, K.I.; Copenhagen, D.R. Modulation of neuronal function by intracellular pH. *Neurosci. Res.* **1996**, *24*, 109–116. [CrossRef]

46. Mahadevan, V.; Woodin, M.A. Regulation of neuronal chloride homeostasis by neuromodulators. *J. Physiol.* **2016**, *594*, 2593–2605. [CrossRef] [PubMed]

47. Côme, E.; Marques, X.; Poncer, J.C.; Lévi, S. KCC2 membrane diffusion tunes neuronal chloride homeostasis. *Neuropharmacology* **2020**, *169*, 107571. [CrossRef]

48. Yousuf, M.S.; Zubkow, K.; Tenorio, G.; Kerr, B. The chloride co-transporters, NKCC1 and KCC2, in experimental autoimmune encephalomyelitis (EAE). *Neuroscience* **2017**, *344*, 178–186. [CrossRef]

49. Zhang, S.; Zhou, J.; Zhang, Y.; Liu, T.; Friedel, P.; Zhuo, W.; Somasekharan, S.; Roy, K.; Zhang, L.; Liu, Y.; et al. The structural basis of function and regulation of neuronal cotransporters NKCC1 and KCC2. *Commun. Biol.* **2021**, *4*, 226. [CrossRef] [PubMed]

50. Medina, I.; Friedel, P.; Rivera, C.; Kahle, K.T.; Kourdougli, N.; Uvarov, P.; Pellegrino, C. Current view on the functional regulation of the neuronal K(+)-Cl(-) cotransporter KCC2. *Front. Cell. Neurosci.* **2014**, *8*, 27. [CrossRef]

51. Friedel, P.; Ludwig, A.; Pellegrino, C.; Agez, M.; Jawhari, A.; Rivera, C.; Medina, I. A Novel View on the Role of Intracellular Tails in Surface Delivery of the Potassium-Chloride Cotransporter KCC2. *eNeuro* **2017**, *4*. [CrossRef] [PubMed]

52. Deitmer, J.W.; Rose, C.R. pH regulation and proton signalling by glial cells. *Prog. Neurobiol.* **1996**, *48*, 73–103. [CrossRef]

53. Traynelis, S.F.; Cull-Candy, S.G. Proton inhibition of N-methyl-D-aspartate receptors in cerebellar neurons. *Nature* **1990**, *345*, 347–350. [CrossRef] [PubMed]

54. DeVries, S.H. Exocytosed protons feedback to suppress the Ca2+ current in mammalian cone photoreceptors. *Neuron* **2001**, *32*, 1107–1117. [CrossRef]

55. Ruminot, I.; Gutiérrez, R.; Peña-Münzenmayer, G.; Añazco, C.; Sotelo-Hitschfeld, T.; Lerchundi, R.; Niemeyer, M.I.; Shull, G.E.; Barros, L.F. NBCe1 mediates the acute stimulation of astrocytic glycolysis by extracellular K+. *J. Neurosci.* **2011**, *31*, 14264–14271. [CrossRef]

56. Theparambil, S.M.; Ruminot, I.; Schneider, H.P.; Shull, G.E.; Deitmer, J.W. The electrogenic sodium bicarbonate cotransporter NBCe1 is a high-affinity bicarbonate carrier in cortical astrocytes. *J. Neurosci.* **2014**, *34*, 1148–1157. [CrossRef]

57. Brookes, N.; Turner, R.J. K(+)-induced alkalinization in mouse cerebral astrocytes mediated by reversal of electrogenic Na(+)-HCO3- cotransport. *Am. J. Physiol.* **1994**, *267*. [CrossRef]

58. Bevensee, M.O.; Apkon, M.; Boron, W.F. Intracellular pH regulation in cultured astrocytes from rat hippocampus. II. Electrogenic Na/HCO3 cotransport. *J. Gen. Physiol.* **1997**, *110*, 467–483. [CrossRef]

59. Ruffin, V.A.; Salameh, A.I.; Boron, W.F.; Parker, M.D. Intracellular pH regulation by acid-base transporters in mammalian neurons. *Front. Physiol.* **2014**, *5*, 43. [CrossRef]

60. Hentschke, M.; Wiemann, M.; Hentschke, S.; Kurth, I.; Hermans-Borgmeyer, I.; Seidenbecher, T.; Jentsch, T.J.; Gal, A.; Hübner, C.A. Mice with a targeted disruption of the Cl-/HCO3- exchanger AE3 display a reduced seizure threshold. *Mol. Cell. Biol.* **2006**, *26*, 182–191. [CrossRef] [PubMed]

61. Ivanov, A.; Mukhtarov, M.; Bregestovski, P.; Zilberter, Y. Lactate Effectively Covers Energy Demands during Neuronal Network Activity in Neonatal Hippocampal Slices. *Front. Neuroenergetics* **2011**, *3*, 2. [CrossRef] [PubMed]

62. Bonnet, U.; Bingmann, D.; Speckmann, E.J.; Wiemann, M. Small intraneuronal acidification via short-chain monocarboxylates: First evidence of an inhibitory action on over-excited human neocortical neurons. *Life Sci.* **2018**, *204*, 65–70. [CrossRef] [PubMed]

63. Trapp, S.; Lückermann, M.; Kaila, K.; Ballanyi, K. Acidosis of hippocampal neurones mediated by a plasmalemmal Ca2+/H+ pump. *Neuroreport* **1996**, *7*, 2000–2004. [CrossRef]

64. Chao, A.C.; Dix, J.A.; Sellers, M.C.; Verkman, A.S. Fluorescence measurement of chloride transport in monolayer cultured cells. Mechanisms of chloride transport in fibroblasts. *Biophys. J.* **1989**, *56*, 1071–1081. [CrossRef]

65. Krapf, R.; Berry, C.A.; Verkman, A.S. Estimation of intracellular chloride activity in isolated perfused rabbit proximal convoluted tubules using a fluorescent indicator. *Biophys. J.* **1988**, *53*, 955–962. [CrossRef]

International Journal of
Molecular Sciences

Article

Calcium Imaging Reveals Fast Tuning Dynamics of Hippocampal Place Cells and CA1 Population Activity during Free Exploration Task in Mice

Vladimir P. Sotskov [1],*, Nikita A. Pospelov [1], Viktor V. Plusnin [2,3] and Konstantin V. Anokhin [1,4],*

1   Institute for Advanced Brain Studies, Lomonosov Moscow State University, 119991 Moscow, Russia; nik-pos@yandex.ru
2   National Research Center "Kurchatov Institute", 123098 Moscow, Russia; witkax@mail.ru
3   Department of NBIC-Technologies, Moscow Institute of Physics and Technology, 141700 Dolgoprudny, Russia
4   P.K. Anokhin Institute of Normal Physiology RAS, 125315 Moscow, Russia
*   Correspondence: vsotskov@list.ru (V.P.S.); k.anokhin@gmail.com (K.V.A.)

Abstract: Hippocampal place cells are a well-known object in neuroscience, but their place field formation in the first moments of navigating in a novel environment remains an ill-defined process. To address these dynamics, we performed in vivo imaging of neuronal activity in the CA1 field of the mouse hippocampus using genetically encoded green calcium indicators, including the novel NCaMP7 and FGCaMP7, designed specifically for in vivo calcium imaging. Mice were injected with a viral vector encoding calcium sensor, head-mounted with an NVista HD miniscope, and allowed to explore a completely novel environment (circular track surrounded by visual cues) without any reinforcement stimuli, in order to avoid potential interference from reward-related behavior. First, we calculated the average time required for each CA1 cell to acquire its place field. We found that 25% of CA1 place fields were formed at the first arrival in the corresponding place, while the average tuning latency for all place fields in a novel environment equaled 247 s. After 24 h, when the environment was familiar to the animals, place fields formed faster, independent of retention of cognitive maps during this session. No cumulation of selectivity score was observed between these two sessions. Using dimensionality reduction, we demonstrated that the population activity of rapidly tuned CA1 place cells allowed the reconstruction of the geometry of the navigated circular maze; the distribution of reconstruction error between the mice was consistent with the distribution of the average place field selectivity score in them. Our data thus show that neuronal activity recorded with genetically encoded calcium sensors revealed fast behavior-dependent plasticity in the mouse hippocampus, resulting in the rapid formation of place fields and population activity that allowed the reconstruction of the geometry of the navigated maze.

Keywords: Ca^{2+} indicators; calcium in vivo imaging; place cells; place fields; cognitive maps

Citation: Sotskov, V.P.; Pospelov, N.A.; Plusnin, V.V.; Anokhin, K.V. Calcium Imaging Reveals Fast Tuning Dynamics of Hippocampal Place Cells and CA1 Population Activity during Free Exploration Task in Mice. *Int. J. Mol. Sci.* **2022**, *23*, 638. https://doi.org/10.3390/ijms23020638

Academic Editors: Piotr D. Bregestovski and Carlo Matera

Received: 1 December 2021
Accepted: 4 January 2022
Published: 7 January 2022

1. Introduction

It is well known that neurons in the CA1 field of the hippocampus form a representation (also referred to as a cognitive map) of a novel context, while animals explore a novel environment [1,2]. The long-term dynamics of such cognitive maps have been well explored in studies [3–5], revealing that the place code can be stable for weeks, though subserved by a drifting population of CA1 neurons. Moreover, it is known that multiple cognitive maps can coexist in the hippocampus in a stable manner and switch between different navigating sessions and even within the same navigating session [6,7].

However, the short-term dynamics of place field emergence and initial tuning are still ill-defined. In particular, it is unclear whether place fields are established at the first moment at which the animal arrives in a novel place, or whether several repeated visits are necessary for place cells to become tuned. A recent study of head-restrained mice in a virtual navigation task demonstrated "immediate" place cells that appear and fire in a stable manner from the first lap in a novel virtual environment [8]. However, it is unclear how rapidly the place codes emerge in real conditions of animal free navigation.

Many of the previous studies on place cell registration used rewarded approaches with pre-trained animals [7,9]. However, since goal-directed behavior may confound the factor of spatial navigation [10,11], we used a reward-free task where mice were allowed to explore a completely novel environment in the shape of an elevated circular track with proximal and distal visual cues (Figure 1).

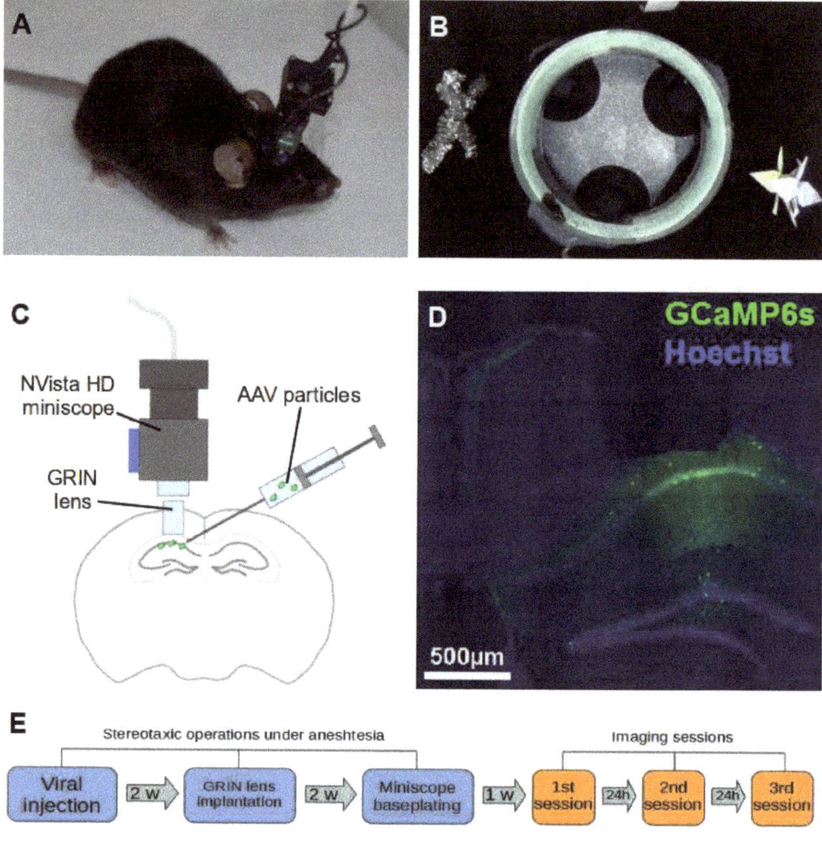

Figure 1. (**A,B**) Mouse with an attached NVista HD miniscope exploring the circular track. (**C**) A scheme of calcium sensor injection and GRIN lens implantation. (**D**) A coronal brain section with a footprint of a GRIN lens and calcium sensor expression. (**E**) The timeline of surgical preparations and imaging.

To record place cell activity, we used head-mounted NVista HD miniscopes [12], which are capable of capturing calcium signals from hundreds of neurons in freely moving animals. To image cells, we used a set of genetically encoded calcium indicators: both conventional GCaMP6s and GCaMP7f, as well as novel ones, NCaMP7 and FGCaMP7. NCaMP7 is a new calcium indicator with enhanced brightness, containing a mNeonGreen fluorescent protein, while FGCaMP7 is a novel calcium sensor based on fungi calmodulin with lowered affinity to the intracellular environment and designed specifically for in vivo miniscopic calcium imaging [13,14]. In these papers, we described in detail their dynamic parameters, such as mean amplitude, and rise and decay times of typical calcium transients. Importantly, we showed that there is no notable difference in such parameters between these new calcium indicators and a conventional one, GCaMP6s. Moreover, the applicability of these sensors for the analysis of in vivo neural functions in awake mice was directly demonstrated by calcium imaging of hippocampal place cells.

Using these tools, we set out to investigate the dynamics of the initial place field formation in the mouse hippocampus, not only at the level of individual place cells, but also at the level of a whole imaged CA1 population. The activity of large populations of neurons is often well embraced by low-dimensional dynamics [15–17]. This makes it possible to describe computations performed by groups of cells using the dynamics of a small number of underlying "latent factors", each of which corresponds to a separate pattern of neuronal coactivation. However, it is not possible to observe the latent factors directly, because they are often related to the initial variables in a non-obvious and non-linear way. In this paper, we utilize the tools from manifold learning to construct an underlying low-dimensional "neural manifold" from rapidly tuned place cell activity and to explore its representational power.

2. Materials and Methods

2.1. Animals and Surgical Procedures

Nine C57Bl/6J mice aged from 2 to 3 months at the beginning of the experiment were used for this study. All surgical protocols were described in detail in our previous papers [13,14,18,19]. First, a viral vector encoding one of the calcium sensors (GCaMP6s/GCaMP7f/NCaMP7/FGCaMP7) was delivered to the CA1 field of the hippocampus of the mice. Animals were anesthetized with a zoletil–xylazine mixture (40 and 5 mg/kg, respectively) and fixed in a stereotaxic holder (Stoelting Inc., Wood Dale, IL, USA). Then, a circular 2-mm-diameter craniotomy was made (Bregma: -1.9 mm AP, -1.4 mm ML), and 500 nL of AAV viral particles (AAV-DJ-CAG-GCaMP6s, AAV-DJ-CAG-GCaMP7f, AAV-DJ-CAG-NCaMP7 or AAV-DJ-CAG-FGCaMP7) was injected to a depth of 1.25 mm from the brain surface. Injections were performed through a glass micropipette with a 50 μm tip diameter (Drummond Scientific Comp., Broomall, PA, USA) by UltraMicroPump with a Micro4 Controller (WPI Inc., Sarasota, FL, USA) at a rate of 100 nL/min. After the injection, all exposed surfaces of the brain tissue were sealed with KWIK-SIL silicone adhesive (WPI Inc.). Two weeks later, the animals were anesthetized and fixed in the stereotaxis again, the silicone cap was removed, and the dura mater was perforated and gently removed from the craniotomy site. Then, a column of cortex tissue superficial to the hippocampus was gently aspirated by a blunt needle tip connected to a vacuum source and the hippocampus was exposed and washed with sterile saline. After this, a 1.0-mm-diameter GRIN lens probe (Inscopix Inc., Palo Alto, CA, USA) was lowered slowly to a depth of 1.1 mm while constantly washing the craniotomy site with sterile saline. Next, all the exposed brain tissue was sealed with KWIK-SIL, and the lens probe was fixed to the skull surface with dental acrylic (Stoelting Inc.). After another two weeks, the animals were checked for fluorescent calcium signal under light anesthesia ($\frac{1}{2}$ of the dose described above). The mice were fixed in the stereotaxis, and an NVista HD miniature microscope (Inscopix Inc.) was lowered upon the GRIN lens probe and the optimal field of view was chosen. Then, a baseplate for chronic imaging was affixed to the skull surface with dental acrylic.

2.2. Miniscope Imaging in Freely Behaving Mice

Finally, after a one-week recovery period, awake mice with an attached NVista HD miniscope were placed for 15 min into a custom-made circular O-shaped track (50 cm diameter, 5 cm width, with 5 cm height borders) with proximal (different border material) and distal (placed on a surrounding curtain 20 cm apart from the track) visual cues. Mice were allowed to explore the environment in arbitrary directions and were not forced to move. The imaging session was repeated for 8 of 9 mice on the next day after 24 h and for 3 mice on the third day after 48 h from the first imaging session. The neural activity was recorded at 20 frames per second at resolution 1440 × 1080 px with an NVista HD miniscope. Screenshots and video samples of raw calcium signal can be seen in Figure A1 of Appendix A and in Supplementary Videos S1–S3. The video of mouse behavior was captured with a Sony HDR CX-405 (Sony Corp., Tokyo, Japan) camera at 25 frames per second.

At the end of the experiments, animals were perfused transcardially with 1% paraformaldehyde in 0.1 mM $CaCl_2$, and then brains were extracted and postfixed for 24 h in the same solution. Thin (50 µm) floating sections were prepared with a Leica 1200VT (Leica Microsystems GmbH, Wetzlar, Germany) vibratome, stained with Hoechst dye (Hoechst AG, Frankfurt, Germany) and imaged with an Olympus FluoView 1000 (Olympus Corp., Tokyo, Japan) confocal microscope with a UMPlanFLN 10× NA 0.30 W objective. All sections were inspected and checked for consistency of calcium sensor expression site and GRIN lens implantation locations to the field CA1 of the hippocampus. Samples of such sections can be seen in Figure 1D and in Figure A1 of Appendix A.

2.3. Neural and Behavioral Data Processing

Image processing was performed with the NoRMCorre [20] and MIN1PIPE [21] pipelines and custom MATLAB and Python scripts. First, all movies were downsampled spatially by a factor 2 to increase the computation speed. Then, the NoRMCorre routine was applied to spatially align movies and to correct motion artifacts. Next, the MIN1PIPE routine was applied to corrected movies, and locations and activity traces of putative cell units were extracted and manually inspected (Figure 2A,B). Then, significant calcium events were detected in the activity traces.

Detection of calcium events was performed with a custom routine, which was described in our previous work [18]. First, a threshold of 4 or 5 median absolute deviations (MADs) was applied to extracted neural traces for animals injected with slow (GCaMP6s, NCaMP7 and FGCaMP7) or fast calcium sensors (GCaMP7f), respectively. Then, neighborhoods of each upward threshold crossing were fitted with a typical calcium event model function with fast rise and slow decay (Figure 2C). This model utilizes precise spiking time t_0, rise time t_{on}, decay time t_{off} and spiking amplitude A as the parameters to be optimized. The lower limit of t_{off} was set to 200 ms for mice injected with the GCaMP7f indicator and 500 ms for the other ones. In case of acceptable fit (goodness of fit ≥ 0.8), a calcium event was scored and the fit was locally subtracted from the original trace in order to let subsequent events be fitted and scored (Figure 2D).

Data for each session were processed separately; matching of cells and traces across sessions was performed with the CellReg routine [22] with default parameters (maximum angle of 30 degrees, maximum translation of 14 microns, registration threshold P_same of 0.5). Exact amounts of matched cells across sessions and their contours can be seen in Table A2 and in Figure A2 of Appendix A. Positions of animals were extracted from behavioral video recording with the open-source Bonsai visual programming media [23]. All obtained time series were synchronized and aligned to the beginning of the imaging session.

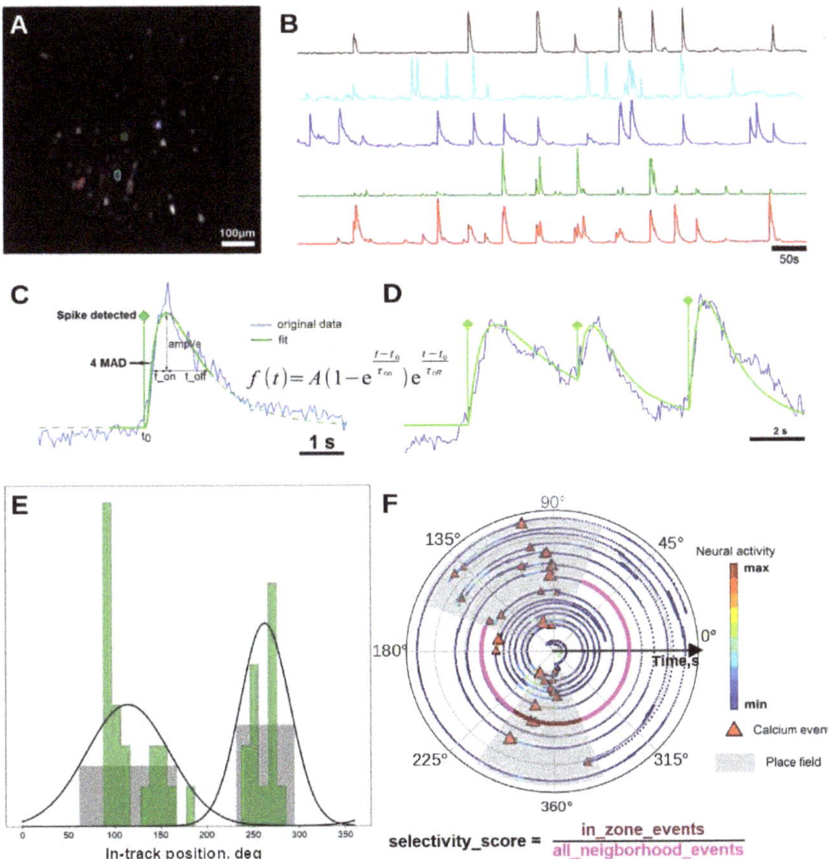

Figure 2. (**A,B**) Selected neuron locations (**A**) and their traces (**B**) extracted by MIN1PIPE routine. (**C**) Scheme of spike detection. Whenever the trace reaches the threshold value, the fitting procedure starts in a range, denoted by the solid green line. Fitting curve *f(t)* is set to a composition of rise and decay exponential factors and spiking amplitude *A*, where t_0, t_{on} and t_{off} are fitting parameters. (**D**) Scheme of multiple spike detection. In case of tolerable goodness of fit (not less than 0.8), the event is scored, and the fitting curve is subtracted from the original trace, allowing next peaks to be scored. (**E**) Spatial event distribution for putative place field detection. Solid black line denotes Gaussian mixture fit of the distribution and grey zones are putative place fields. (**F**) An example of a detected place field based on the distribution on the left. Trajectory of the mouse is unfolded in the axial direction for better readability and colored with respect to raw activity of the correspondent place cell. To check the consistency of putative place field, we calculated selectivity score each time the animal attended place field zone (in the depicted case, selectivity score equals 1).

2.4. Place Cell Detection

Given some uncertainty that exists in the procedure of place cell detection [24] and that the commonly used approach based on spatial information [25] requires a high running speed of animals and does not consider the repeatability of place cell firing each time the animal visits the place field, we used a conservative approach, where we checked both the spatial and temporal persistence of place cell firing. The entire track space was divided into 20 sectorial bins sized 5×7.5 cm. For each cell, an overall number of calcium events was calculated for each bin and the distribution of calcium events was smoothed with a Gaussian kernel (sigma = 1.25), normalized and thresholded by a value of 0.5. Then, centroids of all subthreshold peaks were considered putative place field centers. Width of place fields was evaluated as width of subthreshold peaks + 1 bin from each side. Place fields wider than a half of the track were excluded from further analysis. Next, for each place field, we checked whether the correspondent cell fired in all epochs when the animal attended a given place field. For this purpose, for each attendance of a given place field, we calculated the selectivity score as a ratio of the number of events within the attendance epoch over the total number of events within this epoch and adjacent epochs between visits (Figure 2F). For each place field, we smoothed the selectivity score sequence with a Gaussian filter (sigma = 1). Only epochs of attendance with smoothed selectivity scores more than 0.5 were considered relevant; place cells without at least three subsequent relevant epochs were discarded from the analysis. The tuning latency for each place field was calculated as the number of epochs (or as time in seconds) before the beginning of the first sequence of three or more relevant epochs. We considered that the cognitive map retained between sessions in case of significant similarity of the distribution of place field location shifts between sessions to a normal distribution around zero ($p < 0.05$, Chi-square test).

2.5. Dimensionality Reduction

In this paper, we used the Laplacian eigenmaps method, which constructs a discrete approximation to a continuous low-dimensional representation that naturally arises from the geometry of the manifold [26]. Before dimensionality reduction, the calcium data for each animal were presented in the form of the matrix $D_{N \times T}$, where N is the number of selected cells and T is the total duration of recording. Therefore, D_{ij} stands for the calcium activity of the neuron i at the timeframe j. Only neurons with 5 or more calcium events during the recording were considered.

We built a similarity graph G based on the data in the original high-dimensional space. Each vector V_t was considered a point in $\mathbb{R}^N$. Thus, each column of the matrix $D_{N \times T}$ was treated as a single N-dimensional vector of neural activity at a certain timeframe. Hence, we obtained T multidimensional vectors in $\mathbb{R}^N$, representing neural activity in different moments in time: $\{V_t\}, t \in [0, T]$. To simplify the graph construction procedure and reduce the computation time, we applied mean filtering to the initial time series of calcium activity with window size $w = 2, 4, 8$. We made sure that the choice of the window size did not qualitatively affect our results. This is because the calcium signal hardly changes during the time corresponding to the window sizes of 40, 80, or 160 ms, respectively. All results here were obtained for window size $w = 8$. Thus, the effective signal length and number of nodes in G was $T_{eff} = T/w$.

For each of the T_{eff} similarity graph nodes, exactly k nearest neighbors were calculated using Euclidean distance. The number of nearest neighbors k was the only free parameter in our dimensionality reduction procedure. It tended to be chosen as small as possible (however, its value should have ensured the integrity of the resulting graph). Keeping k small is motivated by the local linearity assumption: if k becomes sufficiently large, the Euclidean distance may not reflect the proximity relations between data points because of the possible nonlinear curvature of the underlying manifold. The situation is also complicated by the dimensionality curse in the initial space [27].

The resulting graph adjacency matrix was explicitly made symmetric to preclude directed edge formation. This means that the "nearest neighbor" relation is made mutual: if some node i is connected to a node j, the reverse is also true. It should be noted that, due to the symmetrization procedure, the number of nearest neighbors k sets only the minimal number of edges for a given node, but not a precise one. The similarity graph was ensured to be connected. If it had more than one connected component after the construction procedure, the largest one was taken (if the total share of discarded points did not exceed 5%; otherwise, dimensionality reduction for a given k was considered unsuccessful). The time moments corresponding to the excluded points were not considered further in the analysis.

Once a similarity graph was constructed, it was presented in the form of an adjacency matrix $A = \{a_{ij}\}$ with matrix elements $a_{ij} = a_{ji}$ taking non-negative values. The absence of self-loops implies the vanishing of the diagonal elements: $a_{ii} = 0$. The matrix elements were considered binary: $a_{ij} = 1$, if the nodes i and $j \neq i$ are connected, and $a_{ij} = 0$ otherwise. The next step was constructing a discrete Laplacian—a symmetric, positive semidefinite matrix that can be considered a diffusion operator on the graph G. The spectral decomposition of the graph Laplacian matrix can be used to optimally embed the graph in a low-dimensional space [26]. According to the Laplacian eigenmaps algorithm, we considered the following generalized eigenvalue problem:

$$\mathbf{Lv} = \lambda D\mathbf{v} \tag{1}$$

Here, D is the degree matrix of the network, whose elements are defined as $d_{ij} = deg(n_i)$ if $i = j$ and $d_{ij} = 0$ otherwise, where $deg(n_i)$ is the degree of the node i: $deg(n_i) = \sum_j a_{ij}$. L stands for the graph Laplacian matrix, which is defined as

$$L = D - A \tag{2}$$

The algorithm utilizes first $m + 1$ solutions $\{\mathbf{f}_i\}, i \in [0, m]$ of the generalized eigenvalue problem (1) (ordered by the associated eigenvalues $\{\lambda_i\}$ in the ascending order) to construct an optimal embedding of the graph in $\mathbb{R}^m$. To be precise, the k-th component of an eigenvector $\mathbf{u}_i$ defines an i-th coordinate of a low-dimensional embedding for a data point $\mathbf{v}_k \in \mathbb{R}^N$. Since the dimensionality of $\mathbf{u}_i$ is equal to the number of nodes in the graph, the first m non-trivial eigenvectors are enough to construct a m-dimensional embedding for each data point.

The first eigenvector $\mathbf{v}_0$ corresponding to $\lambda_0 = 0$ was left out because its components were constant. It is known that a Laplacian matrix of a graph with c connected components has c zero eigenvalues [28]. However, we restricted ourselves to the case of connected graphs, which was ensured by the construction procedure. Hereinafter, we assume that, for our graph, the problem (1) has a single zero eigenvalue.

To measure the reconstruction error of the track geometry, we calculated the residual variance (RV) [29] between real mouse coordinates and points in the latent space. The residual variance was calculated as $RV = 1 - \rho^2(D_h, D_l)$, where ρ defines the Pearson correlation, and elements of $T_{eff} \times T_{eff}$ matrices D_h and D_l denote the pairwise Euclidean distances calculated over the original trajectory points over the low-dimensional embedding points, respectively.

3. Results

In the previous studies [13,14,18,30], we developed a non-rewarded paradigm, where mice with a head-mounted NVista HD miniscope explored a custom-made O-shaped circular track surrounded by curtains with distinctive visual cues (Figure 1B). Mice demonstrated vigorous exploratory behavior, making on average 19 laps across the track during a 15 min imaging session. It should be noted that since mice were able to arbitrarily change the moving direction, the number of laps varied significantly from session to session, and some of the laps were not full.

The mice were transfected with AAV vectors carrying different calcium indicators (namely GCaMP6s, GCaMP7f, NCaMP7 and FGCaMP7; detailed information can be seen in Table A1 of Appendix A). All mice underwent identical surgical, imaging and behavioral protocols. Mice explored the track at one, two or three consequent sessions, and the first time the context was absolutely novel for them. We isolated neuron locations, calcium traces and detected place-selective cells. For this, we selected candidate cells by the presence of distinct peaks in the overall (across entire session time) spatial distribution of calcium events of a given cell, and then checked if this cell fired or not each time the animal entered its putative place field. Only candidate cells with stable firing statistics throughout the session were considered place-selective (for more details, see Methods). We allowed each cell to have multiple place fields; however, the majority (85% on average) of place cells had single place fields.

On the first day, all animals demonstrated a uniform distribution of place fields across the track, without any distinct fluctuations in the vicinity of visual cues or other locations in the track. We matched cell activity across days and monitored the place selectivity of the same cell on all days. It turned out that, on the second day, 3 of 8 imaged mice had cognitive maps similar to the first day (Figure 3). In the other five mice, cognitive maps were not preserved, which may have been partially due to cells that were not active on the first day.

3.1. Selectivity Score and Tuning Latency

To assess the spatial selectivity of place cells to their fields, we used the selectivity score: for each place field, it was calculated whenever the animal attended the place field as the ratio of the number of in-field calcium events of the place cell over the total number of calcium events across the current lap (which may be not full). A selectivity score of 1 corresponded to a cell that fired exclusively in its place field and a score of 0 took place when the place cell did not fire at its place field during this visit. The time and the number of attendances when the smoothed selectivity score hits the threshold value of 0.5 are considered as the tuning latency or time of specialization. Since each mouse had its own trajectory and some mice explored the track faster than others, the number of visits appeared to be a more universal parameter for the estimation of the tuning dynamics of individual place cells rather than time itself. The distributions of tuning latency both in the time and number-of-visits domains are shown in Figure 4A. On the first day, a notable percentage (25.1%) of place fields were established at the very first attendance, while an average place field was formed at the 7th attendance. In the time domain, 23.1% of fields appeared within the first minute in the environment, while the average tuning latency equaled 247 s. On the second and third days, the average tuning latency of place fields decreased to values of 193 and 159 s, respectively, values that correspond to the 5th attendance of the place field. The improvement of tuning latency nominated in visits on the 2nd day was found significant (Figure 4C).

3.2. Selectivity Score Dynamics within Session and across Days

Regarding the selectivity score itself, its evolution was distributed in a similar manner (Figure 4B) across all animals, characterized by strong decay of the rate of cells with longer tuning latency. The mean selectivity score significantly increased from the first to the last attendance on each day of the experiment, but the between-days difference in selectivity score at the first and the last attendance appeared to be not significant (Figure 4D,E). Moreover, no significant difference in selectivity score improvement was observed on the 2nd day between mice with a retained cognitive map versus mice in which the map was not retained. Since we did not observe any cumulation of average selectivity score between days, we searched for it at the level of individual place cells. Importantly, the improvement in tuning latency appeared to be independent of the place field shift between sessions, i.e., cells that preserved their place field did not improve their tuning latency better than cells whose place field shifted (Figure 4F–H).

Taken together, these data suggest that the selectivity of place cell firing rose faster with each new day of the experiment, but without any significant cumulation. The retaining or remapping of the spatial representation in mice does not correlate with significantly higher selectivity scores or faster tuning dynamics. However, since these results are based only on individual place cell firing statistics, and since not only place cells can contribute to spatial coding [4,31], we performed a population analysis to confirm our results.

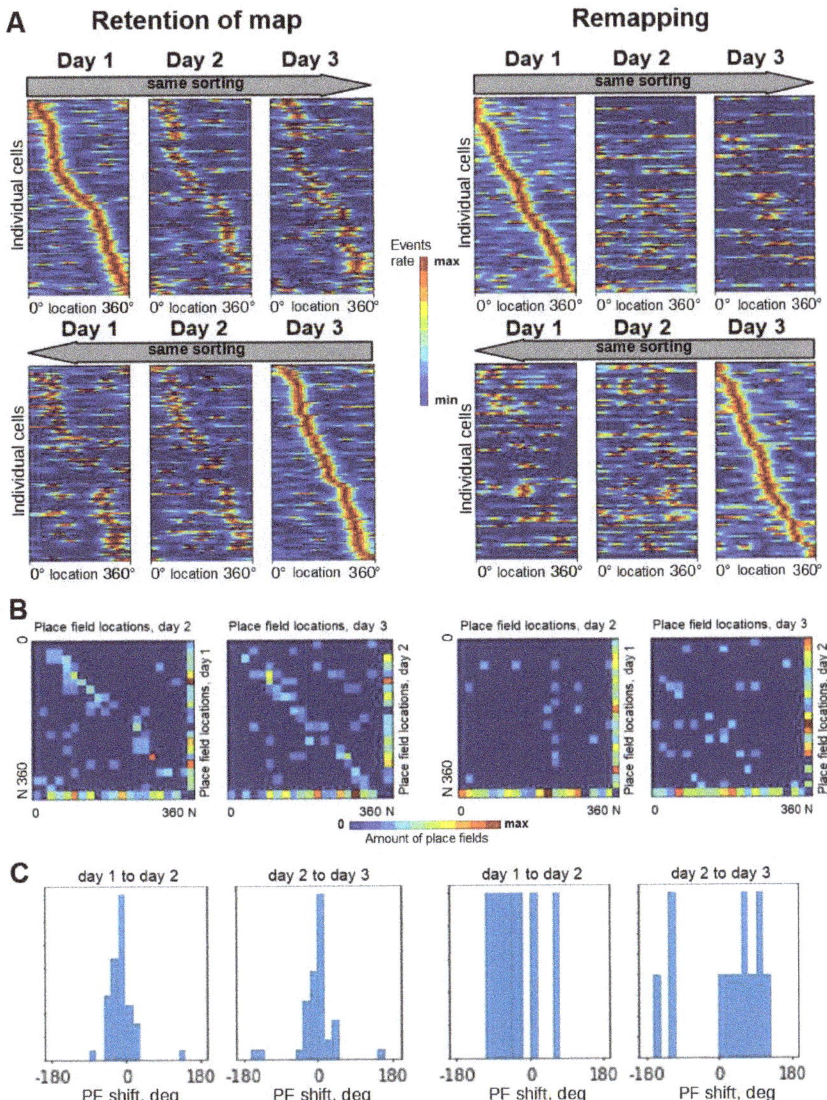

Figure 3. (**A**) Distribution of place fields across days in a case of retained representation (left) and in a case of remapping (right). Each line in a block corresponds to overall activation rate across the track space of each individual place cell in three consequent sessions. Upper row: cells sorted by their peak firing locations on the 1st day. Lower row: cells sorted by their peak firing locations on the 3rd day. (**B**) Heat maps of place cells that changed their place field locations between days. Multiple place fields of the same place cell are scored as fractions. N, not a place cell. Between-day transitions with the retention of the map result in heat maps with distinct diagonals, while remapping transitions do not. (**C**) Distribution of place field shifts for single-place-field place cells (between 1st and 2nd days and between 2nd and 3rd days). Between-day transitions with the retention of the map show sharp peaks in such distributions while remapping transitions show uniform distributions.

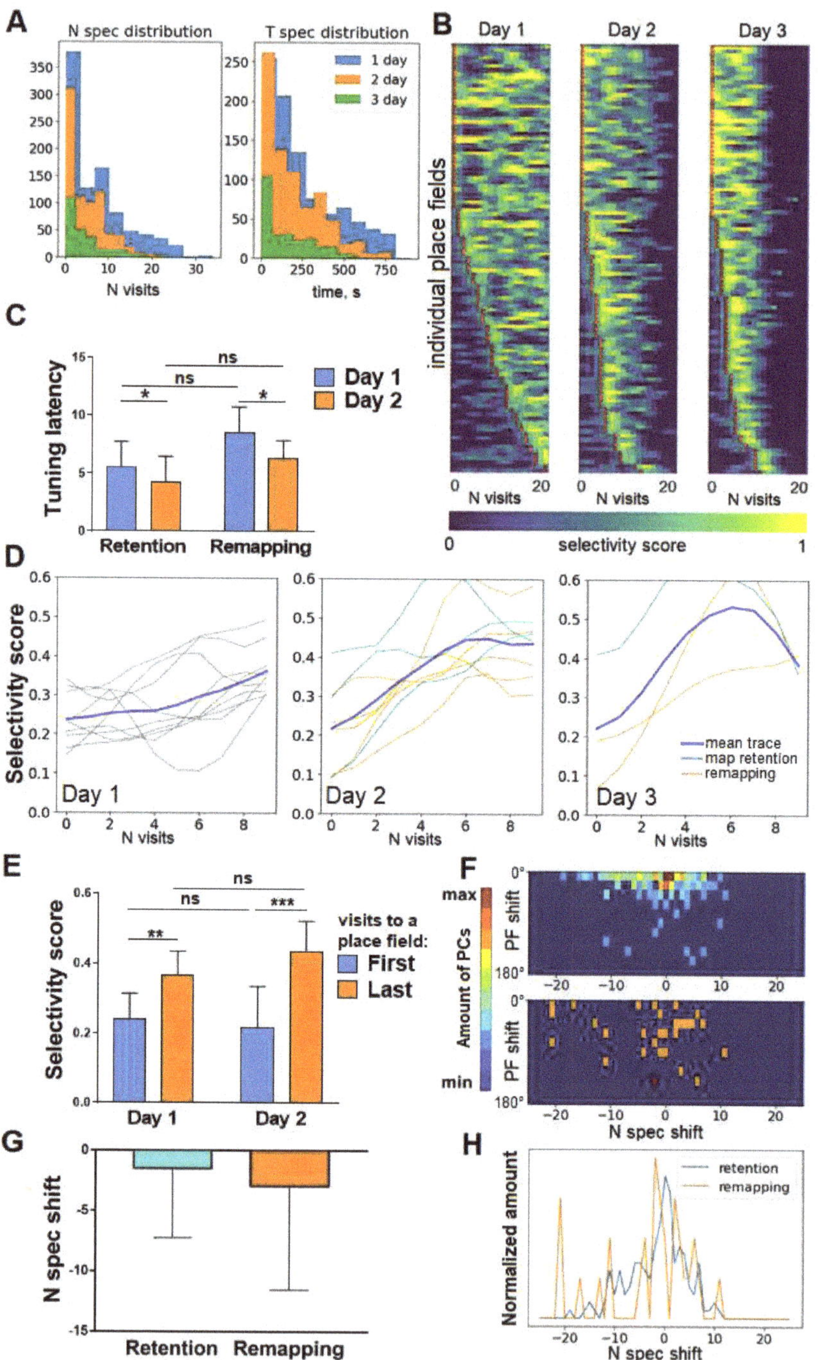

Figure 4. (**A**). Distribution of tuning latency both in the number-of-visits domain and in the time domain. (**B**). Sample of selectivity score distribution in one individual mouse on three consequent days (place fields are sorted independently for each session). Red triangles denote the number of visits where the specialization occurred (n_spec). (**C**) Tuning latency significantly decreases on the second day, while no

difference is observed between retained-map and remapped mice. Two-way ANOVA, * $p = 0.0488$ factor day, ns—not significant ($p > 0.05$). (**D**) Evolution of mean selectivity score across days in all mice. Mean scores for each animal are represented in thin lines. (**E**) Mean selectivity score significantly rises within the first and the second day from the first to the last visit to a place field, while no difference is observed between starting or ending selectivity score on the 1st versus that on the 2nd day. Two-way ANOVA and post hoc Bonferroni test, ** $p = 0.0261$, *** $p = 0.0006$, ns—$p > 0.05$. (**F**) Scatter plot of individual cell variance between 2nd and 1st days in n_spec (n_spec shift) versus place field shift for retained-map mice (above) and remapping mice (below). Only one-field place cells were taken into account. (**G**) Shifts in n_spec do not differ significantly between retained-map and remapped mice. Unpaired Student's t-test, $p = 0.2544$. (**H**) Distribution of n_spec shifts across all one-field place cells for mice with the retention of the map and for mice with remapping.

3.3. Nonlinear Dimensionality Reduction Reveals Track Geometry from Multidimensional Place Cell Activity

We performed a populational analysis of the neural data of the first six mice with a sufficient (>200) number of detected cells on the 1st day and reduced the dimensionality of the data with Laplacian eigenmaps (see Methods). The first two axes of the low-dimensional space coincided with the coordinates of the mouse in the physical environment that it was exploring (with the accuracy of rotation by a fixed angle: Figure 5A,B). It is important to note that the algorithm did not receive any information about the real position of the mouse as an input. This result could not be reproduced with PCA, indicating the nonlinear nature of the problem (Figure 5C). The best result was achieved using low-energy Laplacian eigenmodes of the similarity graph of neuronal activity vectors (Figure 5D).

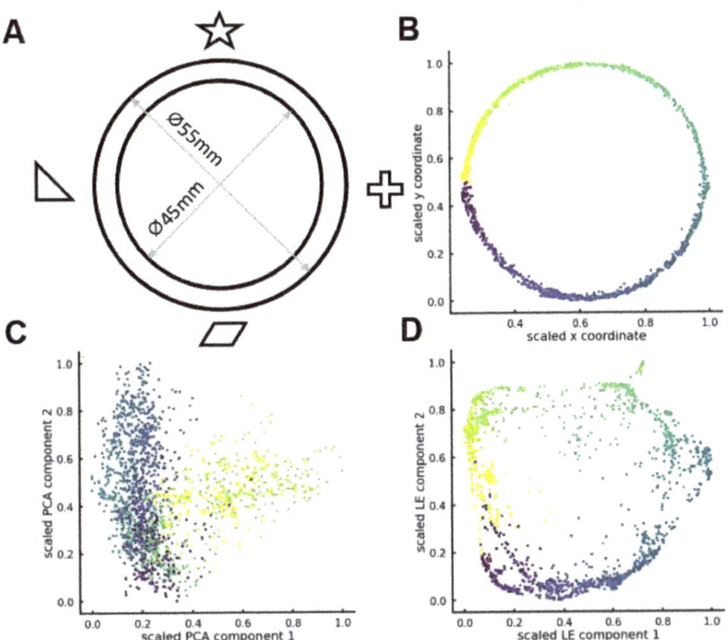

Figure 5. (**A**) Round track with visual cues, (**B**) pointwise track representation, (**C**) first two axes of PCA embedding, (**D**) first two axes of LE embedding (eigenvectors of the graph Laplacian).

3.4. Dependence of Geometry Coding Quality on Population Size

Next, we estimated the quality of the decoding of the space. The reconstruction error of the track space in the embedding (residual variance, RV) decreased with the number of registered cells (Figure 6A). We attribute this to the fact that nonlinear dimensionality reduction is able to distinguish "population" variables from the aggregate activity of many cells.

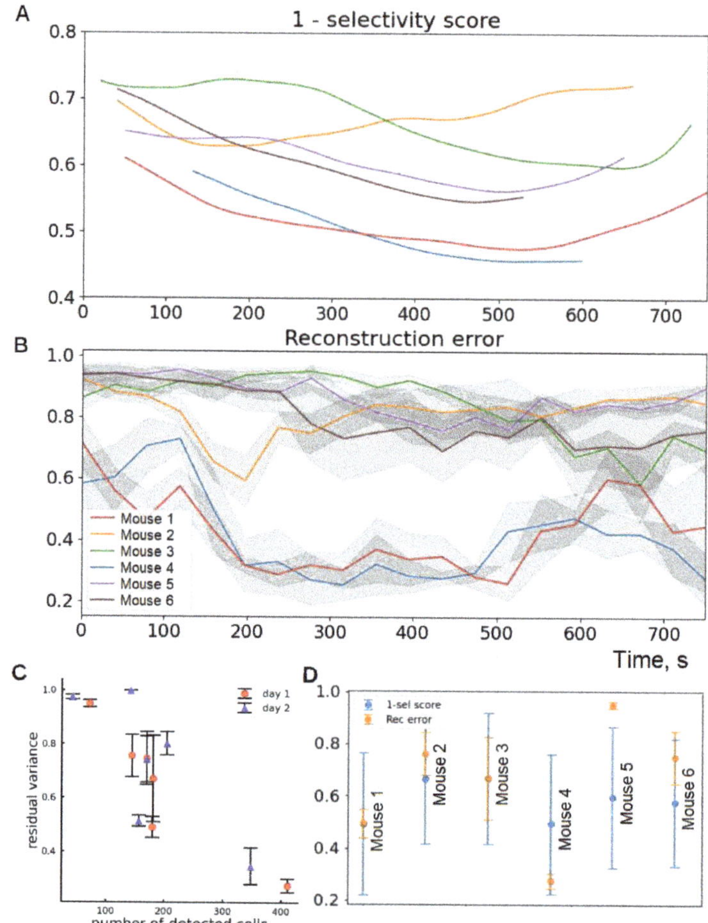

Figure 6. (**A**) Negated mean selectivity score plus one (unselectivity score), which was interpolated, smoothed with running average filter (window length 250 s) and plotted in the time domain. (**B**) Evolution of reconstruction error within the timeline of the 1st day session. Dimensionality reduction was performed for a sliding time window of length 250 s. Shadings represent standard deviations for LE with different graph construction parameters. NC stands for the number of cells registered. (**C**) Residual variance of the embedding depends on the number of detected cells. (**D**) Distribution of mean 1-selectivity score and of mean reconstruction error across different animals. These distributions demonstrate cosine similarity of 0.968 ± 0.047.

3.5. Representation Quality Dynamics over Time

The residual variance of track representation decreased over time as the mice became familiar with the environment explored (Figure 6 B). This effect was stronger the more cells were recorded for a given animal. We attribute this to the gradual formation of a population code in which information about the environment is distributed among many neurons. We compared these data with the negated average selectivity score for each mouse. The mousewise distributions of mean reconstruction error and of mean selectivity score demonstrate cosine similarity of 0.968 (Figure 6D). This result shows the consistency of different measures of quality of spatial coding.

4. Discussion

The main finding of this study is that place representations are promptly formed in mice during free exploration in a completely novel environment. This is consistent with the previous study by Muller et al. [32], where it was shown that place-selective cell firing began within a few minutes in rats exploring a novel environment. Moreover, recent data [8] show a similar distribution of place field forming times, nominated in laps, for mice navigating in a regular manner in a virtual environment. Here, we extend these results to a natural environment with completely free navigation conditions, where mice were capable of arbitrarily choosing the moving direction.

It is known that several types of spatial representation remapping may occur between different navigating sessions, including full, partial and rate remapping [33,34]. However, the retention of the cognitive maps was also observed [3,35]. The ratio of animals with preserved cognitive maps on the 2nd day in our experiments appeared to be consistent with the previous study [7], where it was shown that the global remapping is a stochastic process and several distinct cognitive maps can coexist in the same animal. Importantly, multiple visual cues, distal and proximal, were used in this study, which also is consistent with our experimental design. This could serve as an additional verifier that the mice trained in our paradigm demonstrated normal parameters of spatial representations even in the absence of reinforcement and goal-directed behavior.

The selectivity score dynamics that we observed suggest that the strength of place coding of the imaged CA1 population increases within each session, but without significant cumulation on further sessions. Nevertheless, we observed a reduction in average tuning latency on the second day, when the environment was familiar to the animals. This can be interpreted in the following manner: the strength of place coding starts from similar levels each session but increases faster in a familiar environment than in a novel one. Such dynamics can be associated with a gradual improvement in place coding between trials [36]. However, in this case, one could expect a more robust rise in the selectivity score in mice with retained representations or at least in place cells preserving their place fields, but we have not observed any significant difference. This implies that faster tuning of cognitive maps may be conditioned by some mechanisms not at the level of individual place cells but at the level of the whole CA1 population. Additional research should be done to clarify this question.

It is known that not only place cells may contribute to spatial coding [4,31,37]. By means of a population analysis, we demonstrated that a population of all registered cells as a whole can encode the space of the environment and that the quality of such encoding complements the average selectivity score. This can provide a basis for the estimation of the exact contribution of non-place cells to spatial code by excluding the activity of place cells from population activity, which will be a subject of further analysis.

We have not considered the direction specificity of place cells. It is known that, in one-dimensional tracks, there are direction-specific place fields and one cell can have place fields that are specific in different directions [38,39]. However, given the complete arbitrariness of the animals' trajectories in our paradigm, it is difficult to take the direction into account due to unequal statistics of directions. Given this, we targeted our procedure of place cell detection to omnidirectional place cells, taking into account only continuous statistics of selective cell firing during sequential animals' entrances in the place field in any direction.

One could expect some improvement in spatial code stability evoked by edge, border and object-specific cells [40]. There are no distinct edges and borders within our behavior paradigm, though some cells may be specifically tuned at distal and proximal visual cues around the circular track. However, the evaluation of the exact contribution of such putative cue-specific cells to spatial coding, as well as their precise identification, are obstructed by the overlap of their activity with the activity of "regular" place cells, since cues are integral parts of the environment. Additional modifications to the experimental setup will be required to isolate the contribution of cue-specific cells in further studies.

Due to the dispersion in the number of cells detected across different animals, we checked the consistency of our statistical comparison by excluding mice #8–9 with the lowest number of detected neurons (see Table A1 of Appendix A). As a result, the p-value of the factor day for the comparison of t_spec decreased from a value of $p = 0.0488$ to a value of $p = 0.0436$ (Figure 4C), and the p-values of the comparison of selectivity score between the first and the second visit to a field changed from values of $p = 0.0261$ and $p = 0.0006$ to values $p = 0.0153$ and $p = 0.0003$ for the first and the second session, respectively (Figure 4E). This change did not alter the statistical significance of the results, and therefore we retained mice #8 and 9 in the analysis.

According to [41,42], place codes can be modulated by the exploratory behavior of animals. We did not find any clear behavioral triggers or environmental cues that could activate place cell tuning. Given the relatively high (25%) fraction of immediate early-tuned (at the first visit to a field) place fields, special attention should be paid to the precise registration of all behavior parameters from the first seconds of the mouse's entry into the environment. Nevertheless, at the level of discrete behavior acts, no behavior acts can be completed within such time, which poses a question about some internal states of the network activity leading to the rapid emergence of observed specializations.

Such mechanisms were suggested in studies by Dragoi and Tonegawa [43,44], which reported a possibility that some pre-existing representations supply immediate place coding in a novel environment. However, the contribution of such representations to overall place coding may be accomplished by hippocampal replays between consequent sessions, which may lead to the engagement of so-called "slowly-firing cells" of higher plasticity [45]. Taken together, these approaches may explain the further distinction and completion of novel and familiar environments.

Despite the clear geometric and physical meaning of the LE algorithm utilized in our study, this approach has several drawbacks, which are common for many manifold learning methods. In particular, this method does not scale well with increasing amounts of accumulated data, because its implementation relies on the spectral decomposition of the affinity matrix or related operators, such as the Laplacian or the transition matrix. In general, this leads to a computational complexity of $O(n^a)$, $2.4 < a < 3$, depending on the particular implementation. There is also no natural method for constructing a low-dimensional embedding for the neural activity vector, which was not represented in the initial data and, therefore, did not contribute to the formation of the graph. For this purpose, one has to construct special approximate nonlinear operators [46], which is not always possible. This limits the possibility of using the constructed embedding to analyze new data. To address this issue, we are working on new robust neural network-based methods for the dimensionality reduction of calcium signal data.

5. Conclusions

We have estimated the basic parameters of place cell selectivity within an imaging session at the first and second days of circular maze exploration. On the first day, the mean tuning latency of all place fields in all mice equaled 247 s. On average, place specialization was attained at the seventh visit of an animal to a place field, while 25.1% of place fields were established at their first attendance. On the second day, 3 of 8 mice demonstrated retention of their spatial representation, while 5 of 8 mice did not. In both cases, tuning latency on the second day was significantly lower than on the first day. On each day, the mean selectivity score significantly rose within the session. However, no cumulation was observed on the second day, and the initial and ultimate selectivity scores did not differ significantly between the first and the second day. Moreover, no difference in selectivity score or tuning latency dynamics was detected between the mice that had map retention or underwent remapping, neither at the level of individual cells nor at the level of average values.

Additionally, our nonlinear dimensionality reduction performed on CA1 neuronal activity data revealed the geometry of the environment explored by the mice. The reconstruction error for the six most informative mice on the first day of exploration corresponded to the negated mean selectivity score of these mice.

Taken together, these results reveal the fast emergence and tuning dynamics of place cell codes and demonstrate the applicability of novel calcium indicators NCaMP7 and FGCaMP7 for the light-controlled analysis of neural functions in behaving mice.

Supplementary Materials: The following supporting information can be downloaded at: https://www.mdpi.com/article/10.3390/ijms23020638/s1.

Author Contributions: V.P.S. and K.V.A. designed the experiment; V.P.S. and V.V.P. carried out surgical preparations, in vivo calcium imaging and place cell detection; V.P.S. preprocessed calcium imaging data and behavior data and analyzed selectivity score and tuning dynamics at the level of individual place cells; V.V.P. performed statistical analysis; N.A.P. carried out dimensionality reduction analysis; V.P.S., N.A.P. and K.V.A. wrote the manuscript. All authors revised the manuscript. All authors have read and agreed to the published version of the manuscript.

Funding: This work was supported by the Russian Ministry of Science and Higher Education Project № 075-15-2020-801. The analysis of miniscope calcium activity recordings from the mouse hippocampal neurons during new memory acquisition was supported by the Russian Science Foundation Project № 20-15-00283.

Institutional Review Board Statement: All methods for animal care and all experimental protocols were approved by the National Research Center "Kurchatov Institute" Committee on Animal Care (NG-1/109PR of 13 February 2020) and were in accordance with the Russian Federation Order Requirements N 267 M3 and the National Institutes of Health Guide for the Care and Use of Laboratory Animals.

Data Availability Statement: The data that support the findings of this study and any custom written code are available from the corresponding author upon reasonable request.

Acknowledgments: We kindly acknowledge Fedor V. Subach and Natalia V. Barykina for providing the AAV particles for this study. We are grateful to Sergey K. Nechaev, Alexander S. Gorsky and Ksenia A. Toropova for fruitful discussions. We thank the reviewers for their thorough analysis and quality-improving comments on the manuscript.

Conflicts of Interest: The authors declare no conflict of interest.

Appendix A. Extended Data Tables and Figures

Table A1. Mousewise statistics for each registered session. 1PF Cells, single-field place cells. N spec and T spec, the average number of visit to a place field and correspondent time in seconds, since which this place field is considered tuned. First visit and Last visit sel.sc., the average values of selectivity score at the first and at the last attendance of a place field.

Mouse	Session	Calcium Sensor	Cells	Place Cells	Place Fields	1PF Cells	N Spec	T Spec	First Visit sel.sc.	Last Visit sel.sc.
	1		207	59	63	55	8.63	217.8	0.23	0.38
Mouse 1	2	GCaMP6s	235	24	28	21	8.79	219.2	0.10	0.30
	3		233	30	34	26	5.59	177.8	0.06	0.38
	1		263	90	101	79	10.10	237.3	0.20	0.30
Mouse 2	2	GCaMP6s	297	88	102	74	5.69	146.5	0.23	0.38
	3		228	80	92	68	7.37	196.4	0.19	0.41
	1		320	121	135	107	4.83	170.4	0.34	0.45
Mouse 3	2	GCaMP6s	315	106	114	98	2.21	152.2	0.41	0.44
	3		301	113	127	99	2.14	127.0	0.41	0.36
Mouse 4	1	GCaMP6s	562	257	298	218	3.90	256.4	0.32	0.49
	2		487	247	294	202	4.07	246.6	0.30	0.49
Mouse 5	1	NCaMP7	283	74	82	66	7.99	252.1	0.20	0.35
	2		317	86	96	77	6.58	186.0	0.09	0.46
Mouse 6	1	NCaMP7	200	26	30	22	6.63	245.1	0.15	0.33
	2		189	11	13	9	4.62	208.8	0.09	0.58
Mouse 7	1	GCaMP7f	292	130	174	87	11.26	305.6	0.16	0.31
	2		239	72	81	63	6.70	153.9	0.21	0.35
Mouse 8	1	FGCaMP7	95	10	10	10	6.10	144.4	0.31	0.31
	2		123	35	38	32	5.68	102.4	0.30	0.47
Mouse 9	1	NCaMP7	103	12	19	7	7.26	271.5	0.23	0.35

Table A2. Cell matching and remapping rate statistics between sessions. We considered that the cognitive map retained between sessions in case of the significant similarity of the distribution of place field location shifts between sessions to a normal distribution around zero ($p < 0.05$, Chi-square test).

Mouse	Sessions Compared	Cells Matched	Retention p-Value	Retention of Map
	1 vs. 2	157	0.886	no
Mouse 1	2 vs. 3	179	0.657	no
	1 vs. 3	153	0.839	no
	1 vs. 2	127	0.839	no
Mouse 2	2 vs. 3	138	0.552	no
	1 vs. 3	116	0.006	yes
	1 vs. 2	229	<0.001	yes
Mouse 3	2 vs. 3	217	<0.001	yes
	1 vs. 3	116	<0.001	yes
Mouse 4	1 vs. 2	396	<0.001	yes
Mouse 5	1 vs. 2	208	<0.001	yes
Mouse 6	1 vs. 2	120	0.456	no
Mouse 7	1 vs. 2	69	0.457	no
Mouse 8	1 vs. 2	77	0.237	no

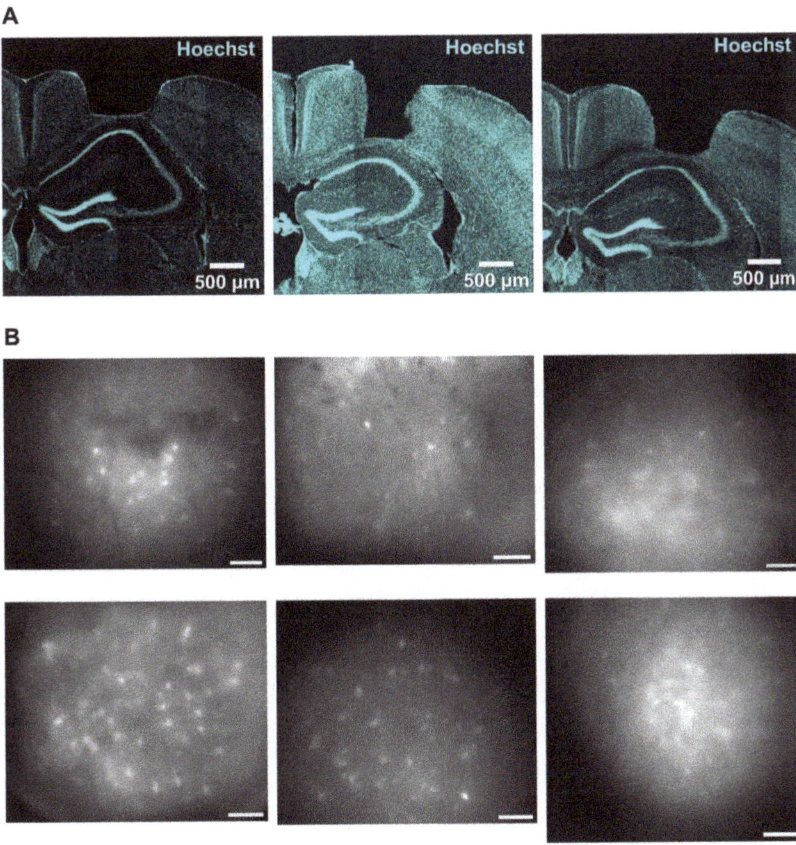

Figure A1. (**A**) Samples of histology sections indicating GRIN lens implantation locations stained with Hoechst. (**B**) Samples of raw NVista HD video screenshots. Scale bars, 100 μm.

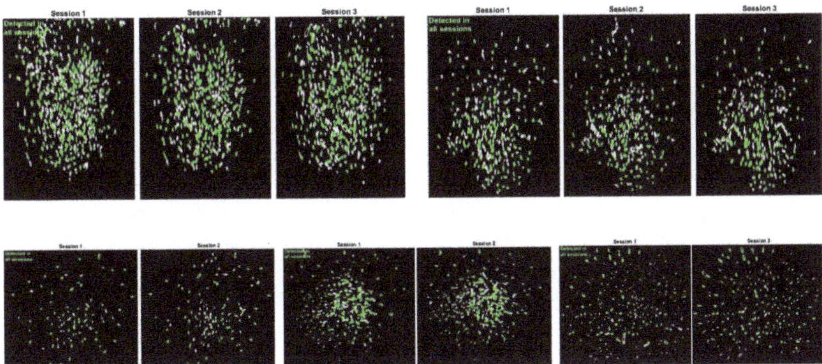

Figure A2. Samples of cell matching between sessions. Matched population is marked in green.

References

1. O'Keefe, J.; Dostrovsky, J. The hippocampus as a spatial map. Preliminary evidence from unit activity in the freely-moving rat. *Brain Res.* **1971**, *4*, 171–175. [CrossRef]
2. O'Keefe, J. Place units in the hippocampus of the freely moving rat. *Exp. Neurol.* **1976**, *51*, 78–109. [CrossRef]
3. Ziv, Y.; Burns, L.; Cocker, E.; Hamel, E.; Ghosh, K.; Kitch, L.; El Gamal, A.; Schnitzer, M. Long-term dynamics of CA1 hippocampal place codes. *Nat. Neurosci.* **2013**, *16*, 264–266. [CrossRef]
4. Rubin, A.; Geva, N.; Sheintuch, L.; Ziv, Y. Hippocampal ensemble dynamics timestamp events in long-term memory. *Elife* **2015**, *4*, e12247. [CrossRef]
5. Kinsky, N.; Sullivan, D.; Mau, W.; Hasselmo, M.; Eichenbaum, H. Hippocampal Place Fields Maintain a Coherent and Flexible Map across Long Timescales. *Curr. Biol.* **2018**, *28*, 3578–3588. [CrossRef]
6. Barnes, C.; Suster, M.; Shen, J.; McNaughton, B. Multistability of cognitive maps in the hippocampus of old rats. *Nature* **1997**, *388*, 272–275. [CrossRef]
7. Sheintuch, L.; Geva, N.; Baumer, H.; Rechavi, Y.; Rubin, A.; Ziv, Y. Multiple Maps of the Same Spatial Context Can Stably Coexist in the Mouse Hippocampus. *Curr. Biol.* **2020**, *30*, 1467–1476. [CrossRef] [PubMed]
8. Dong, C.; Madar, A.; Sheffield, M. Distinct place cell dynamics in CA1 and CA3 encode experience in new environments. *Nat. Commun.* **2021**, *12*, 2977. [CrossRef]
9. Mehta, M.; Barnes, C.; McNaughton, B. Experience-dependent, asymmetric expansion of hippocampal place fields. *Proc. Natl. Acad. Sci. USA* **1997**, *94*, 8918–8921. [CrossRef] [PubMed]
10. Aoki, Y.; Igata, H.; Ikegaya, Y.; Sasaki, T. The Integration of Goal-Directed Signals onto Spatial Maps of Hippocampal Place Cells. *Cell Rep.* **2019**, *27*, 1516–1527. [CrossRef]
11. Hok, V.; Lenck-Santini, P.; Roux, S.; Save, E.; Muller, R.; Poucet, B. Goal-related activity in hippocampal place cells. *J. Neurosci.* **2007**, *27*, 472–482. [CrossRef] [PubMed]
12. Ghosh, K.; Burns, L.; Cocker, E.; Nimmerjahn, A.; Ziv, Y.; Gamal, A.; Schnitzer, M. Miniaturized integration of a fluorescence microscope. *Nat. Methods* **2011**, *8*, 871–878. [CrossRef]
13. Barykina, N.; Sotskov, V.; Gruzdeva, A.; Wu, Y.; Portugues, R.; Subach, O.; Chefanova, E.; Plusnin, V.; Ivashkina, O.; Anokhin, K.; et al. FGCaMP7, an Improved Version of Fungi-Based Ratiometric Calcium Indicator for In Vivo Visualization of Neuronal Activity. *Int. J. Mol. Sci.* **2020**, *21*, 3012. [CrossRef]
14. Subach, O.; Sotskov, V.; Plusnin, V.; Gruzdeva, A.; Barykina, N.; Ivashkina, O.; Anokhin, K.; Nikolaeva, A.; Korzhenevskiy, D.; Vlaskina, A.; et al. Novel Genetically Encoded Bright Positive Calcium Indicator NCaMP7 Based on the mNeonGreen Fluorescent Protein. *Int. J. Mol. Sci.* **2020**, *21*, 1644. [CrossRef]
15. Gallego, J.; Perich, M.; Miller, L.; Solla, S. Neural Manifolds for the Control of Movement. *Neuron* **2017**, *94*, 978–984. [CrossRef]
16. Gallego, J.; Perich, M.; Naufel, S.; Ethier, C.; Solla, S.; Miller, L. Cortical population activity within a preserved neural manifold underlies multiple motor behaviors. *Nat. Commun.* **2018**, *9*, 4233. [CrossRef]
17. Yu, B.; Cunningham, J.; Santhanam, G.; Ryu, S.; Shenoy, K.; Sahani, M. Gaussian-process factor analysis for low-dimensional single-trial analysis of neural population activity. *J. Neurophysiol.* **2009**, *102*, 614–635. [CrossRef]
18. Barykina, N.; Subach, O.; Doronin, D.; Sotskov, V.; Roshchina, M.; Kunitsyna, T.; Malyshev, A.; Smirnov, I.; Azieva, A.; Sokolov, I.; et al. A new design for a green calcium indicator with a smaller size and a reduced number of calcium-binding sites. *Sci. Rep.* **2016**, *6*, 34447. [CrossRef]
19. Barykina, N.; Doronin, D.; Subach, O.; Sotskov, V.; Plusnin, V.; Ivleva, O.; Gruzdeva, A.; Kunitsyna, T.; Ivashkina, O.; Lazutkin, A.; et al. NTnC-like genetically encoded calcium indicator with a positive and enhanced response and fast kinetics. *Sci. Rep.* **2018**, *8*, 15233. [CrossRef] [PubMed]
20. Pnevmatikakis, E.; Giovannucci, A. NoRMCorre: An online algorithm for piecewise rigid motion correction of calcium imaging data. *J. Neurosci. Methods* **2017**, *291*, 83–94. [CrossRef]
21. Lu, J.; Li, C.; Singh-Alvarado, J.; Zhou, Z.; Fröhlich, F.; Mooney, R.; Wang, F. MIN1PIPE: A Miniscope 1-Photon-Based Calcium Imaging Signal Extraction Pipeline. *Cell Rep.* **2018**, *23*, 3673–3684. [CrossRef]
22. Sheintuch, L.; Rubin, A.; Brande-Eilat, N.; Geva, N.; Sadeh, N.; Pinchasof, O.; Ziv, Y. Tracking the Same Neurons across Multiple Days in Ca2+ Imaging Data. *Cell Rep.* **2017**, *21*, 1102–1115. [CrossRef]
23. Lopes, G.; Bonacchi, N.; Frazão, J.; Neto, J.; Atallah, B.; Soares, S.; Moreira, L.; Matias, S.; Itskov, P.; Correia, P.; et al. Bonsai: An event-based framework for processing and controlling data streams. *Front. Neuroinform.* **2015**, *9*, 7. [CrossRef]
24. Grijseels, D.; Shaw, K.; Barry, C.; Hall, C. Choice of method of place cell classification determines the population of cells identified. *PLoS Comput. Biol.* **2021**, *17*, e1008835. [CrossRef]
25. Markus, E.; Barnes, C.; McNaughton, B.; Gladden, V.; Skaggs, W. Spatial information content and reliability of hippocampal CA1 neurons: effects of visual input. *Hippocampus* **1994**, *4*, 410–421. [CrossRef]
26. Belkin, M.; Niyogi, P. Laplacian eigenmaps for dimensionality reduction and data representation. *Neural Comput.* **2003**, *15*, 1373–1396. [CrossRef]
27. Altman, N.; Krzywinski, M. The curse(s) of dimensionality. *Nat. Methods* **2018**, *15*, 399–400. [CrossRef]
28. Chung, F. *Spectral Graph Theory*; American Mathematical Society: Philadelphia, PA, USA, 1997.
29. Tenenbaum, J.; De Silva, V.; Langford, J. A global geometric framework for nonlinear dimensionality reduction. *Science* **2000**, *290*, 2319–2323. [CrossRef]

30. Sotskov, V.; Plusnin, V.; Pospelov, N.; Anokhin, K. The Rapid Formation of CA1 Hippocampal Cognitive Map in Mice Exploring a Novel Environment. In *Advances in Cognitive Research, Artificial Intelligence and Neuroinformatics. Intercognsci 2020. Advances in Intelligent Systems and Computing*; Velichkovsky, B., Balaban, P., Ushakov, V., Eds.; Springer: Cham, Switzerland, 2021; pp. 452–457.

31. Meshulam, L.; Gauthier, J.L.; Brody, C.D.; Tank, D.; Bialek, W. Collective Behavior of Place and Non-place Neurons in the Hippocampal Network. *Neuron* **2017**, *96*, 1178–1191. [CrossRef]

32. Muller, R.; Kubie, J.; Ranck, J. Spatial firing patterns of hippocampal complex-spike cells in a fixed environment. *J. Neurosci.* **1987**, *7*, 1935–1950. [CrossRef] [PubMed]

33. Latuske, P.; Kornienko, O.; Kohler, L.; Allen, K. Hippocampal Remapping and Its Entorhinal Origin. *Front. Behav. Neurosci.* **2018**, *11*, 253. [CrossRef]

34. Lee, J.Q.; LeDuke, D.O.; Chua, K.; McDonald, R.J.; Sutherland, R.J. Relocating cued goals induces population remapping in CA1 related to memory performance in a two-platform water task in rats. *Hippocampus* **2018** *28*, 431–440. [CrossRef]

35. Mankin, E.A.; Sparks, F.T.; Slayyeh, B.; Sutherland, R.J.; Leutgeb, S.; Leutgeb, J.K. Neuronal code for extended time in the hippocampus. *Proc. Natl. Acad. Sci. USA* **2012**, *109*, 19462–19467. [CrossRef]

36. Karlsson, M.; Frank, L. Network dynamics underlying the formation of sparse, informative representations in the hippocampus. *J. Neurosci.* **2008**, *28*, 14271–14281. [CrossRef]

37. Chang, C.; Guo, W.; Zhang, J.; Newman, J.; Sun, S.; Wilson, M. Behavioral clusters revealed by end-to-end decoding from microendoscopic imaging. *bioRxiv* **2021**. [CrossRef]

38. Muller, R.; Bostock, E.; Taube, J.; Kubie, J. On the directional firing properties of hippocampal place cells. *J. Neurosci.* **1994**, *14*, 7235–7251. [CrossRef]

39. McNaughton, B.; Barnes, C.; O'Keefe, J. The contributions of position, direction, and velocity to single unit activity in the hippocampus of freely-moving rats. *Exp. Brain Res.* **1983**, *52*, 41–49. [CrossRef]

40. Scaplen, K.M.; Gulati, A.A.; Heimer-McGinn, V.L.; Burwell, R.D. Objects and landmarks: hippocampal place cells respond differently to manipulations of visual cues depending on size, perspective, and experience. *Hippocampus* **2014**, *24*, 1287–1299. [CrossRef]

41. Kentros, C.; Agnihotri, N.; Streater, S.; Hawkins, R.; Kandel, E. Increased attention to spatial context increases both place field stability and spatial memory. *Neuron* **2004**, *42*, 283–295. [CrossRef]

42. Frank, L.; Brown, E.; Wilson, M. Trajectory encoding in the hippocampus and entorhinal cortex. *Neuron* **2000**, *27*, 169–178. [CrossRef]

43. Dragoi, G.; Tonegawa, S. Preplay of future place cell sequences by hippocampal cellular assemblies. *Nature* **2011**, *469*, 397–401. [CrossRef]

44. Dragoi, G.; Tonegawa, S. Distinct preplay of multiple novel spatial experiences in the rat. *Proc. Natl. Acad. Sci. USA* **2013**, *110*, 9100–9105. [CrossRef] [PubMed]

45. Grosmark, A.; Buzsáki, G. Diversity in neural firing dynamics supports both rigid and learned hippocampal sequences. *Science* **2016**, *351*, 1440–1443. [CrossRef] [PubMed]

46. Bengio, Y.; Paiement, J.-F.; Vincent, P.; Delalleau, O.; Le Roux, N.; Ouimet, M. Out-of-sample extensions for LLE, Isomap, MDS, Eigenmaps, and Spectral Clustering. In Proceedings of the 16th International Conference on Neural Information Processing Systems (NIPS'03), Whistler, BC, Canada, 9–11 December 2003; MIT Press: Cambridge, MA, USA, 200; pp. 177–184.

International Journal of
Molecular Sciences

Article

Reversible Photocontrol of Dopaminergic Transmission in Wild-Type Animals

Carlo Matera [1,2,3], Pablo Calvé [4,†], Verònica Casadó-Anguera [5,†], Rosalba Sortino [1,2,†], Alexandre M. J. Gomila [1,2], Estefanía Moreno [5], Thomas Gener [4], Cristina Delgado-Sallent [4], Pau Nebot [4], Davide Costazza [1], Sara Conde-Berriozabal [6], Mercè Masana [6], Jordi Hernando [7], Vicent Casadó [5], M. Victoria Puig [4] and Pau Gorostiza [1,2,8,*]

[1] Institute for Bioengineering of Catalonia (IBEC), the Barcelona Institute for Science and Technology, 08028 Barcelona, Spain
[2] Biomedical Research Networking Center in Bioengineering, Biomaterials and Nanomedicine (CIBER-BBN), 28029 Madrid, Spain
[3] Department of Pharmaceutical Sciences, University of Milan, 20133 Milan, Italy
[4] Hospital del Mar Medical Research Institute (IMIM), Barcelona Biomedical Research Park, 08003 Barcelona, Spain
[5] Department of Biochemistry and Molecular Biomedicine, Faculty of Biology, Institute of Biomedicine, University of Barcelona, 08028 Barcelona, Spain
[6] Department of Biomedical Sciences, Faculty of Medicine and Health Sciences, Institute of Neuroscience, University of Barcelona, IDIBAPS, CIBERNED, 08036 Barcelona, Spain
[7] Department of Chemistry, Autonomous University of Barcelona (UAB), 08193 Cerdanyola del Vallès, Spain
[8] Catalan Institution for Research and Advanced Studies (ICREA), 08010 Barcelona, Spain
* Correspondence: pau@icrea.cat
† These authors contributed equally to this work.

Citation: Matera, C.; Calvé, P.; Casadó-Anguera, V.; Sortino, R.; Gomila, A.M.J.; Moreno, E.; Gener, T.; Delgado-Sallent, C.; Nebot, P.; Costazza, D.; et al. Reversible Photocontrol of Dopaminergic Transmission in Wild-Type Animals. *Int. J. Mol. Sci.* **2022**, *23*, 10114. https://doi.org/10.3390/ijms231710114

Academic Editor: Giuseppe Di Giovanni

Received: 3 August 2022
Accepted: 31 August 2022
Published: 4 September 2022

Publisher's Note: MDPI stays neutral with regard to jurisdictional claims in published maps and institutional affiliations.

Abstract: Understanding the dopaminergic system is a priority in neurobiology and neuropharmacology. Dopamine receptors are involved in the modulation of fundamental physiological functions, and dysregulation of dopaminergic transmission is associated with major neurological disorders. However, the available tools to dissect the endogenous dopaminergic circuits have limited specificity, reversibility, resolution, or require genetic manipulation. Here, we introduce azodopa, a novel photoswitchable ligand that enables reversible spatiotemporal control of dopaminergic transmission. We demonstrate that azodopa activates D_1-like receptors in vitro in a light-dependent manner. Moreover, it enables reversibly photocontrolling zebrafish motility on a timescale of seconds and allows separating the retinal component of dopaminergic neurotransmission. Azodopa increases the overall neural activity in the cortex of anesthetized mice and displays illumination-dependent activity in individual cells. Azodopa is the first photoswitchable dopamine agonist with demonstrated efficacy in wild-type animals and opens the way to remotely controlling dopaminergic neurotransmission for fundamental and therapeutic purposes.

Keywords: azobenzene; behavior; brainwave; dopamine; GPCR; in vivo electrophysiology; optogenetics; optopharmacology; photochromism; photopharmacology; photoswitch; zebrafish

1. Introduction

Dopamine receptors (DARs) are members of the class A G protein-coupled receptor (GPCR) family and are prominent in the vertebrate central nervous system (CNS). Their primary endogenous ligand is the catecholaminergic neurotransmitter dopamine, a metabolite of the amino acid tyrosine. Among the many neuromodulators used by the mammalian brain to regulate circuit function and plasticity, dopamine (Figure 1a) stands out as one of the most behaviorally powerful [1,2]. Dopaminergic neurons are critically involved in diverse vital CNS functions, including voluntary movement, feeding, reward, motivation, sleep, attention, memory, and cognition. The extracellular concentration of dopamine oscillates following day/night cycles and plays important physiological roles in the regulation of olfaction, retinal function [3], and circadian rhythms [4]. Disentangling these diverse components of DAR signaling and dopaminergic transmission is an unmet need both of basic research and medicine, because their abnormal function leads to complex medical conditions such as Parkinson's and Huntington's diseases, schizophrenia, attention deficit hyperactivity disorder, Tourette's syndrome, drug abuse, and addiction [1]. To date, five subtypes of DARs have been cloned: D_1, D_2, D_3, D_4, and D_5. Based on their coupling to either $G_{\alpha s,olf}$ proteins or $G_{\alpha i/o}$ proteins, which, respectively, stimulate or inhibit the production of the second messenger cAMP, DARs are classified as D_1-like receptors (D_1, D_5) or D_2-like receptors (D_2, D_3, D_4). However, both classes are known to signal through multiple pathways. Targeting these receptors using specific agonists and antagonists allows to modulate dopaminergic transmission and dopamine-dependent functions. Indeed, hundreds of compounds interfering with the dopaminergic system have been developed, and many of them are clinically used to treat various disorders. They also constitute pharmacological tools to study the role of dopamine in synaptic and neural circuits [5,6] as well as the mechanisms underlying dopamine-related debilitating conditions [7]. However, conventional ligands cannot differentiate among specific neuronal sub-populations in heterogeneous brain regions where multiple neuronal subtypes exist, thus potentially activating DARs that mediate distinct or even opposing physiological functions [2]. For this reason, the lack of circuit selectivity is a confounding element in basic research and likely cause of the poor safety as well as efficacy of many dopaminergic drugs. Hence, methods to activate DARs noninvasively with high spatiotemporal resolution are required both for research and therapeutic purposes.

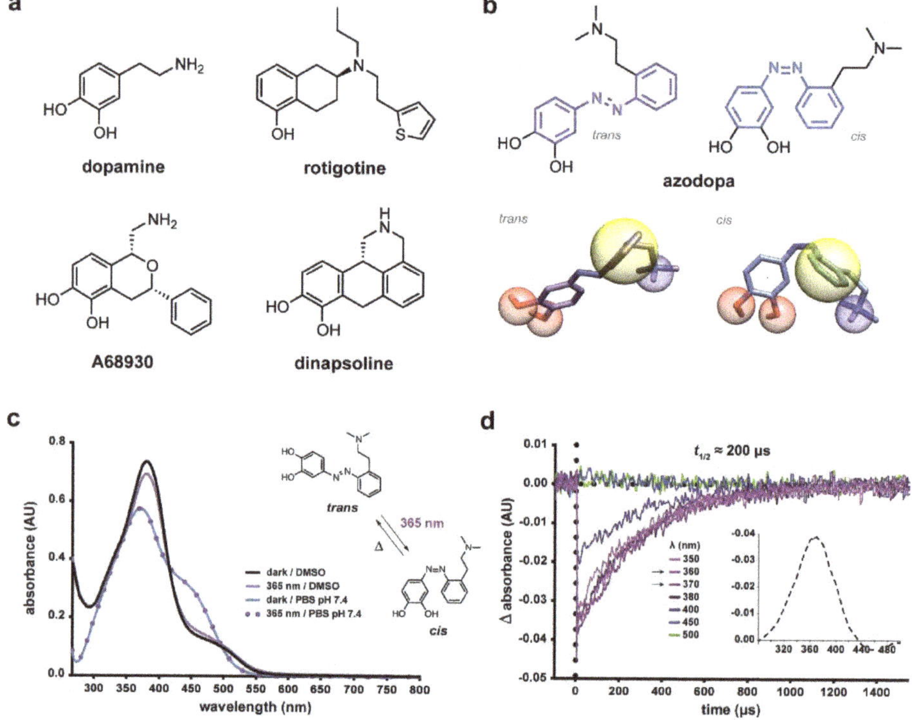

Figure 1. Design, structure, and photochromism of azodopa. (**a**) Chemical structure of dopamine and representative (semi)rigidified derivatives: rotigotine (nonselective agonist), A68930 (D_1 agonist), and dinapsoline (D_1 agonist). (**b**) 2D and 3D chemical structure of the photochromic dopamine ligand

azodopa (*trans* and *cis* isomers). Essential pharmacophoric features for D_1 receptor binding are highlighted: blue spheres represent cationic site points and *H*-bond donors, i.e., the protonated amino function that can form a salt bridge and a hydrogen bond with Asp103 and Ser107, respectively; red spheres represent *H*-bond acceptors, i.e., the hydroxyl groups of the catechol ring that can interact with Ser198, Ser199, and Ser202; yellow spheres represent hydrophobic elements, i.e., the aromatic ring that can form π–π interactions with Phe288 and Phe289. The mutual position and orientation of such pharmacophoric features in the receptor-bound conformation should affect binding affinity and efficacy of the ligands. (**c**) Photochromic behavior of azodopa (50 μM) studied with steady-state spectroscopy in aqueous (PBS, pH 7.4) and organic (DMSO) solutions. As lifetimes of *cis*-hydroxyazobenzenes are very short in polar protic solvents, no changes in the absorption spectrum of aqueous solutions of azodopa could be observed after illumination with 365 nm light (3 min). (**d**) Photochromic behavior of azodopa (30 μM) investigated by transient absorption spectroscopy in water (only representative traces are shown for the sake of clarity; see Figure S7 for the full experiment). Transient absorption time traces were measured at different wavelengths upon excitation of *trans*-azodopa with a 5 ns pulsed laser at λ = 355 nm (3 mJ/pulse energy) and 25 °C. Thermal relaxation half-life of the *cis* isomer (200 μs) was estimated by applying an exponential one-phase decay model (GraphPad Prism 6). Inset: Transient absorption spectrum of *trans*-azodopa upon pulsed irradiation at λ = 355 nm recorded at t = 0 μs. X-values represent wavelength (nm), Y-values represent ΔA (arbitrary units, AU).

Pursuing a traditional pharmacological approach would hardly pay off in such a physiological scenario, because a drug generally affects its target in multiple CNS regions at once and its effect is slowly reversible. Light is an unparalleled input signal to noninvasively manipulate biological systems in precisely designated patterns, and photopharmacology [8–10], which relies on molecular photoswitches to regulate bioactive compounds, has already been successful in GPCRs [11–16], ion channels [17,18], and enzymes [19,20]. The possibility of using light as an external stimulus to manipulate specific populations of dopaminergic neurons has generated enormous interest in neurobiology since the invention of optogenetics [21,22], in particular to elucidate the basis of complex behavioral and cognitive processes [23,24]. However, the achievements of optogenetics rely on the overexpression of exogenous proteins that lack critical aspects of endogenous GPCR signaling, including their native ligand binding sites, downstream molecular interactions, and other elements that can affect receptor dynamics. Thus, current optogenetic tools may provide partial, or sometimes inaccurate, insights into biological processes. In addition, the application of genetic manipulation techniques to human subjects is still importantly hampered by safety (possible immune responses against the gene transfer), regulatory, and economic issues. Complementary approaches to address some of these limitations have been proposed. Trauner, Isacoff, and collaborators developed light-gated DARs through a combined chemical–genetic method in which a weakly photoswitchable ligand was tethered via a maleimide–thiol conjugation to a genetically engineered cysteine residue at the target receptor in order to improve photocontrol [25]. They also used a genetically targeted membrane anchor to tether a dopaminergic ligand via SNAP-tag labeling [15]. Unlike the exogenous optogenetics tools described above, light-gated receptors bear a single-point mutation and provide a nearly physiological study system, but they still require gene delivery and overexpression. Genetics-free methods have also been described. For instance, Etchenique, Yuste, and collaborators developed a caged dopamine compound based on ruthenium–bipyridine chemistry and used it to activate dendritic spines with two-photon excitation [26]. More recently, Gmeiner et al. described two caged DAR antagonists that can serve as valuable tools for light-controlled blocking of D_2/D_3 receptors [27]. However, notwithstanding its virtues, uncaging is an irreversible chemical process, while a photopharmacological modulation based on reversible, byproduct-free molecular photoswitches has important advantages in vivo. Noteworthy in this regard is the work published by König and collaborators, who developed a set of photochromic small molecules by incorporation of dithienylethenes and fulgides into known dopamine receptor ligands [28]. Two of those compounds, named **29** and **52**, showed an interesting isomer-dependent (open vs.

closed form) efficacy at activating D_2 receptors, although neither in situ photoswitching nor in vivo validation of their effects were reported. We present here the design, synthesis, and photopharmacological characterization of the first chemical tool that enables the reversible photocontrol of native dopamine receptors in wild-type animals. Notably, it can be used to separate the retinal component of dopaminergic neurotransmission in zebrafish and manipulate brain waves in mice.

2. Results

2.1. Design Strategy and Synthesis

Aryl azo compounds, especially azobenzenes, have emerged as the photoswitches of choice in photopharmacology because of their physical and chemical properties, which make them especially suitable for biological applications [29,30]. An elegant strategy for the incorporation of an azobenzene into a bioactive ligand [11] relies on the isosteric replacement of the two-atom linker between the two aromatic rings with a diazene unit (–N=N–) [11,20,31–33], which entails minimal perturbation of pharmacophore and drug-like properties, thus accounting for the success of this so-called azologization approach [31]. However, to the best of our knowledge, this strategy is not applicable to any of the known dopamine agonists, since the scaffold of an isomerizable aryl azo compound is not directly conceivable in their structure.

In the quest for a freely diffusible drug-like dopaminergic photoswitch, we noticed that most agonists (especially to D_1 receptors, see Figure 1a) are rigid or semi-rigid, conformationally restrained structures in which essential pharmacophoric features are held in their mutual position [34]. Indeed, two main routes can be identified in the early development of dopamine agonists, namely, the rigidification of the dopamine molecule and the dissection of apomorphine, one of the first potent dopamine agonists to be found [35]. Dopamine rigidification led to the discovery of potent agonists, whereas the apomorphine de-rigidification generally reduced its efficacy [36,37]. It stands then to reason that governing the geometry of such structures would enable the control of their biological effects. This could be achieved by "building" upon a semi-rigid and photoisomerizable molecular frame the structural elements required for DAR activation and using light as external control signal. This approach can be likened to a "photoswitch decoration", in which pharmacophore groups are introduced into the structure of a molecular photoswitch to design a light-controlled bioactive compound. Common features in the dopamine agonist pharmacophore model are the following: (I) a cationic site point (an amino group) that forms a salt bridge with an aspartic acid residue in the receptor's third transmembrane helix (TM3); (II) one or two hydrogen bond acceptor/donor sites (e.g., hydroxy groups), that interact with serine residues in TM5; and (III) an aromatic ring system that takes part in π–π interactions with hydrophobic residues in TM6 [38–40]. On these grounds, we devised a photoswitchable dopamine agonist by decorating an azobenzene molecule with two hydroxyl substituents (II) on one phenyl ring, and an amino group (I) connected through a short linker to the other phenyl ring (III) (Figure 1b). This molecule, that we named azodopa, carries the main DAR binding determinants and enables to change their relative position upon photoisomerization (Figure 1b). Azodopa was synthesized in two steps via an azo-coupling reaction between commercially available 2-(2-(dimethylamino)ethyl)aniline and 1,2-dihydroxybenzene (see supplementary Scheme S1 and Supplementary Materials, SM, for detailed synthetic procedures and physicochemical characterization).

2.2. Photochemical Characterization

An essential condition for azodopa to function as a photoswitchable dopamine ligand is that it photoisomerizes between two different configurations (*trans* and *cis*). We first characterized it by steady-state UV–Vis absorption spectroscopy (Figure 1c), which revealed an absorption maximum at 370 nm in aqueous solution (PBS, pH 7.4) and at 380 nm in DMSO. It is known that the thermal *cis*-to-*trans* isomerization of 4-hydroxyazobenzenes follows a very fast kinetics in polar protic solvents via solvent-assisted proton transfer tautomerization, whereas it proceeds more slowly in aprotic and nonpolar solvents [41]. In agreement, no changes in the absorption spectrum of azodopa could be detected with a conventional UV–Vis spectrophotometer upon illumination at 365 nm in aqueous buffer because of the short lifetime of its *cis* isomer in this medium, likely because of the presence of a non-negligible concentration of azonium cation at physiological pH and/or the formation of hydrogen-bonded complexes with the solvents that accelerate the thermal *cis*-to-*trans* isomerization via tautomerization, and only a small change could be recorded in a polar but aprotic solvent such as DMSO, where the tautomerization is still possible, although less pronounced (Figure 1c). No measurements were performed in nonpolar solvents (e.g., toluene) because of the very poor solubility of our compound in this kind of media. We next determined the optimal isomerization wavelength and lifetime of *cis*-azodopa by means of transient-absorption spectroscopy with ns-time resolution. Upon pulsed excitation of azodopa in aqueous media with UV and violet light, an instantaneous and remarkable decrease of the absorption signal was detected, which can be ascribed to the depletion of the *trans* ground electronic state due to the photoinduced formation of the corresponding *cis* isomer (Figure 1d) [41,42]. The initial absorption value was quickly recovered because of the fast thermal *cis*-to-*trans* back-isomerization which restores the initial concentration of the *trans* isomer with an estimated half-life of about 200 µs (Figure 1d). Thus, since conversion to the *trans* form occurs almost immediately after turning the light off, we performed all biological experiments under continuous illumination. Overall, our absorption spectroscopy studies showed that azodopa undergoes *trans*-to-*cis* photoisomerization upon illumination with UV and violet light and it spontaneously reverts to its full *trans* isomer in a fraction of a millisecond once the light is switched off. Other mechanisms, such as excited-state intramolecular proton transfer, could also play a role. In any case, azodopa should allow fast regulation of DAR activity using a single illumination source to induce *trans*-to-*cis* isomerization.

2.3. In Vitro Pharmacology

We next tested the effects of azodopa on D_1-like receptors, for which freely diffusible photoswitchable agonists have not been reported. First, we evaluated the binding affinity of azodopa in mammalian D_1-like and D_2-like receptors using competitive radioligand binding experiments performed either in the dark or under illumination at 365 nm (see SM for details) [43,44]. Azodopa displayed a good capacity to bind to D_1-like and D_2-like receptors, with higher affinity in the *trans* configuration in both cases. In particular, for D_1-like receptors, we calculated an almost fourfold decrease in affinity at 365 nm (K_d^{dark} = 600 nM, $K_d^{365\ nm}$ = 2200 nM) (supplementary Figure S8 and Table S1). Since the activation of D_1-like receptors promotes cyclic adenosine monophosphate (cAMP) formation and phosphorylation by ERK1/2, we also investigated the ability of azodopa to behave as a D_1-like agonist by studying cAMP accumulation (Figure 2a) and ERK1/2 phosphorylation (Figure 2b) in cells overexpressing human D_1 receptors, both in the dark and under continuous illumination at 365 nm (to compare the "full *trans*" vs. the "*cis*-enriched" form, respectively). Intracellular cAMP accumulation was measured in HEK-293T cells transiently transfected with D_1 receptors in a time-resolved fluorescence resonance energy transfer (TR-FRET) assay. We found that azodopa induced cAMP accumulation in a dose- and light-dependent manner (Figure 2a). In particular, we observed substantial differences between the two forms at 5 µM and 10 µM, with the *trans* form being significantly more effective at inducing cAMP production than the *cis*-enriched form at the same concentrations. When cells were co-treated with the D_1-like antagonist SKF83566 (300 nM), the effect of azodopa was largely reduced or even disappeared, indicating that azodopa activates D_1 receptors. In cAMP assays performed in non-transfected HEK-293T cells used as negative control, azodopa (10 µM), dopamine (1 µM), and SKF38393 (300 nM) did not produce any effect (Figure S9). ERK1/2 phosphorylation was measured in HEK-293T cells transiently transfected with D_1 by Western blot analysis using phospho-ERK1/2 antibody. The application of azodopa (10 µM) promoted ERK1/2 phosphorylation to an extent that was significantly different between the two conditions (dark vs. light). The full *trans* form displayed an efficacy 3.6-fold greater than the *cis*-enriched form (Figure 2b), in agreement with the observations in cAMP accumulation assays. Again, co-treatment of the cells with SKF83566 abolished azodopa responses, showing that the effects observed were mediated by D_1. Although we did not thoroughly characterize the pharmacological activity profile, azodopa displayed D_1-mediated photoactivity (blocked by SKF83566) also in zebrafish (see later).

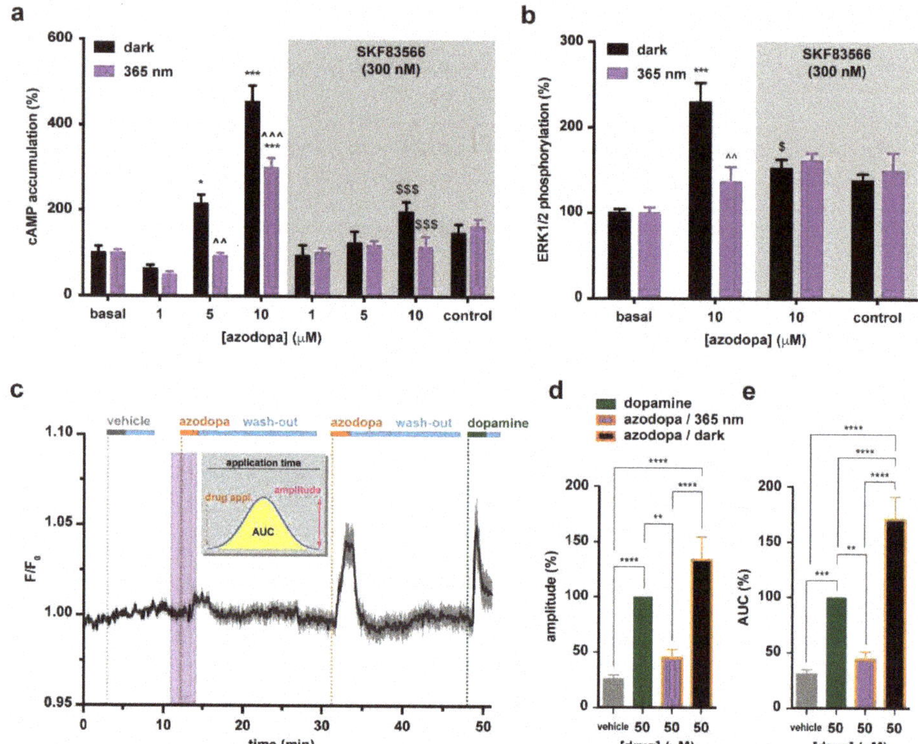

Figure 2. In vitro pharmacological characterization of azodopa. (**a**) Effect on D_1-mediated adenylyl cyclase activation. cAMP accumulation experiments in HEK-293T cells transiently transfected with D_1 and treated with different concentrations of azodopa, in the dark (black bars) or under illumination (purple bars), in the presence (gray area) or not (white area) of a D_1-like receptor antagonist (SKF83566). Values are represented in percentage vs. basal levels of cAMP. Data are mean $\pm$ S.E.M. (6 experiments performed in quadruplicate). Statistical differences were analyzed by two-way ANOVA followed by Tukey's post hoc test (*** $p < 0.001$ vs. basal; * $p < 0.05$ vs. basal; ^^^ $p < 0.001$ vs. dark; ^^ $p < 0.01$ vs. dark; \$\$\$ $p < 0.001$ vs. controls non-pretreated with the antagonist). (**b**) Effect on D_1-mediated ERK1/2 activation. ERK1/2 phosphorylation was determined in HEK-293T cells transiently transfected with D_1 and treated with different concentrations of azodopa, in the dark (black bars) or under illumination (purple bars), in the presence (gray area) or not (white area) of a D_1-like antagonist (SKF83566). Values are represented in percentage vs. basal levels of ERK1/2 phosphorylation. Data are mean $\pm$ S.E.M. (3 or 4 experiments performed in triplicate or quadruplicate). Statistical differences were analyzed by two-way ANOVA followed by Tukey's post hoc test (*** $p < 0.001$ vs. basal; ^^ $p < 0.01$ vs. dark; \$ $p < 0.05$ vs. controls non-pretreated with the antagonist). (**c–e**) Effect on D_1-mediated intracellular calcium release compared to dopamine. (**c**) Real-time calcium imaging response (averaged traces, black line, n = 24 cells) in HEK-293T cells co-expressing D_1 receptors and R-GECO1 as calcium indicator. Traces were recorded upon direct application of azodopa (50 µM, orange bars) in the dark (white area) and under illumination (purple area). Shadow

represents "± S.E.M.". Gray and green bars indicate the application of vehicle (control) and dopamine (reference agonist), respectively. Light blue bars indicate wash-out periods. See example frames and raw data traces of individual cells in supplementary Figure S10, and supplementary Video S1 for the entire movie. Two values of the calcium responses generated by azodopa were calculated (Origin 8 software) and compared: the peak amplitude $\Delta F/F_0$ (**d**), calculated as the difference between the maximal and the minimal intensity of each response (**** $p < 0.0001$ for vehicle vs. dopamine; **** $p < 0.0001$ for vehicle vs. azodopa/dark; **** $p < 0.0001$ for azodopa/365 nm vs. azodopa/dark; ** $p = 0.035$ for dopamine vs. azodopa/365 nm), and the area under the curve (*AUC*) (**e**), calculated as the integral over the entire application time of vehicle or drugs (**** $p < 0.0001$ vehicle vs. dopamine; **** $p < 0.0001$ for vehicle vs. azodopa/dark; **** $p < 0.0001$ for dopamine vs. azodopa dark; **** $p < 0.0001$ for azodopa/365 nm vs. azodopa/dark; *** $p = 0.001$ for vehicle vs. dopamine; ** $p = 0.0025$ for dopamine vs. azodopa/365 nm). Data are mean ± S.E.M. (n = 40 cells from 3 independent experiments). Data were normalized over the maximum response obtained with the saturating concentration of dopamine (50 µM) and were analyzed by one-way ANOVA followed by Tukey's post hoc test for statistical significance. All statistical analyses (panels (**a,b,d,e**)) were performed with GraphPad Prism 6.

It is known that D_1 receptors are also linked to other second messenger systems. These include receptor-mediated activation of phospholipase C (G_q coupling) to generate inositol 1,4,5-trisphosphate (IP_3) which participates in phosphoinositide turnover and calcium-regulated signaling pathways in the brain [45]. IP_3 receptors are mainly located in the endoplasmic reticular membrane where IP_3 can mobilize Ca^{2+} from intracellular stores [45]. Thus, as a complementary method to characterize azodopa in vitro, we performed Ca^{2+}-imaging experiments in HEK-293T cells co-expressing D_1 receptors and R-GECO1 as calcium indicator, and used dopamine as a control (Figure 2c–e, supplementary Figures S10 and S11, supplementary Video S1). Azodopa (50 µM) was applied to the cells both in the dark and under illumination with UV light. Dopamine (50 µM) was tested as reference agonist. Robust increases of intracellular calcium were observed by the application of azodopa in the dark (pure *trans* isomer), whereas only weak increases were recorded when azodopa was applied under illumination (Figure 2c and Figure S10). Calcium responses were quantified and compared by peak amplitude and area under the curve (AUC). The analysis of these parameters showed that *trans*-azodopa stimulates the release of intracellular calcium (similar to dopamine) and this effect is abolished under illumination. In this experiment, *trans*-azodopa displayed significantly higher efficacy than dopamine (Figure 2d,e). No responses were observed in control experiments in HEK cells (n = 25) not expressing D_1, thus confirming that the calcium oscillations were elicited by a specific interaction at this receptor (Figure S11). Moreover, we verified that *trans*-azodopa fully recovers its efficacy after long (60 min) exposure to 365 nm light, demonstrating that its effects are reversibly photodependent and are not due to an irreversible photodegradation of the molecule (Figures S20 and S21). Our results in calcium imaging experiments suggest that *trans*-azodopa activates the G_q/phospholipase C pathway, in addition to the canonical G_s/adenylyl cyclase pathway already investigated. Such intriguing behavior has been described also for other dopamine agonists [45].

Taken as a whole, the results from our in vitro experiments illustrated in Figure 2 show that *trans*-azodopa is a full agonist at D_1-like receptors and that its effects can be partially or completely switched off with light. The reduction of azodopa efficacy under illumination can be attributed to the photoisomerization process which decreases the partial concentration of the *trans* isomer at the target receptor and/or disrupts the ligand interaction(s) at the binding pocket.

2.4. Behavioral Effects in Zebrafish Larvae

The promising photopharmacological profile of azodopa prompted to use it in vivo to modulate dopaminergic neurotransmission. For that purpose, we designed a behavioral assay to record and quantify the locomotor activity of living animals as a function of drug concentration and illumination. It is known that dopamine plays a pivotal role in motor control in humans, as it activates striatal direct pathway neurons that directly project to the output nuclei of the basal ganglia through D_1 receptors, whereas it inhibits striatal indirect pathway neurons that project to the external pallidum through D_2 receptors [46]. We chose zebrafish larvae (*Danio rerio*) as animal model for our experiments because of their transparency, which facilitates the delivery of light, and their morphological, genetic, and behavioral similarity to higher vertebrates [10,47–52]. Indeed, a functional nervous system is established after only 4–5 days of embryonic development in zebrafish, enabling them to perform complex behaviors such as swimming and exploratory activity. In particular, the major dopaminergic pathways in mammals are also represented in the zebrafish brain, and homologous receptors for most of the mammalian subtypes have been identified in these animals. Humans and zebrafish share 100% of the amino acids in the binding site for D_1 and D_3 receptors, and 85–95% for D_2 and D_4 receptors, and generally similar effects are observed for dopaminergic ligands in zebrafish and in mammals [51]. As a rule of thumb, dopamine receptor agonists increase the locomotor activity, whereas antagonists decrease it [53]. However, disentangling the action of drugs on the multiple components of dopaminergic transmission (including different brain regions [2], the regulation of retinal function [3], and of circadian rhythms [4]) constitutes an unmet need of pharmacology and medicine. Therefore, we set to test azodopa in vivo and to take advantage of photocontrolling dopaminergic responses.

Zebrafish larval movements were tracked using a ZebraBox device for automated behavioral recording. Zebrafish larvae at 6 days post-fertilization (6 dpf) were randomly divided into control (vehicle) and treatment groups (with azodopa added to water). Each individual was placed in a separate well of a 96-well plate. Our setup allowed exposing the animal to controlled cycles of dark and 365 nm UV illumination, using the following protocol: dark (20 min, for adaptation), UV light (30 s), dark (20 min), and then four cycles of UV light (30 s) and dark (5 min) (see SM for details and supplementary Video S2). In order to identify alterations in behavior, we measured multiple properties of locomotor activity to determine the activity level [54]. In particular, we focused on fast movements and measured swimming distances and duration of fast swimming, defined as speed ≥ 6 mm·s^{-1}.

We first studied a wide concentration range (spanning from 1 nM to 1 mM azodopa) to determine if behavioral effects could be observed without signs of acute toxicity. No significant differences in locomotor activity were observed between the control group and the treatment groups up to a concentration of 10 µM, neither in the dark nor upon illumination (Figure S12). A strong increase in swimming activity was recorded at 1 mM, but the effect ceased after 30 min, possibly because the fish were exhausted. We observed the most interesting alterations of the behavioral profile at 100 µM (Figures 3a–d and S12). The changes in the swimming activity over time for this group and the control group are plotted in Figure 3a (full experiment represented in Figure S12; see also Video S2). We detected a progressive increase in activity for 30–40 min and relatively small photoresponses (e.g., at 20 min in Figure 3a) which is often due to the slow uptake of the drug in the fish [14]. After this period, animals treated with 100 µM azodopa displayed a high swimming activity in the dark that was sustained for the whole duration of the experiment, and that was abolished during each period of UV illumination (30 s bouts between minute 40 and 60 in Figure 3a). These results agree with the intracellular signaling photoresponses observed in vitro and confirm a *trans*-on/*cis*-off dopamine agonist profile. An averaged time course (between 40–41.5 min, integrated every 5 s) of the swimming activity is magnified in Figure 3b. Representative trajectories of individual fish in wells containing the vehicle or 100 µM azodopa in the dark (40–40.5 min) and under illumination (40.5–41 min) are shown in Figure 3c, where green and red lines indicate slow (<6 mm·s^{-1}) and fast (≥ 6 mm·s^{-1}) swimming periods, respectively (see Figure S13 for the entire plate and SM for experimental details and analysis). Figure 3d shows the quantification and statistical analysis of the total distances swum by the control group and the treatment group (100 µM azodopa) during the last four cycles, namely, once the maximum effect of the drug was reached and maintained (30 s pre-illumination, 30 s illumination, and 30 s post-illumination periods).

Fish treated with azodopa covered a distance three times longer than the control group during the same dark periods, and this effect was blocked under illumination. Interestingly, we also noticed a reduction in swimming activity during the first 30 s of dark after each illumination pulse in the treatment group, likely because the fish needed some time to recover the maximum level of activity. The differences between the total distances swum by the treatment group during the 30 s periods of dark (pre- or post-illumination) and the 30 s periods of illumination were also statistically significant (Figure 3d).

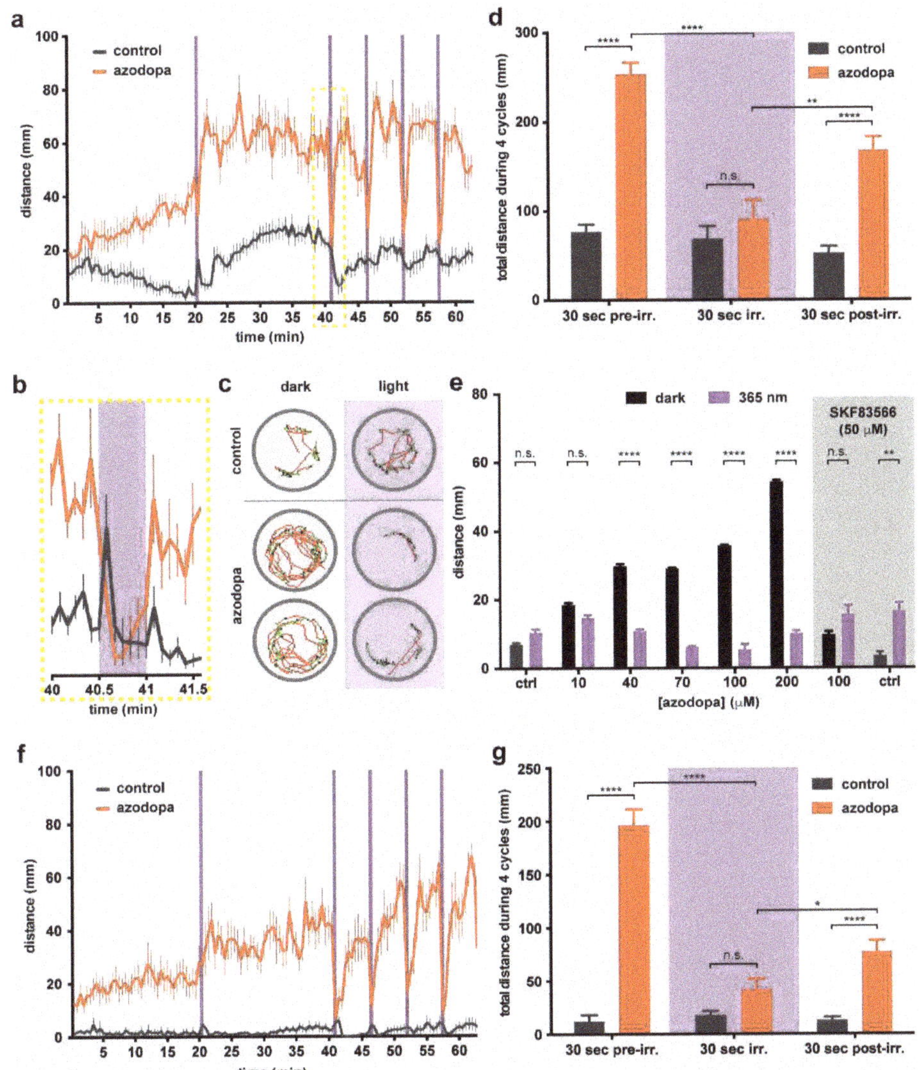

Figure 3. Behavioral effects of azodopa in zebrafish. (a–e) Experiments with normal zebrafish. **(a)** Swimming activity (distance/time) in larvae exposed to vehicle (control, gray line) or 100 µM azodopa (treatment, orange line) in the dark (white areas) or under illumination with 365 nm light (purple bars). Data are mean ± S.E.M. (n = 11–12 individuals/group). **(b)** Representative time frame (40–41.5 min) of the swimming activity integrated every 5 s, showing how the effect of azodopa can be completely shut down upon illumination. The spike of activity observed for the control group

upon illumination represents the startle response to the light stimulus. (**c**) Exemplary trajectories of individual larvae in one well containing the vehicle and two wells containing 100 μM azodopa in the dark (40–40.5 min) and under illumination (40.5–41 min). Green lines and red lines indicate slow and fast swimming periods, respectively. The remarkable and light-dependent difference in behavior between untreated and azodopa-treated fish can be appreciated by observing these trajectories and the supplementary Video S2. (**d**) Quantification of the total distances swum by the control group (vehicle) and the treatment group (100 μM azodopa) during 4 consecutive cycles of illumination (30 s) and dark (30 s before and after illumination). Data are mean ± S.E.M. (n = 11–12 individuals/group) and were analyzed by two-way ANOVA followed by Tukey's post hoc test (**** $p < 0.0001$; ** $p = 0.0037$). (**e**) Dose–response study of azodopa (white area) and effect of a co-administered D_1-like antagonist (gray box). Different groups of larvae were exposed to increasing concentrations of azodopa. For quantification, the average distance swum by each group during the last 4 consecutive dark–light cycles (30 s integration) was considered. The graph shows that *trans*-azodopa (black bars) increases the fish locomotor activity in a dose-dependent manner, but its effects are abolished by the co-administration of a potent and selective D_1-like antagonist (SKF83566, 50 μM). Data are mean ± S.E.M. (n = 12 individuals/group) and were analyzed by two-way ANOVA followed by Sidak's post hoc test (**** $p < 0.0001$; ** $p = 0.0063$). (**f,g**) Experiments with blinded zebrafish. (**f**) Swimming activity (distance/time) in larvae exposed to vehicle (control, gray line) or 100 μM azodopa (treatment, orange line) in the dark (white areas) or under illumination with 365 nm light (purple bars). Data are mean ± S.E.M. (n = 12 individuals/group). (**g**) Quantification of the total distances swum by the control group (vehicle) and the treatment group (100 μM azodopa) during 4 consecutive cycles of illumination (30 s) and dark (30 s before and after illumination). Data are mean ± S.E.M. (n = 12 individuals/group) and were analyzed by two-way ANOVA followed by Tukey's post hoc test (**** $p < 0.0001$; * $p = 0.0232$). All statistical analyses (panels (**d,e,g**)) were performed with GraphPad Prism 6.

Next, we sought to verify whether the effect of azodopa on zebrafish locomotor activity is dose-dependent by testing a range of concentrations around 100 μM. For our analysis, we averaged the distance swum by each group during the last four consecutive dark–light cycles (30 s integration). We observed that azodopa increased the swimming activity in the dark in a dose-dependent manner, while the effect under 365 nm illumination was smaller at all concentrations and yielded a weak dose dependence (Figure 3e; see also Figure S14 for a representation of the activity profile at all concentrations). Moreover, we observed that co-application of the D_1-like antagonist SKF83566 (50 μM) abolished the behavioral effects produced by azodopa at 100 μM in the dark, and restored zebrafish activity to control levels (Figures 3e and S15). These experiments allow to exclude the participation of adrenergic receptors in the photoresponses in vivo [14], and support the hypothesis that they are mediated by D_1-like rather than D_2-like receptors (SKF83566 is thousand-fold more potent in the formers). However, they cannot rule out the involvement of 5-HT_2 receptors, as the antagonist binds them with only 20-fold weaker potency.

To distinguish between the contribution of visual responses to (1) the changes in fish locomotion and (2) the dopaminergic modulation with azodopa, we repeated the experiments with blinded zebrafish larvae. The zebrafish retina contains four different cone photoreceptor subtypes (UV, S, M, L), each one defined by the expression of specific opsins that confer a particular wavelength-sensitivity. UV cones express SWS1, an opsin with peak sensitivity in the UV range ($\lambda_{max} = 354$ nm) [55]; therefore, the removal of functional UV cones can be used to suppress UV-dependent behaviors [56]. Blinded zebrafish larvae were obtained via a noninvasive blinding technique that induces photoreceptor apoptosis while preserving the rest of the retina (see SM for details) [57]. They were remarkably inactive and unresponsive to illumination (Figure 3f,g). The activity profile of azodopa-treated blinded zebrafish was qualitatively similar to the one observed with azodopa-treated normal zebrafish (Figure 3a,f respectively). Azodopa (100 μM) produced a remarkable increase of the swimming activity of blind larvae for the entire experiment during the dark periods, and this effect was abolished upon illumination. Quantification of the total distances swum by the control group and the treatment group (100 μM azodopa) during the last four dark/light cycles confirmed our observations: the locomotion of blinded animals could be significantly photoswitched with azodopa. The overall decrease in swimming activity of blinded zebrafish (Figure 3f,g) compared to normal zebrafish (Figure 3a–e) is due to the induced blindness, which reduces their exploratory tendencies. This phenomenon is more pronounced in the control-blinded group, which is almost immobile, and makes the effect of azodopa on locomotion appear even stronger: azodopa-treated blinded larvae covered a total distance 16 times longer (during pre-illumination periods) and 6 times longer (during post-illumination periods) than the one swum by control animals (Figure 3g).

The fact that azodopa can elicit behavioral photoresponses in blinded zebrafish and that they have similar magnitude to those in normal fish show that retinal photoreceptors are not directly involved in the observed change in locomotion upon illumination. Instead, photoresponses must be attributed to other dopaminergic circuits in the CNS (present both in blinded and normal larvae) that are effectively placed under the control of light with azodopa. Interestingly, the time course of photoresponses does display differences between blinded and normal larvae, and these must be ascribed to the presence of visual inputs in normal animals. The most outstanding one is the recovery of locomotion after turning off UV light in the presence of azodopa, which is significantly faster in normal zebrafish larvae than in blinded ones (see four cycles in Figure 3a,f, and quantification in supplementary Figure S16). Thus, azodopa enables time-resolved behavioral experiments that contain unique information about the dopaminergic modulation of retinal function [3], and that will be further investigated with spatially-resolved neuronal activity maps.

Overall, behavioral responses in zebrafish larvae agree with our previous findings in vitro and confirm that azodopa is a reversible photoswitchable dopaminergic agonist displaying dose-dependent locomotor effects. In addition, we demonstrated that zebrafish larvae, previously exposed to azodopa and dark–light cycles, recovered normal swimming behavior after washout (see supplementary Figure S22 and SM), and were still alive about 48 h after the experiment. The robust photocontrol of behavior and the apparent absence of acute toxicity encouraged us to test the potential of azodopa in a mammalian model.

2.5. Electrophysiological Recordings in Anesthetized Mice

We studied whether azodopa modulates neural activity in the cortex of mice. For such experiments, we developed a custom setup that combined in vivo electrophysiological recordings and the possibility to illuminate with 365 nm LEDs. The Open Ephys data acquisition system was used to record neural activity via an octrode (four two-wire stereotrode array) inserted in the superficial layers of the secondary motor cortex (M2) with a large craniotomy that allowed exposure of cortical tissue (Figure 4a).

We first investigated the effects of azodopa in the absence of light in two mice anesthetized with isoflurane (Mouse 1 and Mouse 2, Figure 4b–e). In each animal, we recorded spiking activity of individual neurons (single unit activity, SUA) and local field potentials (LFPs) in M2 before and after the administration of *trans*-azodopa. We quantified mean firing rates (spikes per second) and LFP power from 1 to 10 Hz. Transient oscillatory signals observed in LFPs reflect the summed synchronized activity of neural networks and are also called neural oscillations. Neural oscillations between 1 and 10 Hz have been associated with cognitive processing during alertness and slow waves during slow wave sleep and anesthesia (i.e., UP and DOWN states). The experiments started after cortical activity was stable for at least 10 min. Then, we recorded baseline neural activity for 10 more minutes. During this period, slow fluctuations of neural activity were observed both at the single-neuron and LFP levels that were associated with UP and DOWN states typically observed during anesthesia (Figure 4b). Later, 10 µl of *trans*-azodopa at 3 µM concentration were administered manually on the surface of the cortex with a standard pipette. Azodopa increased neural activity few seconds after its administration. The effects were transient in many neurons and the general activity remained elevated for at least 5 min (Figure 4c). There was a boost in the firing rate of individual neurons that no longer followed the UP and DOWN cycles. In addition, more neurons could be identified after *trans*-azodopa administration (21 neurons during baseline and 33 neurons after azodopa in mouse 1; see also supplementary Figure S17). In fact, in mouse 2, we isolated 23 neurons during baseline and only 10 neurons after *trans*-azodopa because spiking activity was so elevated that spikes from different neurons co-occurred altering the shape of the waveforms and thus prevented their classification. Corresponding with the increased spiking activity of neurons, the power of oscillatory activities augmented, although moderately compared to changes in firing rate (Figure 4c, bottom panel). Statistical analyses confirmed significant increases of firing rate and 1–10 Hz power when combining the two animals (Figure 4d,e).

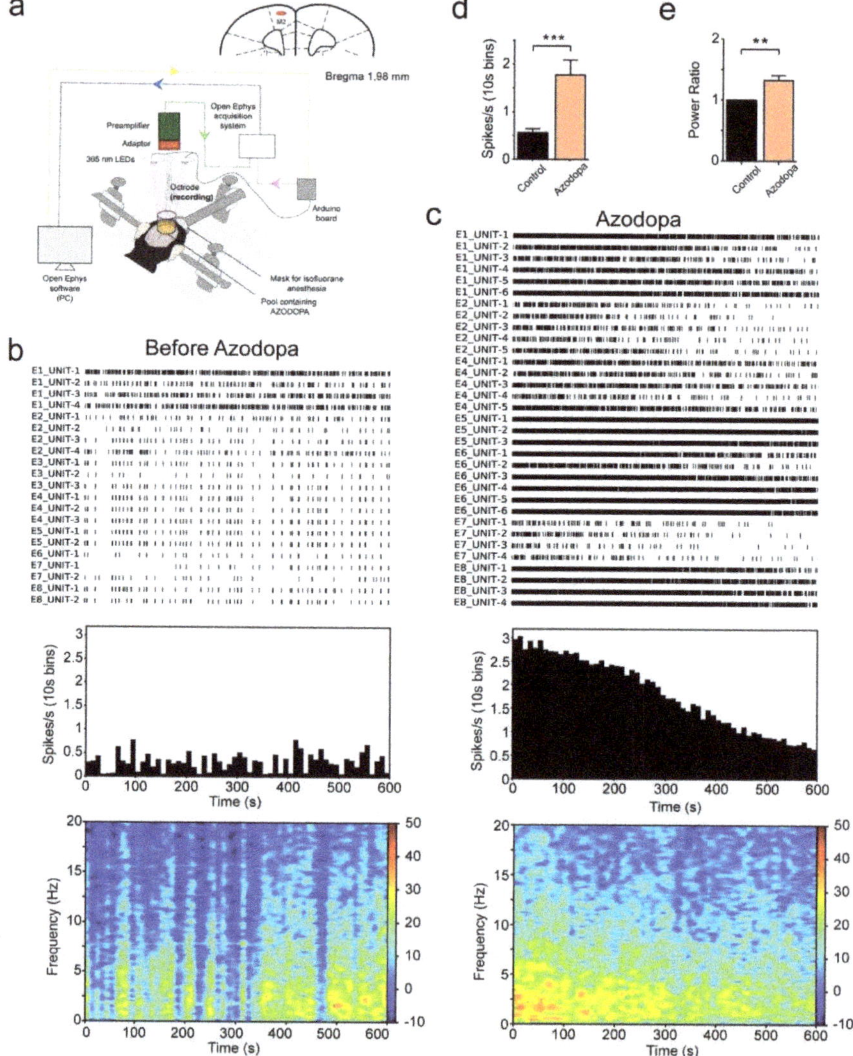

Figure 4. Effect of cortical administration of *trans*-azodopa on electrophysiological recordings in anesthetized mice. (**a**) Animals were anesthetized with isoflurane and placed in a stereotaxic apparatus. A craniotomy was drilled above the secondary motor cortex (M2) and an octrode was inserted in the superficial layers. Analogic signals were bandpass filtered and digitized by a preamplifier, amplified by an Open Ephys data acquisition system (green arrow), and finally visualized and recorded in a PC (blue arrow). Neural activity was recorded during baseline conditions and after

administration of *trans*-azodopa on the cortical surface (3 µM concentration in 10 µL volume). See the Supplementary Materials for further details of the setup and supplementary Figures S17–S19 for the effect of illumination on azodopa. (**b**) Neural activity during baseline conditions in Mouse 1. Raster plot of spiking activity in the cortex of one mouse for 10 min. Each row depicts the spiking activity of a single neuron (unit), each tick representing an action potential. We used arrays of 8 electrodes (octrodes) in each animal, from which several units could be recorded. Neurons are labeled by their electrode number (E1 to E8). Firing rates were stable and followed the UP and DOWN slow fluctuations typical of anesthesia. The quantification of firing rates and average time–frequency spectrogram of the power of neural oscillations (n = 8 electrodes) are shown below. (**c**) Neural activity after the administration of 3 µM *trans*-azodopa in Mouse 1 (zero indicates the time of administration). Azodopa boosted the firing rate of neurons and increased the power of neural oscillations. (**d**) Azodopa increased spiking activity of cortical neurons. Mean firing rate of neurons before and after the administration of *trans*-azodopa. Data are mean ± S.E.M. (n = 44 neurons during baseline vs. 43 neurons after azodopa in two mice) and were analyzed with an unpaired *T*-test (*** p = 0.0002). (**e**) Azodopa increased the power (1–10 Hz) of neural oscillations in the two animals. Due to the large differences in the baseline power of the two mice, we normalized the power to its baseline for visualization purposes only. Data are mean ± S.E.M. (n = 16 channels per condition from two mice) and were analyzed with a paired *T*-test (** p = 0.0011).

We next investigated the effects of azodopa under 365 nm illumination in two more mice anesthetized with isoflurane (Mouse 3 and Mouse 4, supplementary Figures S17–S19). In each mouse, we first injected 0.5 mL of saline and later 0.5 mL 3 µM azodopa in consecutive experiments. Solutions were administered with an infusion pump over the course of one minute to avoid environmental noise and to allow electrophysiological recordings during the administration. Moreover, drug administration was conducted under illumination so that azodopa penetrated cortical tissue in an inactive form. Then, we illuminated the motor cortex with five consecutive cycles of one minute of darkness (OFF) and one minute of light (ON) (supplementary Figure S17). Here, we focused on differential spiking activity during light and dark cycles because azodopa displayed a stronger effect than on the power of neural oscillations in the previous experiments (Figure 4).

After the injection of saline, the mean firing rate of neurons remained similar to baseline levels, and neural activity slightly increased during the light cycles compared to the dark cycles (supplementary Figure S17). In the presence of azodopa, neural activity increased as in direct application experiments (Figure 4) and some neurons showed rapid and reversible changes in their firing patterns during the ON and OFF light cycles. Interestingly, some of these neurons decreased their spiking activity during the light periods (DARK-ON) and others increased it (LIGHT-ON; supplementary Figures S18 and S19). Although these heterogeneous effects of light cancelled each other out in the global average (supplementary Figure S17c), manual pooling of single unit recordings yielded significant differences in neural activity between dark and light conditions (supplementary Figure S19). Since dopamine receptors are expressed both in excitatory and inhibitory neurons of the frontal cortex, we hypothesize that DARK-ON neurons may correspond to those expressing D_1-like receptors, which decrease their firing rates when azodopa is deactivated by light, and LIGHT-ON neurons may be cells affected indirectly by network effects driven by DARK-ON neurons.

Overall, our first in vivo studies in anesthetized mice indicate that *trans*-azodopa exerts excitatory actions on brain cortical microcircuits, as reflected by the increased spiking and oscillatory activities. The effects of azodopa were transient; the firing rate of many neurons was increased for few minutes, while in others the elevated firing lasted more than 10 min. This could be due to the diffusion (and therefore dilution) of azodopa within cortical tissue over time, and possibly to its metabolic washout and reuptake by synaptic terminals [58]. The increase of cortical activity by *trans*-azodopa is consistent with our results from functional studies in cell cultures and suggests that *trans*-azodopa may also act as a dopamine agonist in the mouse brain [59]. The general excitatory action of azodopa was modulated by UV light in individual neurons, some of which revealed a decrease in spiking activity while others increased it. These heterogeneous responses to D_1-like receptor activation in wild-type animals bear physiological relevance and are currently being studied.

3. Discussion

Understanding the dopaminergic system dynamics is a central question in neurobiology and neuropharmacology. In fact, DARs are involved in the modulation of fundamental physiological functions such as voluntary movement, motivation, cognition, emotion, reward, and neuroendocrine secretion, among others, and a dysregulation of the dopaminergic transmission is unavoidably associated with major psychiatric and neurological disorders. However, the available techniques to dissect neuronal circuits and their role in pathological conditions have several shortcomings. Electrical stimulation lacks cellular specificity and conventional pharmacological manipulation lacks temporal and spatial resolution [60]. Optogenetic tools allow the modulation of specific neural circuit elements with millisecond precision, but are limited by non-uniform expression of the optogenetic actuators and generation of non-physiological patterns of activity throughout the targeted population of neurons. Here, we introduce a novel photopharmacological agonist that enables reversible spatiotemporal control of intact dopaminergic pathways in vivo. We show that azodopa triggers DAR-mediated cAMP accumulation and ERK1/2 phosphorylation as well as phospholipase C activation in its *trans* configuration, and its efficacy can be reduced or switched off under illumination. Accordingly, azodopa allows the reversible photocontrol of zebrafish motility on a timescale of seconds, increasing the swimming activity exclusively in the *trans* active state. Azodopa can be bath-applied, does not require microinjection or genetic manipulation, and is compatible with high-throughput behavioral screening in wild-type or transgenic fish, and with other treatments including pharmacological ones. For example, the inactivity of blinded larvae under control conditions can be risen to levels comparable to normal (untreated) animals by adding *trans*-azodopa in the water without illumination. In addition, locomotion is reduced to control levels upon photoisomerization to the *cis* form, which suggests that azodopa might be used to interfere with extracellular dopamine/melatonin cycles in the retina involved in circadian rhythms [3]. Furthermore, the intriguing observation that DAR activation by azodopa *cis-trans* isomerization (dark relaxation) produces faster behavioral responses in normal fish than in blinded ones offers new opportunities to interrogate the dopaminergic modulation of retinal circuits with spatiotemporal, pharmacological, and cell-type specificity [61]. Our results thus complement recently reported genetic–photopharmacological agonists [15] and photopharmacological antagonists [25], and open the way to dissect dopaminergic neurotransmission in intact animals. Characterizing in detail the pharmacological profile and safety of azodopa was not the aim of this work, but experiments with zebrafish in the presence of a selective antagonist suggested that D_1-like receptors are the main mediators of the photocontrolled behavior.

trans-Azodopa also exerts excitatory actions on brain cortical microcircuits at firing rates and frequency bands relevant for behavior in mice [62]. In agreement with our in vitro studies, azodopa induced a general increase in neural excitability, although the time course and light dependence of the responses was heterogeneous in individual neurons. We identified cortical neurons with light-ON and dark-ON patterns, which is expected when administering in the intact brain an agonist of dopamine receptors that are expressed both in excitatory and inhibitory neurons. These responses must be characterized further, but it must be noted that isoflurane anesthesia produces profound inhibitory effects on brain activity. The activity of neural networks may be more difficult to modulate under anesthesia than during alertness, which prompts to evaluate the effects of azodopa under light/dark regimes in awake mice.

From a photochromic point of view, azodopa displays a short half-life of thermal relaxation, which is useful in neurobiology since a single wavelength of light allows to rapidly toggle the photoswitch between its two isomers. However, azodopa requires UV light for deactivation, which is normally undesired in photopharmacology because of poor tissue penetration and potential phototoxicity. Moreover, azodopa is active in the most thermodynamically stable configuration (*trans*). Although there are clinical conditions that might take advantage of a dark-active drug that can be deactivated on demand (e.g., to reduce levodopa-induced dyskinesia in Parkinson's disease) [63], light-activatable compounds are generally preferred for research and therapeutic purposes. Hence, new dopaminergic photoswitches must be developed that are active in the less thermodynamically stable configuration [64] and photoisomerize with visible [65] or infrared light [66,67] in order to unleash their full potential as photopharmacological tools.

In summary, azodopa is a photochromic activator of endogenous dopamine receptors that does not require genetic manipulation, and is the first photoswitchable dopaminergic agonist with demonstrated efficacy in vivo in intact wild-type animals, including mammals. This ligand allows analyzing the different components of the dopaminergic circuitry and is a breakthrough in developing new photoswitchable drugs potentially useful to manage neurological conditions, including movement disorders and addiction.

Supplementary Materials: The following supporting information can be downloaded at: https://www.mdpi.com/article/10.3390/ijms231710114/s1, References [68,69] are cited in the supplementary materials.

Author Contributions: Conceptualization: C.M. and P.G.; formal analysis: C.M., P.C., V.C.-A., R.S., A.M.J.G., E.M., T.G., C.D.-S., P.N. and J.H.; investigation: C.M., P.C., V.C.-A., R.S., A.M.J.G., E.M., T.G., C.D.-S., P.N., D.C., S.C.-B., M.M. and J.H.; writing—original draft preparation: C.M. and P.G.; writing—review and editing: C.M., V.C., M.V.P. and P.G.; visualization: C.M., P.C., V.C.-A. and R.S.; supervision: V.C., M.V.P. and P.G.; project administration: V.C., M.V.P. and P.G.; funding acquisition: V.C., M.V.P. and P.G. All authors have read and agreed to the published version of the manuscript.

Funding: This research was funded by EU Horizon 2020 Framework Programme for Research and Innovation: Human Brain Project WaveScalES, Specific Grant Agreement 2 No. 785907 and Specific Grant Agreement 3 No. 945539; NEUROPA project, Grant Agreement No. 863214; DEEPER project ICT-36-2020-101016787; European Union Regional Development Fund within the framework of the ERDF Operational Program of Catalonia 2014-2020: CECH project; Ministry of Science and Innovation DEEP RED grant PID2019-111493RB-I00 funded by MCIN/AEI/10.13039/501100011033; Ministerio de Economía y Competitividad and European Regional Development Funds: Grant no. SAF2017-87629-R to V.C., SAF2016-80726-R to M.V.P., SAF2017-88076-R to M.M.; AGAUR/Generalitat de Catalunya: CERCA Programme; Generalitat de Catalunya: Grant no. 2017-SGR-1442 to P.G., 2017-SGR-1497 to V.C., 2017-SGR-210 to M.V.P., and 2017-SGR-00465 to J.H.; Fundaluce and "la Caixa" foundations: ID 100010434, grant agreement LCF/PR/HR19/52160010.

Institutional Review Board Statement: All experiments on animals were conducted in compliance with EU directive 2010/63/EU and Spanish guidelines (Laws 32/2007, 6/2013 and Real Decreto 53/2013) and were authorized by the Barcelona Biomedical Research Park (PRBB) Animal Research Ethics Committee and the local government (code #9891).

Data Availability Statement: The datasets generated during and/or analyzed during the current study are available from the corresponding author on reasonable request.

Acknowledgments: The authors wish to thank Ewa Błasiak (Jagiellonian University, Kraków) for donating the plasmid encoding the human D1 receptor, Fabio Riefolo (IBEC Barcelona) for his help with HPLC analyses, and Jordi Ortiz (Autonomous University of Barcelona) for useful discussions. Molecular graphics and analyses were performed with the UCSF Chimera package. Chimera is developed by the Resource for Biocomputing, Visualization, and Informatics at the University of California, San Francisco (supported by NIGMS P41-GM103311). Mass spectrometry was performed at the IRB Barcelona Mass Spectrometry Core Facility, which actively participates in the BMBS European COST Action BM 1403 and is a member of Proteored, PRB2-ISCIII, supported by grant PRB2 (IPT13/0001—ISCIIISGEFI/FEDER).

Conflicts of Interest: The authors declare no conflict of interest.

References

1. Beaulieu, J.-M.; Gainetdinov, R.R. The physiology, signaling, and pharmacology of dopamine receptors. *Pharmacol. Rev.* **2011**, *63*, 182–217. [CrossRef] [PubMed]
2. Tritsch, N.X.; Sabatini, B.L. Dopaminergic modulation of synaptic transmission in cortex and striatum. *Neuron* **2012**, *76*, 33–50. [CrossRef] [PubMed]
3. Firsov, M.L.; Astakhova, L.A. The Role of Dopamine in Controlling Retinal Photoreceptor Function in Vertebrates. *Neurosci. Behav. Phys.* **2015**, *46*, 138–145. [CrossRef]
4. Korshunov, K.S.; Blakemore, L.J.; Trombley, P.Q. Dopamine: A Modulator of Circadian Rhythms in the Central Nervous System. *Front. Cell. Neurosci.* **2017**, *11*, 91. [CrossRef] [PubMed]
5. Puig, M.V.; Miller, E.K. The role of prefrontal dopamine D1 receptors in the neural mechanisms of associative learning. *Neuron* **2012**, *74*, 874–886. [CrossRef]
6. Puig, M.V.; Miller, E.K. Neural Substrates of Dopamine D2 Receptor Modulated Executive Functions in the Monkey Prefrontal Cortex. *Cereb. Cortex* **2015**, *25*, 2980–2987. [CrossRef] [PubMed]
7. Beaulieu, J.-M.; Espinoza, S.; Gainetdinov, R.R. Dopamine receptors—IUPHAR Review 13. *Br. J. Pharmacol.* **2015**, *172*, 1–23. [CrossRef]
8. Lerch, M.M.; Hansen, M.J.; van Dam, G.M.; Szymanski, W.; Feringa, B.L. Emerging Targets in Photopharmacology. *Angew. Chem. Int. Ed. Engl.* **2016**, *55*, 10978–10999. [CrossRef] [PubMed]
9. Hüll, K.; Morstein, J.; Trauner, D. In Vivo Photopharmacology. *Chem. Rev.* **2018**, *118*, 10710–10747. [CrossRef]
10. Pianowski, Z.L. *Molecular Photoswitches: Synthesis, Properties, and Applications*; VCH: Hoboken, NJ, USA, 2021.
11. Pittolo, S.; Gómez-Santacana, X.; Eckelt, K.; Rovira, X.; Dalton, J.; Goudet, C.; Pin, J.-P.; Llobet, A.; Giraldo, J.; Llebaria, A.; et al. An allosteric modulator to control endogenous G protein-coupled receptors with light. *Nat. Chem. Biol.* **2014**, *10*, 813–815. [CrossRef]
12. Agnetta, L.; Bermudez, M.; Riefolo, F.; Matera, C.; Claro, E.; Messerer, R.; Littmann, T.; Wolber, G.; Holzgrabe, U.; Decker, M. Fluorination of Photoswitchable Muscarinic Agonists Tunes Receptor Pharmacology and Photochromic Properties. *J. Med. Chem.* **2019**, *62*, 3009–3020. [CrossRef] [PubMed]
13. Riefolo, F.; Matera, C.; Garrido-Charles, A.; Gomila, A.M.J.; Sortino, R.; Agnetta, L.; Claro, E.; Masgrau, R.; Holzgrabe, U.; Batlle, M.; et al. Optical Control of Cardiac Function with a Photoswitchable Muscarinic Agonist. *J. Am. Chem. Soc.* **2019**, *141*, 7628–7636. [CrossRef] [PubMed]
14. Prischich, D.; Gomila, A.M.J.; Milla-Navarro, S.; Sangüesa, G.; Diez-Alarcia, R.; Preda, B.; Matera, C.; Batlle, M.; Ramírez, L.; Giralt, E.; et al. Adrenergic Modulation with Photochromic Ligands. *Angew. Chem. Int. Ed. Engl.* **2021**, *60*, 3625–3631. [CrossRef] [PubMed]
15. Donthamsetti, P.; Winter, N.; Hoagland, A.; Stanley, C.; Visel, M.; Lammel, S.; Trauner, D.; Isacoff, E. Cell specific photoswitchable agonist for reversible control of endogenous dopamine receptors. *Nat. Commun.* **2021**, *12*, 4775. [CrossRef] [PubMed]
16. Berizzi, A.E.; Goudet, C. Strategies and considerations of G-protein-coupled receptor photopharmacology. *Adv. Pharm.* **2020**, *88*, 143–172.
17. Bregestovski, P.; Maleeva, G.; Gorostiza, P. Light-induced regulation of ligand-gated channel activity. *Br. J. Pharmacol.* **2017**, *172*, 5870. [CrossRef] [PubMed]
18. Izquierdo-Serra, M.; Bautista-Barrufet, A.; Trapero, A.; Garrido-Charles, A.; Díaz-Tahoces, A.; Camarero, N.; Pittolo, S.; Valbuena, S.; Pérez-Jiménez, A.; Gay, M.; et al. Optical control of endogenous receptors and cellular excitability using targeted covalent photoswitches. *Nat. Commun.* **2016**, *7*, 12221. [CrossRef]
19. Szymanski, W.; Ourailidou, M.E.; Velema, W.A.; Dekker, F.J.; Feringa, B.L. Light-Controlled Histone Deacetylase (HDAC) Inhibitors: Towards Photopharmacological Chemotherapy. *Chem. Eur. J.* **2015**, *21*, 16517–16524. [CrossRef]
20. Matera, C.; Gomila, A.M.J.; Camarero, N.; Libergoli, M.; Soler, C.; Gorostiza, P. Photoswitchable Antimetabolite for Targeted Photoactivated Chemotherapy. *J. Am. Chem. Soc.* **2018**, *140*, 15764–15773. [CrossRef]
21. Steinberg, E.E.; Janak, P.H. Establishing causality for dopamine in neural function and behavior with optogenetics. *Brain Res.* **2013**, *1511*, 46–64. [CrossRef]
22. Stauffer, W.R.; Lak, A.; Yang, A.; Borel, M.; Paulsen, O.; Boyden, E.S.; Schultz, W. Dopamine Neuron-Specific Optogenetic Stimulation in Rhesus Macaques. *Cell* **2016**, *166*, 1564–1571.e6. [CrossRef] [PubMed]
23. Chang, C.Y.; Esber, G.R.; Marrero-Garcia, Y.; Yau, H.-J.; Bonci, A.; Schoenbaum, G. Brief optogenetic inhibition of dopamine neurons mimics endogenous negative reward prediction errors. *Nat. Neurosci.* **2016**, *19*, 111–116. [CrossRef] [PubMed]
24. Hare, B.D.; Shinohara, R.; Liu, R.J.; Pothula, S.; DiLeone, R.J.; Duman, R.S. Optogenetic stimulation of medial prefrontal cortex Drd1 neurons produces rapid and long-lasting antidepressant effects. *Nat. Commun.* **2019**, *10*, 223. [CrossRef] [PubMed]
25. Donthamsetti, P.C.; Winter, N.; Schönberger, M.; Levitz, J.; Stanley, C.; Javitch, J.A.; Isacoff, E.Y.; Trauner, D. Optical Control of Dopamine Receptors Using a Photoswitchable Tethered Inverse Agonist. *J. Am. Chem. Soc.* **2017**, *139*, 18522–18535. [CrossRef] [PubMed]
26. Araya, R.; Andino-Pavlovsky, V.; Yuste, R.; Etchenique, R. Two-photon optical interrogation of individual dendritic spines with caged dopamine. *ACS Chem. Neurosci.* **2013**, *4*, 1163–1167. [CrossRef] [PubMed]
27. Gienger, M.; Hübner, H.; Löber, S.; König, B.; Gmeiner, P. Structure-based development of caged dopamine D2/D3 receptor antagonists. *Sci. Rep.* **2020**, *10*, 829. [CrossRef] [PubMed]

28. Lachmann, D.; Studte, C.; Männel, B.; Hübner, H.; Gmeiner, P.; König, B. Photochromic Dopamine Receptor Ligands Based on Dithienylethenes and Fulgides. *Chem. Eur. J.* **2017**, *23*, 13423–13434. [CrossRef]
29. Szymanski, W.; Beierle, J.M.; Kistemaker, H.A.V.; Velema, W.A.; Feringa, B.L. Reversible photocontrol of biological systems by the incorporation of molecular photoswitches. *Chem. Rev.* **2013**, *113*, 6114–6178. [CrossRef]
30. Broichhagen, J.; Frank, J.A.; Trauner, D. A Roadmap to Success in Photopharmacology. *Acc. Chem. Res.* **2015**, *48*, 1947–1960. [CrossRef]
31. Schoenberger, M.; Damijonaitis, A.; Zhang, Z.; Nagel, D.; Trauner, D. Development of a new photochromic ion channel blocker via azologization of fomocaine. *ACS Chem. Neurosci.* **2014**, *5*, 514–518. [CrossRef]
32. Borowiak, M.; Nahaboo, W.; Reynders, M.; Nekolla, K.; Jalinot, P.; Hasserodt, J.; Rehberg, M.; Delattre, M.; Zahler, S.; Vollmar, A.; et al. Photoswitchable Inhibitors of Microtubule Dynamics Optically Control Mitosis and Cell Death. *Cell* **2015**, *162*, 403–411. [CrossRef] [PubMed]
33. Morstein, J.; Awale, M.; Reymond, J.-L.; Trauner, D. Mapping the Azolog Space Enables the Optical Control of New Biological Targets. *ACS Cent. Sci.* **2019**, *5*, 607–618. [CrossRef] [PubMed]
34. Casagrande, C.; Bertolini, G. Perspectives in the design and application of dopamine receptor agonists. In *Perspectives in Receptor Research*; Elsevier: Amsterdam, The Netherlands, 1996; Volume 24, pp. 67–84.
35. Rodenhuis, N. New, Centrally Acting Dopaminergic Agents with an Improved Oral Bioavailability: Synthesis and Pharmacological Evaluation. Ph.D. Thesis, University of Groningen, Groningen, The Netherlands, 2000.
36. Pettersson, I.; Liljefors, T. Structure-activity relationships for apomorphine congeners. Conformational energies vs. biological activities. *J. Comput. Aided Mol. Des.* **1987**, *1*, 143–152. [CrossRef] [PubMed]
37. Cannon, J.G.; Borgman, R.J.; Aleem, M.A.; Long, J.P. Centrally acting emetics. 7. Hofmann and Emde degradation products of apomorphine. *J. Med. Chem.* **1973**, *16*, 219–224. [CrossRef] [PubMed]
38. Malo, M.; Brive, L.; Luthman, K.; Svensson, P. Investigation of D_2 receptor-agonist interactions using a combination of pharmacophore and receptor homology modeling. *ChemMedChem* **2012**, *7*, 471–82–338. [CrossRef] [PubMed]
39. Malo, M.; Brive, L.; Luthman, K.; Svensson, P. Investigation of D_1 receptor-agonist interactions and D_1/D_2 agonist selectivity using a combination of pharmacophore and receptor homology modeling. *ChemMedChem* **2012**, *7*, 483–94–338. [CrossRef] [PubMed]
40. Kołaczkowski, M.; Bucki, A.; Feder, M.; Pawłowski, M. Ligand-optimized homology models of D_1 and D_2 dopamine receptors: Application for virtual screening. *J. Chem. Inf. Modeling* **2013**, *53*, 638–648. [CrossRef] [PubMed]
41. Garcia-Amorós, J.; Sánchez-Ferrer, A.; Massad, W.A.; Nonell, S.; Velasco, D. Kinetic study of the fast thermal cis-to-trans isomerisation of para-, ortho- and polyhydroxyazobenzenes. *Phys. Chem. Chem. Phys.* **2010**, *12*, 13238–13242. [CrossRef]
42. Izquierdo-Serra, M.; Gascón-Moya, M.; Hirtz, J.J.; Pittolo, S.; Poskanzer, K.E.; Ferrer, È.; Alibés, R.; Busqué, F.; Yuste, R.; Hernando, J.; et al. Two-photon neuronal and astrocytic stimulation with azobenzene-based photoswitches. *J. Am. Chem. Soc.* **2014**, *136*, 8693–8701. [CrossRef]
43. Ferré, S.; Casadó, V.; Devi, L.A.; Filizola, M.; Jockers, R.; Lohse, M.J.; Milligan, G.; Pin, J.-P.; Guitart, X. G protein-coupled receptor oligomerization revisited: Functional and pharmacological perspectives. *Pharmacol. Rev.* **2014**, *66*, 413–434. [CrossRef]
44. Casadó, V.; Ferrada, C.; Bonaventura, J.; Gracia, E.; Mallol, J.; Canela, E.I.; Lluis, C.; Cortés, A.; Franco, R. Useful pharmacological parameters for G-protein-coupled receptor homodimers obtained from competition experiments. Agonist-antagonist binding modulation. *Biochem. Pharmacol.* **2009**, *78*, 1456–1463. [CrossRef] [PubMed]
45. Jin, L.-Q.; Wang, H.-Y.; Friedman, E. Stimulated D1 dopamine receptors couple to multiple Gα proteins in different brain regions. *J. Neurochem.* **2001**, *78*, 981–990. [CrossRef] [PubMed]
46. Chiken, S.; Sato, A.; Ohta, C.; Kurokawa, M.; Arai, S.; Maeshima, J.; Sunayama-Morita, T.; Sasaoka, T.; Nambu, A. Dopamine D1 Receptor-Mediated Transmission Maintains Information Flow through the Cortico-Striato-Entopeduncular Direct Pathway to Release Movements. *Cereb. Cortex* **2015**, *25*, 4885–4897. [CrossRef] [PubMed]
47. Basnet, R.M.; Zizioli, D.; Taweedet, S.; Finazzi, D.; Memo, M. Zebrafish Larvae as a Behavioral Model in Neuropharmacology. *Biomedicines* **2019**, *7*, 23. [CrossRef] [PubMed]
48. Howe, K.; Clark, M.D.; Torroja, C.F.; Torrance, J.; Berthelot, C.; Muffato, M.; Collins, J.E.; Humphray, S.; McLaren, K.; Matthews, L.; et al. The zebrafish reference genome sequence and its relationship to the human genome. *Nature* **2013**, *496*, 498–503. [CrossRef] [PubMed]
49. Kalueff, A.V.; Stewart, A.M.; Gerlai, R. Zebrafish as an emerging model for studying complex brain disorders. *Trends Pharmacol. Sci.* **2014**, *35*, 63–75. [CrossRef] [PubMed]
50. Lange, M.; Norton, W.; Coolen, M.; Chaminade, M.; Merker, S.; Proft, F.; Schmitt, A.; Vernier, P.; Lesch, K.-P.; Bally-Cuif, L. The ADHD-susceptibility gene lphn3.1 modulates dopaminergic neuron formation and locomotor activity during zebrafish development. *Mol. Psychiatry* **2012**, *17*, 946–954. [CrossRef]
51. Ek, F.; Malo, M.; Åberg Andersson, M.; Wedding, C.; Kronborg, J.; Svensson, P.; Waters, S.; Petersson, P.; Olsson, R. Behavioral Analysis of Dopaminergic Activation in Zebrafish and Rats Reveals Similar Phenotypes. *ACS Chem. Neurosci.* **2016**, *7*, 633–646. [CrossRef]
52. Levitas-Djerbi, T.; Appelbaum, L. Modeling sleep and neuropsychiatric disorders in zebrafish. *Curr. Opin. Neurobiol.* **2017**, *44*, 89–93. [CrossRef]

53. Irons, T.D.; Kelly, P.E.; Hunter, D.L.; Macphail, R.C.; Padilla, S. Acute administration of dopaminergic drugs has differential effects on locomotion in larval zebrafish. *Pharmacol. Biochem. Behav.* **2013**, *103*, 792–813. [CrossRef]

54. Ingebretson, J.J.; Masino, M.A. Quantification of locomotor activity in larval zebrafish: Considerations for the design of high-throughput behavioral studies. *Front. Neural Circuits* **2013**, *7*, 109. [CrossRef] [PubMed]

55. Raymond, P.A.; Barthel, L.K.; Curran, G.A. Developmental patterning of rod and cone photoreceptors in embryonic zebrafish. *J. Comp. Neurol.* **1995**, *359*, 537–550. [CrossRef] [PubMed]

56. Guggiana-Nilo, D.A.; Engert, F. Properties of the Visible Light Phototaxis and UV Avoidance Behaviors in the Larval Zebrafish. *Front. Behav. Neurosci.* **2016**, *10*, 160. [CrossRef] [PubMed]

57. Taylor, S.; Chen, J.; Luo, J.; Hitchcock, P. Light-induced photoreceptor degeneration in the retina of the zebrafish. *Methods Mol. Biol.* **2012**, *884*, 247–254. [PubMed]

58. Kaufman, D.M.; Milstein, M.J. Neurotransmitters and Drug Abuse. In *Kaufman's Clinical Neurology for Psychiatrists*; Elsevier: Amsterdam, The Netherlands, 2013; pp. 501–525.

59. Roffman, J.L.; Tanner, A.S.; Eryilmaz, H.; Rodriguez-Thompson, A.; Silverstein, N.J.; Ho, N.F.; Nitenson, A.Z.; Chonde, D.B.; Greve, D.N.; Abi-Dargham, A.; et al. Dopamine D1 signaling organizes network dynamics underlying working memory. *Sci. Adv.* **2016**, *2*, e1501672. [CrossRef]

60. Nieh, E.H.; Kim, S.-Y.; Namburi, P.; Tye, K.M. Optogenetic dissection of neural circuits underlying emotional valence and motivated behaviors. *Brain Res.* **2013**, *1511*, 73–92. [CrossRef]

61. Roy, S.; Field, G.D. Dopaminergic modulation of retinal processing from starlight to sunlight. *J. Pharm. Sci.* **2019**, *140*, 86–93. [CrossRef]

62. Alemany-González, M.; Gener, T.; Nebot, P.; Vilademunt, M.; Dierssen, M.; Puig, M.V. Prefrontal-hippocampal functional connectivity encodes recognition memory and is impaired in intellectual disability. *Proc. Natl. Acad. Sci. USA* **2020**, *117*, 11788–11798. [CrossRef]

63. Aubert, I.; Guigoni, C.; Håkansson, K.; Li, Q.; Dovero, S.; Barthe, N.; Bioulac, B.H.; Gross, C.E.; Fisone, G.; Bloch, B.; et al. Increased D1 dopamine receptor signaling in levodopa-induced dyskinesia. *Ann. Neurol.* **2005**, *57*, 17–26. [CrossRef]

64. Cabré, G.; Garrido-Charles, A.; González-Lafont, À.; Moormann, W.; Langbehn, D.; Egea, D.; Lluch, J.M.; Herges, R.; Alibés, R.; Busqué, F.; et al. Synthetic Photoswitchable Neurotransmitters Based on Bridged Azobenzenes. *Org. Lett.* **2019**, *21*, 3780–3784. [CrossRef] [PubMed]

65. Garrido-Charles, A.; Huet, A.; Matera, C.; Thirumalai, A.; Llebaria, A.; Moser, T.; Gorostiza, P. Fast photoswitchable molecular prosthetics control neuronal activity in the cochlea. *bioRxiv* **2021**. [CrossRef] [PubMed]

66. Pittolo, S.; Lee, H.; Lladó, A.; Tosi, S.; Bosch, M.; Bardia, L.; Gómez-Santacana, X.; Llebaria, A.; Soriano, E.; Colombelli, J.; et al. Reversible silencing of endogenous receptors in intact brain tissue using 2-photon pharmacology. *Proc. Natl. Acad. Sci. USA* **2019**, *116*, 13680–13689. [CrossRef] [PubMed]

67. Cabré, G.; Garrido-Charles, A.; Moreno, M.; Bosch, M.; Porta-de-la-Riva, M.; Krieg, M.; Gascón-Moya, M.; Camarero, N.; Gelabert, R.; Lluch, J.M.; et al. Rationally designed azobenzene photoswitches for efficient two-photon neuronal excitation. *Nat. Commun.* **2019**, *10*, 907–912. [CrossRef] [PubMed]

68. Casadó, V.; Cortés, A.; Ciruela, F.; Mallol, J.; Ferré, S.; Lluis, C.; Canela, E.I.; Franco, R. Old and new ways to calculate the affinity of agonists and antagonists interacting with G-protein-coupled monomeric and dimeric receptors: The receptor-dimer cooperativity index. *Pharmacol. Ther.* **2007**, *116*, 343–354. [CrossRef] [PubMed]

69. Ferrada, C.; Moreno, E.; Casadó, V.; Bongers, G.; Cortés, A.; Mallol, J.; Canela, E.I.; Leurs, R.; Ferré, S.; Lluís, C.; et al. Marked changes in signal transduction upon heteromerization of dopamine D1 and histamine H3 receptors. *Br. J. Pharmacol.* **2009**, *157*, 64–75. [CrossRef] [PubMed]

Review

Photopharmacology of Ion Channels through the Light of the Computational Microscope

Alba Nin-Hill [1], Nicolas Pierre Friedrich Mueller [2,3], Carla Molteni [4], Carme Rovira [1,5] and Mercedes Alfonso-Prieto [2,6,*]

1 Departament de Química Inorgànica i Orgànica (Secció de Química Orgànica) and Institut de Química Teòrica i Computacional (IQTCUB), Universitat de Barcelona, 08028 Barcelona, Spain; albanin@ub.edu (A.N.-H.); c.rovira@ub.edu (C.R.)
2 Institute for Advanced Simulations IAS-5 and Institute of Neuroscience and Medicine INM-9, Computational Biomedicine, Forschungszentrum Jülich, 52425 Jülich, Germany; nic.mueller@fz-juelich.de
3 Faculty of Mathematics and Natural Sciences, Heinrich-Heine-University Düsseldorf, Universitätsstr. 1, 40225 Düsseldorf, Germany
4 Physics Department, King's College London, London WC2R 2LS, UK; carla.molteni@kcl.ac.uk
5 Institució Catalana de Recerca i Estudis Avançats (ICREA), 08020 Barcelona, Spain
6 Cécile and Oskar Vogt Institute for Brain Research, University Hospital Düsseldorf, Medical Faculty, Heinrich Heine University Düsseldorf, 40225 Düsseldorf, Germany
* Correspondence: m.alfonso-prieto@fz-juelich.de

Abstract: The optical control and investigation of neuronal activity can be achieved and carried out with photoswitchable ligands. Such compounds are designed in a modular fashion, combining a known ligand of the target protein and a photochromic group, as well as an additional electrophilic group for tethered ligands. Such a design strategy can be optimized by including structural data. In addition to experimental structures, computational methods (such as homology modeling, molecular docking, molecular dynamics and enhanced sampling techniques) can provide structural insights to guide photoswitch design and to understand the observed light-regulated effects. This review discusses the application of such structure-based computational methods to photoswitchable ligands targeting voltage- and ligand-gated ion channels. Structural mapping may help identify residues near the ligand binding pocket amenable for mutagenesis and covalent attachment. Modeling of the target protein in a complex with the photoswitchable ligand can shed light on the different activities of the two photoswitch isomers and the effect of site-directed mutations on photoswitch binding, as well as ion channel subtype selectivity. The examples presented here show how the integration of computational modeling with experimental data can greatly facilitate photoswitchable ligand design and optimization. Recent advances in structural biology, both experimental and computational, are expected to further strengthen this rational photopharmacology approach.

Keywords: photopharmacology; photoswitchable ligands; ion channels; voltage-gated ion channels; ligand-gated ion channels; homology modeling; molecular docking; molecular dynamics; enhanced sampling

Citation: Nin-Hill, A.; Mueller, N.P.F.; Molteni, C.; Rovira, C.; Alfonso-Prieto, M. Photopharmacology of Ion Channels through the Light of a Computational Microscope. *Int. J. Mol. Sci.* **2021**, *22*, 12072. https://doi.org/10.3390/ijms222112072

Academic Editors: Piotr D. Bregestovski and Carlo Matera

Received: 10 October 2021
Accepted: 2 November 2021
Published: 8 November 2021

Publisher's Note: MDPI stays neutral with regard to jurisdictional claims in published maps and institutional affiliations.

1. Introduction

Photopharmacology (also known as optopharmacology) is a discipline that aims at regulating the activities of biological systems with light. Light-controlled modulation can be accomplished with photoswitchable compounds [1–4]. Such molecules contain a bioactive ligand coupled to a photochromic group that, upon irradiation, causes bond isomerization or formation. For instance, the most commonly used photochromic group, azobenzene [5], isomerizes between *trans* and *cis* configurations (Figure 1a). This results in changes in both length and dipole moment that can affect the shape and chemical complementarity of the photoswitchable ligand with the protein binding pocket. When the two forms of the photoswitchable ligand have different binding preferences and/or differentially regulate protein function, optical control is achieved. Thereby, irradiation with light of the appropriate wavelength can turn protein activity on or off with high temporal and spatial resolutions.

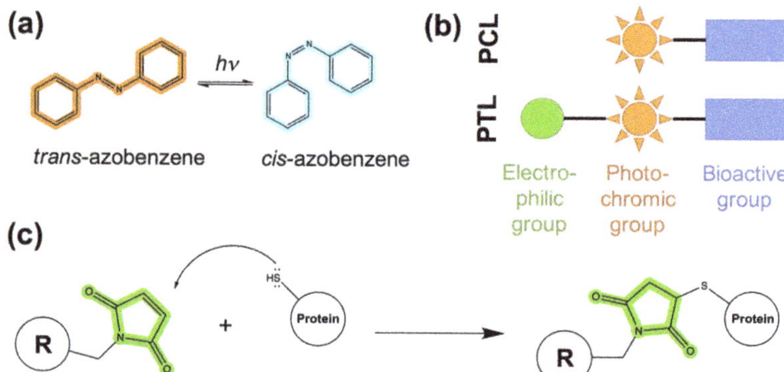

Figure 1. Chemical toolbox for the design of photoswitchable ligands. (**a**) Chemical structure of azobenzene, the most commonly used photochromic group, showing its *trans-cis* photoisomerization. (**b**) Modular design of photoswitchable ligands, either soluble photochromic ligands (PCLs) or photochromic tethered ligands (PTLs). (**c**) Covalent bond formation between the typical tethering protein residue, cysteine (shown here by its sidechain thiol group) and a common electrophile group included in PTLs, maleimide (colored in green); the rest of the PTLs are represented as a substituent R.

Photoswitchable ligands have been widely applied to the field of ion channels, because the picosecond timescale of the photochromic group transition upon irradiation is faster than the timescale of ion flow across the neuronal membrane. Among other applications, photopharmacology has been used to study ion channel properties and kinetics, the regulation of neuronal circuits and the control of animal responses, such as heartbeat, pain, vision and behavior [4]. Two main types of photoswitches have been used: freely diffusible photochromic ligands (PCLs) and photoswitchable tethered ligands (PTLs). Their design is modular (Figure 1b), containing a ligand known to regulate protein function (bioactive group) connected to a photochromic group (e.g., azobenzene). In the case of PTLs, they additionally contain an electrophilic group that binds covalently to an amino acid with nucleophilic properties near the binding site (typically cysteine; see Figure 1c). Although this nucleophilic residue can be naturally present in the target protein, in most cases the reactive cysteine is introduced by site-directed mutagenesis (i.e., optochemical genetics [6]). In addition to PCLs and PTLs, photopharmacological applications to ion channels can also employ photocaged ligands, which contain a protecting group (i.e., the cage) that is cleaved upon light irradiation, resulting in a rapid release of the bioactive molecule (e.g., the neurotransmitter). However, their design has been extensively reviewed in references [4,7–10] and thus will not be considered here.

The first ion channel to be photomodulated was the nicotinic acetylcholine receptor (nAchR). Both PCLs and PTLs were developed [11,12], consisting of a known nAchR ligand linked to a photoswitchable azobenzene group and, for PTLs, also coupled to a benzylic bromide (i.e., the electrophile for Cys tethering). In the *trans* form, the photoswitchable ligand was able to modify the receptor activity, whereas isomerization to *cis* upon UV light irradiation turned off the modulatory effect of the photoswitch. Although at the time molecular biology techniques were still in their infancy, two fortunate coincidences contributed to this first success story. On the one hand, nAChR is highly expressed in the electroplaques of electric eels. On the other, even without sequence knowledge, it was known that treatment with a disulfide reducing agent generated a free cysteine residue that allowed the covalent conjugation of tethered ligands [13]. Despite this remarkable achievement, it was still unexplained why some of the PTLs, designed based on nAChR agonists, acted instead as light-modulated antagonists [11], i.e., why the addition of the azobenzene group was changing the ligand pharmacological properties. More than thirty years later, experimental determination of the first crystal structures of the snail acetylcholine binding protein (AchBP, the soluble counterpart of the ligand binding extracellular domain of nAChR) gave a hint of the molecular basis of this change in activity (Figure 2a). The degree of closure of the so-called loop C over the binding site is correlated with the agonist (closed loop) or antagonist (open loop) activity of the cholinergic ligands [14–17]. This demonstrates that, although a photoswitchable ligand design can be successfully achieved with only ligand structure–activity relationship data, structural knowledge of the target receptor or ion channel can greatly facilitate such a task [2].

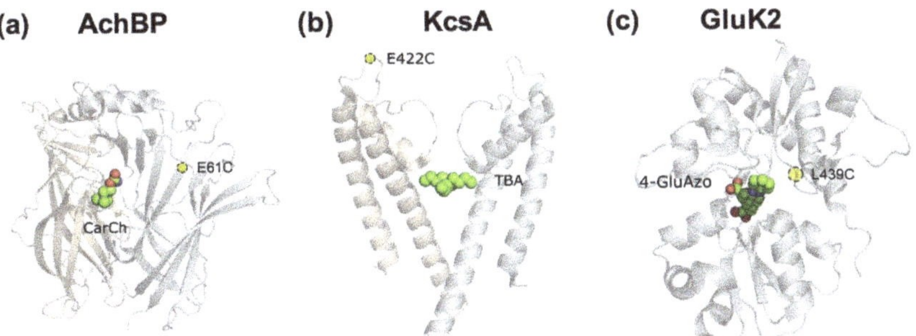

Figure 2. Crystallographic structures used for rational structure-based design of photoswitchable ligands (**a**,**b**) or its validation (**c**). (**a**) Structure of the acetylcholine binding protein (AchBP) bound to carbamylcholine (CarCh), PDB code 1UV6 [15]. (**b**) Structure of the KcsA potassium channel with the pore blocker tetrabutylammonium (TBA) bound in the intracellular site below the selectivity filter, PDB code 2BOB [18]. (**c**) Structure of the ligand binding domain (LBD) of the kainate receptor GluK2 in complex with 4-glutamyl-azobenzene (4-GluAzo), PDB code 4H8I [19]. For the sake of clarity, only one or two subunits of the oligomeric proteins are shown in cartoon representation and colored either in gray or apricot. Ligands are displayed in space-filling representation, with C, O and N atoms colored in green, red and blue, respectively. The position of the cysteine residue serving as covalent attachment point for the PTL is highlighted with a yellow circle.

The next generation of photoswitchable ligands was developed in the early 2000s thanks to the resolution of the crystal structures of a voltage-gated potassium channel and an ionotropic glutamate receptor. Based on the KcsA crystal structure solved in the presence of tetraethylammonium (TEA)-like pore blockers (Figure 2b) [18,20], light-control blockage of the homologous Shaker potassium channel [21,22] was achieved. A Cys mutation was introduced near the extracellular TEA binding site (E422C mutant) and a PTL was synthetized consisting of a Cys-reactive maleimide group, an azobenzene photoswitch and quaternary ammonium (MAQ). *Trans*-MAQ extends from the tethering site to reach the ammonium binding site in the pore, whereas *cis*-MAQ is too short to do so. Similarly, the X-ray structure of the soluble ligand binding domain (LBD) of the kainate receptor [23] was used to design photoswitchable ligands targeting this receptor. Namely, a PCL (consisting of the glutamate agonist and an azobenzene photoswitch, 4-GluAzo) [24] and a PTL (composed of a Cys-reactive maleimide, an azobenzene photoswitch and the glutamate ligand, MAG) [25] were developed. The latter covalently attaches to a light-modulated glutamate receptor (LiGluR) containing an L439C mutation. Moreover, the structure-based PCL design was later validated by solving the crystal structure of 4-GluAzo bound to the GluK2 LBD [19] (Figure 2c). The clamshell-like structure of the LBD showed a closed state, similar to other agonist-bound LBD structures and in contrast with the open state observed in the antagonist-bound structures [23]. Therefore, the design of photoswitchable ligands with the desired activity requires not only information about the protein structure, but also about the dynamical rearrangements occurring after ligand binding [2].

Unfortunately, experimental structures for the (voltage-gated and ligand-gated) ion channels involved in neurotransmission are still scarce, despite recent advances in structural biology tools, in particular cryo-electron microscopy (cryo-EM). To fill this gap, computational modeling can be used to generate structural models for these channels. Moreover, computer-aided drug design techniques (either ligand- or structure-based) can be applied to the design of photoswitchable ligands. Indeed, a recent computational study showed that a large number of bioactive molecules can be susceptible of azologization (i.e., fragment replacement by an isosteric azobenzene group to make the drug photoswitchable) [26]. In other words, a computational microscope, a term coined by the late Klaus Schulten to describe the use of modeling and simulations to study protein function and dynamics [27], can also shed light on the photopharmacology field.

For a systematic review of all the PCLs and PTLs available to date, we refer the reader to several excellent published reviews on photopharmacology or optochemical genetics [1–4,6,10,28–34]. Here we focus on photopharmacological applications targeting ion channels in which structure-based computational methods were applied, in combination with experimental approaches. We start with a short theoretical description of the computational methods used for photopharmacology so far. Then, we present some of the applications published in the literature (Tables S1–S4), where computational methods have been used to rationally design and optimize photoswitchable ligands, explain their observed effect on ion channel activity and/or identify possible tethering positions for Cys mutation. We have classified such studies depending on whether the photoswitchable ligand targets voltage- or ligand-gated ion channels (VGICs and LGICs, respectively). The chemical structures of all the PCLs and PTLs discussed in the text are shown in Figures S1–S4. To the best of our knowledge, most ion channel photopharmacology studies integrating computational methods have been carried out on azobenzene-based photoswitchable ligands.

2. Computational Modeling

Ion channels are oligomeric proteins, in which several subunits assemble to form the functional channel (see Sections 3 and 4 below). The large number of VGIC families, as well as receptor (sub)types for LGICs, gives rise to a large number of possible combinations. Moreover, each of these ion channels can adopt different functional states (open, closed, desensitized or inactivated) and their activities can be regulated by a myriad of ligands (agonists and antagonists, as well as pore blockers and allosteric modulators). Unlike the examples mentioned in the Introduction [21,22,25], an experimental structure of the ion channel of interest (in the relevant functional state and in complex with the ligand used as a bioactive group of the PCL or PTL) may not be available. This structural gap can be filled by structure-based computational approaches, such as those included in Figure 3. In the following, we mention some basic ideas underlying these methodologies; a full description of these computational methods is beyond the scope of this review, and thus we refer the reader to the excellent advanced reviews cited below.

Homology modeling generates structural models of the target protein based on its sequence and the experimental structure of a homologous protein (the so-called template). The quality of the homology model depends on the sequence identity between the target and template proteins, with 35% sequence identity being considered the minimum threshold for homologous membrane proteins to have similar 3D structures [35,36]. Moreover, structural rearrangements occur upon ligand binding and opening/closing of the ion channel pore; thus, it is recommended to choose a template structure not only with the highest sequence identity, but also captured in the appropriate functional state.

Molecular docking aims at predicting protein–ligand interactions (in the case at hand, between the ion channel of interest and the photoswitchable ligand) [37]. Docking can be performed using either an experimental structure or a homology model of the ion channel of interest. If information about the putative ligand binding site is already available, it can be incorporated into the computation protocol to guide the docking (i.e., information-driven docking). This includes the structural information of the bioactive part of the photoswitch bound to the ion channel, as well as mutagenesis data, indicating which residues are likely to be interacting with the bioactive molecule and/or the photoswitch. Otherwise, a blind docking approach, in which all possible binding pockets on the protein surface are explored, can be the method of choice. In most docking protocols, the ligand is considered flexible, whereas the receptor structure can be treated as rigid or flexible. In the latter case, only amino acids surrounding the binding site are usually allowed to move in order to model the induced fit effects. In the case of PTLs, the covalent bond between the tether and the reactive Cys can be modeled by using either a positional constraint (limiting the movement of the electrophile group within a certain sphere from the reactive Cys) or a distance restraint between the two groups. Interestingly, molecular docking (and in general computational modeling) allows one not only to model the photoswitch isomer that preferentially binds to the target protein, but also the other isomer, providing molecular insights into the light-modulated changes in ion channel activity. Moreover, virtual mutations can be introduced in the target protein structure to model the changes in photoswitch binding upon site-directed mutagenesis or when using different ion channel subtypes.

Homology modeling and molecular docking can provide static structures of the target protein in complex with the photoswitch. Additionally, Monte Carlo (MC) or molecular dynamics (MD) simulations can be performed. Therewith, the target protein–photoswitchable ligand complex is embedded in a lipid bilayer mimicking the physiological membrane environment and, upon equilibration/minimization, several configurations of the system are sampled, thus providing a dynamical picture. MD models the physical movements of the system as a function of time by solving Newton's equations of motion, whereas MC generates an ensemble of configurations according to the corresponding Boltzmann distribution. Such simulations allow one to further characterize the conformational rearrangements occurring upon ligand binding and their connection with ion conduction [38–44]. In combination with enhanced sampling methods and/or free energy calculations, MD can also provide an estimate of the ligand affinity (e.g., binding energy differences between the two forms of the photoswitch or between two photoswitchable ligands), as well as molecular insights into the energetic determinants of binding (e.g., to identify the most suitable position for introducing a reactive Cys for PTL covalent attachment).

Nonetheless, such computational methodologies also have limitations. The quality of the homology models may not be sufficient for an accurate prediction of PCL or PTL binding, especially if the sequence identity of the target protein with the template is low (i.e., below 35%) and/or the structural changes occurring upon ligand binding are not similar to those captured in the available experimental structures. Whereas molecular docking can only model small rearrangements in the protein side chains to accommodate the ligand, molecular dynamics combined with enhanced sampling techniques can be attempted to simulate further protein conformational rearrangements. However, it is still challenging to simulate large structural changes in proteins due to the long time scales of these processes and the limited quality of the (protein and ligand) force fields. Therefore, a close interplay of these in silico studies with in vitro and in vivo assays is key for successful photoswitchable ligand design. On the one hand, the experimental data are used to guide the calculations and to validate the computational models. On the other, the computational data provide a molecular explanation of the photoswitch-mediated modulation and help design modifications of the photoswitchable ligand and/or mutations of the ion channel (including Cys mutants for tethering) that can be tested experimentally. In other words, structure-based computational methods provide insights complementary to experiments that link the molecular details of the photoswitchable ligands to the experimentally measured macroscopic effects.

Figure 3 shows a possible workflow for the rational, structure-based designing of photoswitchable ligands, based on the following steps:

(1)   In the most straightforward case, a search in the Protein Data Bank (PDB) yields an experimental structure of the target protein in complex with the bioactive molecule to be used as a basic module of the photoswitchable ligand. A structure of the target protein bound to a similar molecule (in terms of chemical structure and activity) or a structure of a homologous protein–ligand complex can also be used as a surrogate, as demonstrated by the examples mentioned in the Introduction.

(2)   In the absence of an experimental structure of the target protein–bioactive molecule complex, an experimental structure of the apo protein can alternatively be used; ideally, this structure contains the appropriate subunit composition and was captured in the relevant functional state.

(3)   When no experimental structure is available, homology modeling can generate a structural model of the target protein based on the experimental structure of a homologous template protein. When selecting the template structure, one should consider the sequence identity between the target and template and, additionally, other features of the template structure, such as the functional state and the bound ligand(s) .

(4)   Although the (experimental or computational) structure of the protein alone is already informative, carrying out a computational molecular docking of the bioactive molecule can help to further optimize the photoswitchable ligand design. In particular, the predicted binding mode can be used to identify the optimal position to introduce the photochromic group and/or estimate the length of the linker between the different modules of the PCL or PTL, as well as pinpoint potential residues for Cys screening.

(5)   The photoswitchable ligand (PCL or PTL) design follows the modular approach depicted in Figure 1b. As mentioned in steps (1)–(4), structural information on the binding mode of the bioactive module to the target protein can be used to guide such a design.

(6)   In the case of PTLs, their design additionally includes an inspection of the structure of the target protein, either experimental or computational, in order to identify putative tethering positions, i.e., residues near the ligand binding site amenable for cysteine mutagenesis screening.

(7)   Upon design of the photoswitchable ligand, synthesis and experimental testing can already be performed; the latter includes measuring the modulatory effect of the ligand under different light conditions, as well as site-directed mutagenesis (either Cys mutations for PTL covalent attachment or other mutations to confirm the binding site location and PCL/PTL ligand binding mode).

(8)   The observed light-dependent activity (or lack thereof), as well as the effect of mutations, can be rationalized *a posteriori* by performing a molecular docking of the PCL or virtual Cys mutation combined with covalent docking for the PTL. The resulting model of the target protein–photoswitch complex can be inspected to design additional site-directed mutations to validate the predicted PCL/PTL predicted binding mode. Alternatively, molecular docking can be used *a priori* (i.e., before experimental testing) to select the best candidate among several possible photoswitchable ligand designs (for subsequent experimental testing), as well as to explore alternative Cys tethering sites.

(9)   Though the (static) computational models described so far are already useful to understand the molecular basis of light-controlled ion channel modulation, they can additionally be refined by molecular dynamics. Such simulations, alone or in combination with enhanced sampling and free energy techniques, can provide further dynamical and energetic insights into the photoswitch effect, as explained earlier in this section.

(10)   This integrative computational-experimental approach offers a comprehensive understanding of the PCL/PTL effect on ion channel function, including but not limited to the information listed in the last step of the proposed workflow (see Figure 3).

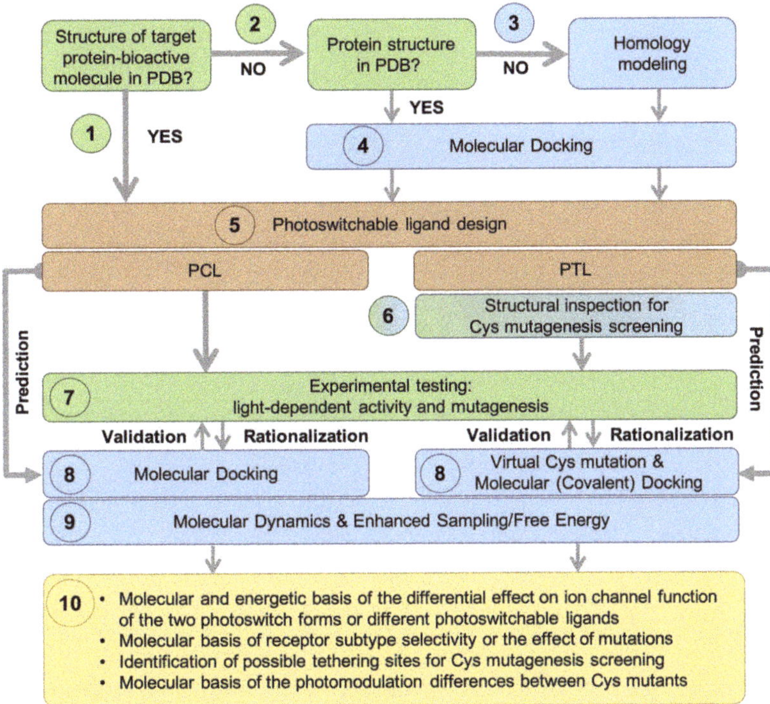

Figure 3. Proposed workflow for rational, structure-based design of photoswitchable ligands. The design step is indicated in orange, the experimental steps in green and the computational steps in blue. Ligand-based data (such as structure–activity relationship data or information about tethered, non-photoswitchable ligands), though not shown here, can also be integrated in the design step. Some of the possible information outcomes of this integrative computational-experimental workflow are listed in the yellow box.

In the following sections, we exemplify the steps of the proposed computational workflow (Figure 3) by discussing published computational modeling and simulation studies of photoswitchable ligands targeting VGICs and LGICs (see also Tables S1–S4).

3. Computational Modeling of Photoswitchable Ligands Targeting Voltage-Gated Ion Channels

VGICs are membrane proteins whose ion conduction pores open and close in response to changes in membrane voltage, intracellular signaling molecules or both [45–47]. This superfamily contains voltage-gated potassium, sodium and calcium channels (Kv, Nav and Cav, respectively), and other members, such as the transient receptor potential (TRP) channels. Two functional domains can be present in this superfamily (Figure 4): the voltage sensing domain (VSD), constituted by four transmembrane helices (S1–S4), and the ion pore domain (PD), formed by two transmembrane helices S5–S6 connected by the pore (P-)loop, which contains the ion selectivity filter. Kv and TRP channels are tetrameric proteins in which each subunit contains a VSD and a PD, whereas in the Nav and Cav channels this tetrameric assembly is encoded in a single gene. Nonetheless, other members of the VGIC superfamily lack a PD (e.g., Kir, HV1 and TPTE channels) or are not tetrameric (such as K2P channels).

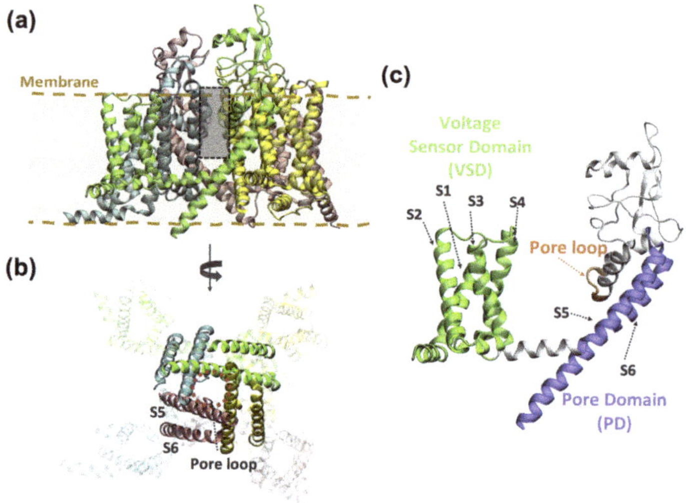

Figure 4. Representative structure of a tetrameric voltage-gated ion channel (VGIC). (**a**) Structure of the voltage-gated sodium channel Nav1.7, PDB code 6J8J [48], with protein subunits colored in green, yellow, pink and cyan, respectively. Only the pore-forming subunits are shown. The binding site of (photoswitchable) pore blockers is indicated with a gray box. (**b**) Alternative view of the same structure from the intracellular side. The helices displayed in solid colors correspond to the pore domain (S5 and S6 helices and the pore loop); the ion-conducting pore is indicated with a dashed red circle. (**c**) Detailed view of one Nav1.7 subunit indicating the two functional domains present in VGICs. The voltage sensor domain (VSD) has four transmembrane helices (S1–S4, colored in green) and the pore domain (PD) is constituted by two transmembrane helices (S5 and S6, in blue), connected by the pore (P-)loop, which contains the ion selectivity filter (in orange).

3.1. Photoswitchable Pore Blockers

A structural inspection of the first crystal structures of the KcsA potassium channel with TEA-like pore blockers [18,20] allowed the design of PTLs that target the homologous Shaker channel [21,22]. The ammonium binding site located in the extracellular vestibule of the pore (Figure 2b) appeared to be at the right distance from residue 422 (15–18 Å) to design a PTL (MAQ, **1** in Figure S1) whose quaternary ammonium group could reach this binding site when the azobenzene is in the *trans* configuration (approx. 17 Å long), but not in *cis* (~10 Å). The introduction of an E422C mutation for tethering and subsequent electrophysiological testing indeed revealed the photomodulation of the PTL-modified Shaker channel. Nonetheless, further structural information of the channel in complex with the photoswitchable ligand could help improve this initial design as well as characterize the molecular determinants of the differential effects of the *trans* and *cis* azobenzene forms.

In this regard, Mourot and coworkers [49] used a crystal structure of the Kv1.2–2.1 chimera in the open state and molecular docking to generate structural models of the Shaker K^+ channel bound to a PCL composed of two quaternary ammonium moieties connected by an azobenzene group (QAQ, **2**). The extended shape of *trans*-QAQ places the two positively charged groups at the right distance to interact with the two quaternary ammonium binding sites inside the pore, one in the extracellular vestibule and the other below the selectivity filter. The latter cannot be occupied by *cis*-QAQ due to its bent shape, explaining why the *cis* form is a less potent pore blocker than the *trans* one. A similar computational approach was used to rationalize the functional effects of FHU-779 (**3**), a PCL composed of an azobenzene-based long tail and the Cav pore blocker diltiazem [50]. In this case, homology modeling was first used to generate a structural model of the Cav1.2 ion pore, based on an open-state structure of the bacterial NavAb sodium channel. Afterwards, molecular docking was performed by Monte Carlo-based minimizations. The computational models again showed that both isomers can be accommodated inside the pore. However, the elongated *trans*-FHU-779 extends along the pore, with the positively charged nitrogen near the selectivity filter, the adjacent benzothiazepine moiety bound to a lateral fenestration and the long photoswitchable tail interacting with the C-terminal region of helix S6. In contrast, the tail of the "folded" *cis* form cannot reach the latter region, explaining the reversible light-dependent block of Cav1.2 by FHU-779.

In order to get further molecular and energetic insights into the binding of pore blocker PCLs, a recent study [51] used MD simulations, together with an enhanced sampling technique (Gaussian accelerated MD) and free energy calculations (based on a molecular mechanics generalized Born surface area or MMGBSA approach). The VGIC studied was the Nav1.4 channel, for which a recent cryo-EM structure in the inactivated state is available, and *p*-diaminoazobenzene (**4**) was used as a simplified model of the aforementioned photoswitchable pore blockers. Interestingly, the simulations revealed that there is more than one binding site for *p*-diaminoazobenzene in the *trans* configuration. *p*-diaminoazobenzene binds to two binding sites compatible with its expected pore blocking activity, one in the central cavity near the selectivity filter and the other near the intracellular gate. In addition, the *trans* form of the PCL appears to bind in a lateral cavity close to the membrane, which includes some residues previously identified as important for the binding of local anesthetics. The occupancy of this third binding site suggests that *p*-diaminoazobenzene could act not only as pore blocker but also have similar effects to local anesthetics.

3.2. Photoswitchable Modulators

Photoswitchable ligands for VGICs are not limited to pore blockers and pore openers [52]. For instance, several PCLs have been reported for TRP channels that act as activators or inhibitors [53–56]. The structural information (either experimental structures or homology models) of TRP channels can then be used to understand the mechanism of ligand-mediated activation or inhibition. In this regard, Lichtenegger and coworkers [57] designed a photoswitchable analog of the endogenous activator diarachidonlyglycerol (OptoDArG, **5**) of the TRPC3 channel. A homology model of the TRPC3 channel was built (based on the cryo-EM structure of the closely related TRPV1) in order to design a mutagenesis screening of the lipid binding cavity. This screening revealed a single glycine residue that connects the binding pocket and the selectivity filter through a lateral fenestration, thus providing clues on how lipid sensing controls ion channel gating. To the best of our knowledge, molecular docking and MD simulations have not been applied yet to study the binding of photoswitchable lipids to TRP channels. However, these two computational techniques have been extensively used to investigate channel modulation by other TRP ligands [58–62] and thus their use could be easily extended to photopharmacological applications.

4. Computational Modeling of Photoswitchable Ligands Targeting Ligand-Gated Ion Channels

LGICs are both ion channels that conduct ions across the neuronal membrane and receptors binding neurotransmitters. There are three main families [47,63,64]: pentameric LGICs (pLGICs), ionotropic glutamate receptors (iGluRs) and ATP-gated purinergic receptor (P2X) ion channels (Figure 5).

pLGICs [65] are composed of five identical or different subunits (homo- and heteropentamers, respectively). They encompass excitatory, cation-selective nicotinic acetylcholine receptors (nAchRs), serotonin or 5-hydroxytryptamine type 3 (5-HT$_3$) receptors and zinc-activated channels (ZAC), as well as inhibitory, anion-selective γ-aminobutyric acid receptors (GABARs) and glycine receptors (GlyRs). In addition, prokaryotic members of the pLGIC family include the *Gloeobacter* ligand-gated ion channel (GLIC), a proton-gated cation-selective channel; the *Erwinia chrysanthemi* ligand-gated ion channel (ELIC), a cation-selective channel activated by small amines, such as GABA; and the *C. elegans* glutamate-gated chloride channel (GluCl). Each subunit of these pentameric receptors (Figure 5a) can be divided into an extracellular domain (ECD, contributing to the neurotransmitter binding site), a transmembrane domain (TMD, formed by four helices, M1-M4, of which M2 lines the ion conduction pore) and, in some cases, an intracellular domain (ICD).

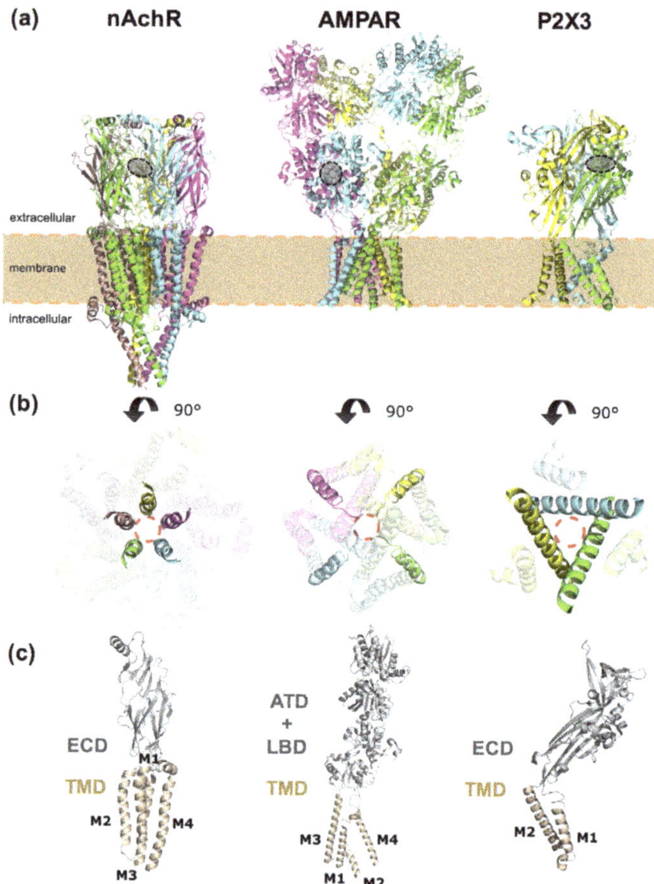

Figure 5. Representative structures of the three ligand-gated ion channel (LGIC) classes. (**a**) Shown from left to right, structures of a pentameric ligand-gated ion channel (pLGIC) based on the nicotinic acetylcholine receptor (nAchR), PDB code 7EKI [66]; an ionotropic glutamate receptor (iGluR), based on the AMPA receptor, PDB code 3KG2 [67]; and an ATP-gated purinergic (P2X) receptor, PDB code 5SVL [68]. Each protein subunit is shown in a different color (yellow, pink, green, cyan and/or magenta). A gray circle indicates the orthosteric binding site of LGICs, where the agonists or antagonists used to design most of the PCLs and PTLs mentioned in the text bind. For the sake of clarity, only one of the symmetric sites of the homomeric receptor is shown (out of the five present in nAchR, four in AMPAR and three in the P2X3 receptor). (**b**) Transmembrane domain (TMD) viewed from the extracellular side. The pore-lining structural elements are displayed in solid colors (i.e., the M2 helix for the nAchR and P2X3 receptors or the M2 helix and the re-entrant pore-loop for AMPAR) and the ion-conducting pore is indicated with a red dashed circle. (**c**) Detailed view of one subunit. The extracellular domains (ECD) of both the nAchR and P2X3 receptors are colored in gray, whereas the transmembrane domain (TMD) is colored in apricot. The AMPAR is displayed with the same color scheme, but the extracellular part of the receptor is divided into an amino terminal domain (ATD) and a ligand binding domain (LBD). Transmembrane helices are labeled from M1 to M4 or from M1 to M2, respectively. The intracellular domain is not shown for the sake of clarity.

iGluRs are LGICs essential for excitatory neurotransmission and can be classified in NMDA, AMPA and kainate receptors. They are tetrameric receptors (Figure 5b), with each subunit containing an extracellular amino terminal domain (ATD), an extracellular ligand binding domain (LBD, containing the glutamate binding site), a TMD (composed by three helices, M1, M3 and M4, as well as a re-entrant pore-loop, M2) and an intracellular carboxy-terminal domain (CTD).

Lastly, P2X receptors conduct mostly cations and are homo- or hetero-trimers (Figure 5c) composed of an ECD, a TMD (with two transmembrane helices, M1–M2, per subunit) and a C-terminal cytosolic tail.

4.1. Nicotinic Acetylcholine Receptors

nAchRs are the first ion channels for which light-modulated ligands were reported [11,12]. Based on a nAchR agonist, two photoswitchable ligands were designed [12]: bis-Q (**6** in Figure S2), a PCL with two quaternary ammonium moieties linked by an azobenzene group, and QBr (**7**), a PTL composed of a quaternary ammonium, azobenzene and benzylic bromide for Cys tethering. Both bis-Q and QBr acted as agonists in the *trans* form, whereas the *cis* form was almost inactive. In contrast, azo-CarCh (**8**) and azo-PTA (**9**) were found to act instead as light-reversible antagonists [11], despite the fact that these two PCLs were also designed based on two known nAChR agonists (carbamylcholine and phenyltrimetylammonium, respectively). This strongly indicates that small changes in the photoswitchable ligand structure can dramatically affect its light-modulated activity and thus structural knowledge of the receptor is needed to improve the ligand design.

Structural information was used by Tochitsky and coworkers [69] to devise light-modulated nAchRs (LinAchRs) that can be activated or inhibited with light, while responding normally to acetylcholine. They used PTLs (MAAch and MAHoCh, **10** and **11** in Figure S2, respectively) consisting of a Cys reactive maleimide group, an azobenzene photoswitch and a ligand head group mimicking known nAChR agonists (acetylcholine and homocholine, respectively). Potential positions to introduce Cys mutations for PTL covalent attachment were identified by inspecting the experimental structure of the soluble AChBP (as a surrogate of the nAChR ECD) in complex with carbamylcholine (a known nAchR agonist) (Figure 2a), as well as a computational model of $\alpha 4\beta 2$ nAchR in complex with MAAch (generated by combining homology modeling and molecular docking). The subsequent Cys screening showed that the E61C mutant was photomodulated by both PTLs, but with the opposite effects. Even though the design of both PTLs was based on nAChR agonists, E61C nAchR was photoactivated by *cis*-MAAch, but photoinhibited by *cis*-MAHoCh. The agonist activity of MAAch was further studied by repeating the aforementioned docking calculation with the homology model of $\alpha 4\beta 2$ nAchR, but adding a positional constraint that restricted the maleimide group to be within a certain radius of the C61 sulfur atom. The obtained docking poses showed that only *cis*-MAAch, but not its *trans* form, can position the bioactive ligand headgroup in the right place. Although a similar calculation was not performed for MAHoCh, the unexpected antagonist activity of this PTL was rationalized based on structural information for other nAChR antagonists. A 'foot-in-the-door' mechanism was proposed [70], by which antagonist binding prevents the complete closure of the ligand binding site (in particular loop C), as required for receptor activation.

Molecular docking was also used to rationalize the differential effect of the *trans* and *cis* forms of a PCL targeting an insect nAchR [71]. AMI-10 (**12**) contains two molecules of imidacloprid, an AchR agonist normally used as an insecticide, linked by an azobenzene group. The PCL turned out to be more effective upon irradiation, with the *cis* form showing a median lethal dose fivefold lower than the *trans*. Using a structure of the sea slug AchBP as a surrogate of the insect nAchR, Xu and coworkers showed that *cis*-AMI-10 can place the two imidacloprid moieties inside the binding pocket, whereas for the *trans* form the second moiety extends outwards, without forming any interaction with the protein. The larger number of protein–ligand interactions for the *cis* isomer thus correlates with its higher insecticide activity. A recent study [72] has reported an alternative design, in which the two imidacloprid moieties are linked by a photoswitchable dithienylethene group (DitIMI) in order to improve solubility compared to azobenzene.

4.2. 5-HT$_3$ Receptors

The 5-HT$_3$ receptor is another important cationic pLGIC, activated by serotonin, which is involved in a series of neurological disorders, from schizophrenia to drug abuse. Pharmacologically, it is the target of several drugs, including antiemetics, which act as antagonists, to alleviate the effects of cancer therapies [73]. To the best of our knowledge, only one study has explored and experimentally characterized azologs of reported antagonists of the 5-HT$_{3A}$ receptor [74], with only one of the investigated compounds retaining antagonist activity with no isomer specificity. Complementarily *trans-cis* switches based on Pro analogs have been used to study the gating mechanism of the 5-HT$_3$R. Although still a controversial mechanism, it has been proposed that the *trans-to-cis* isomerization of a Pro residue located in the loop connecting the M2 and M3 helices at the ECD–TMD interface (Pro8* or Pro281 in the X-ray structure of the mouse 5-HT$_3$R [75]) mediates channel gating. Mutations of this Pro into unnatural amino acid analogs strongly favoring the *trans* isomer resulted in non-functional channels [76], suggesting that, if *trans-cis* isomerization of Pro8* cannot occur, the channel would not open. A follow-up computational study [77] used MD simulations combined with enhanced sampling (metadynamics) to investigate the isomerization of a series of proline analogs (i.e., the ones tested in the aforementioned mutagenesis experiments) using dipeptide model systems in aqueous solution. A comparison of these simulations with the electrophysiology data showed an excellent correlation between the calculated free energy differences between the *cis* and *trans* isomers and the effect of the unnatural mutations on the receptor functional response. However, these simulations were performed on simplified models and thus did not take into account the effects of the receptor environment. These were addressed in subsequent molecular dynamics and metadynamics simulations of a model of the 5-HT$_3$R, built based on an X-ray structure [75] including the extracellular and transmembrane domains, embedded in a 1-palmitoyl-2-oleoyl-sn-glycero-3-phosphocholine (POPC) lipid bilayer (Figure 6a). These simulations showed how the protein environment affects the proline isomerization free energy landscape, which loses symmetry with respect to the case in water [78]. In addition, they provided the molecular details of the network of interactions of the proline potential switch with other residues at the ECD–TMD interface. On the one hand, such interactions select a preferential isomerization path. On the other, Pro isomerization causes the constriction of a ring of negatively charged Asp residues at the top of the pore-lining helix, which might enhance cation attraction and conduction. Altogether, the Pro molecular switch (Figure 6b) appears to behave as the endogenous counterpart of photoswitchable ligands (e.g., Glyght; see Section 4.4). In addition to 5-HT$_3$R, other ion channels [79] seem to control activation by using prolyl isomerization, which is additionally involved in many other biological processes [80,81]. Complementarily, the recent development of light-sensitive unnatural amino acids (UAAs) [82] has opened the way to endow light sensitivity to ion channels directly using these UAA probes.

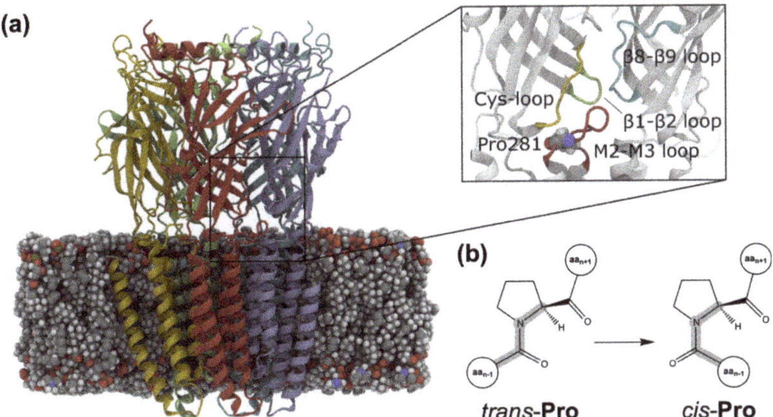

Figure 6. Prolyl isomerization in the serotonin or 5-hydroxytryptamine type 3 (5-HT$_3$) receptor. (**a**) Receptor model embedded in a 1-palmitoyl-2-oleoyl-sn-glycero-3-phosphocholine (POPC) lipid bilayer. The five protein subunits are displayed in red, yellow, green, light blue, and blue, respectively, and the lipid molecules as van der Waals spheres. The inset shows the extracellular domain (ECD)–transmembrane domain (TMD) interface region where the isomerizable Pro281 (in van der Waals spheres) is located, surrounded by the Cys loop (in yellow), the $\beta1$–$\beta2$ loop (green), the $\beta8$–$\beta9$ loop (blue), and the M2–M3 loop (red). Adapted with permission from Crnjar, A.; Comitani, F.; Hester, W.; Molteni, C. Trans–cis proline switches in a pentameric ligand-gated ion channel: how they are affected by and how they affect the biomolecular environment. *J. Phys. Chem. Lett.* **2019**, *10(3)*, 694–700. Copyright 2019 American Chemical Society. (**b**) Schematic representation of the *trans-cis* isomerization of proline.

4.3. GABA$_A$ Receptors

GABA$_A$ receptors are highly pharmacologically relevant pLGICs [83–85], being the targets of benzodiazepines for the treatment of anxiety, insomnia and depression, as well as of clinically used anesthetics (such as etomidate and propofol). The rich pharmacology of GABA$_A$ receptors offers a wide variety of starting options for the design of photoswitchable ligands. Nonetheless, until recently, such efforts were hampered by the lack of experimental structural information on the ligand binding sites [86–88]. The design of photoswitchable ligands would then rely on ligand structure–activity relationship data, as well as mutagenesis and Cys scanning data for the receptor binding sites. PCLs and PTLs were thus generated, targeting the orthosteric (GABA) binding site [89–91] or allosteric binding sites [92–94].

Homology modeling and molecular docking have been used to further characterize the experimentally observed photomodulation at the molecular level. For instance, Lin and coworkers [89] used a homology model of the receptor to map the potential tethering sites for Cys mutation in the orthosteric site and thereby design a light-modulated GABA$_A$R (LiGABA$_A$R), formed by α1(T125C), β2 and γ2S subunits. This model was built based on the experimental structures of AchBP (as a surrogate for the ECD of pLGICs) and the related *Torpedo* nAChR (for the TMD) [95]. Furthermore, molecular docking was performed to rationalize the antagonist effect of one of the proposed PTLs, MAB-0 (**13** in Figure S3), composed of maleimide, azobenzene and 4-hydroxybenzylamine. Although the latter group is not a typical gabaergic agonist/antagonist, in the *trans* form it appears to interact with several aromatic residues in the orthosteric site, thus enabling competitive inhibition against GABA. Instead, the *cis* isomer is not able to place the 4-hydroxybenzylamine near the putative interacting residues, consistent with its lack of effect on channel function. Homology modeling and molecular docking were also used in a recent study [96] to rationalize the different binding preferences of azogabazine (**14**). This PCL is composed of azobenzene and gabazine, a known GABA$_A$R competitive antagonist binding to the orthosteric site. In this case, the homology modeling approach benefited from the resolution of the first cryo-EM structures of heteropentameric GABA$_A$Rs. The model for the murine $\alpha1\beta2\gamma2$L receptor was built using the cryo-EM structure of human $\alpha1\beta3\gamma2$L GABA$_A$R as a template [97]. The *trans*-azogabazine docking poses revealed protein–ligand interactions similar to those observed in cryo-EM structures of other GABA$_A$R–antagonist complexes, explaining its antagonist activity. In addition, the *trans*-azogabazine docking poses were used to design site-directed mutagenesis experiments, which further confirmed the predicted binding mode [96].

The computational methods described above have also been used to investigate the binding of photoswitchable ligands to allosteric sites. Borghese and coworkers [98] used docking to identify possible tethering points in a $\alpha1\beta3\gamma2$ GABA$_A$R for MAP20 (**15**), a PTL composed of methanethiosulfonate, azobenzene and the anesthetic propofol. Capitalizing on a previously published model [99,100] of the parent compound propofol bound to the TMD binding site at the $\beta+\alpha$- interface [101], they manually superimposed the anesthetic part of the PTL and then sampled the rotations of the bond connecting propofol to azobenzene to identify nearby receptor residues amenable for Cys mutation. Experimental testing of the proposed Cys mutants showed that $\beta3$(M283C) and $\alpha1$(V227C) are the two Cys mutants displaying the largest photomodulation of GABA-induced currents upon treatment with MAP20. Subsequent structural modeling of MAP20 conjugated at these positions showed that the PTL can reach one or two propofol $\beta+\alpha$- binding sites depending on the tethering subunit ($\alpha1$(V227C) or $\beta3$(M283C), respectively). The predicted number of occupied binding sites is in line with the photomodulation of $\alpha1\beta3$(M283C)$\gamma2$ GABA$_A$R being larger than for $\alpha1$(V227C)$\beta3\gamma2$. Moreover, the computational models were validated retrospectively by comparison with a cryo-EM structure of GABA$_A$R bound to propofol [102].

Homology modeling and molecular docking have been used to rationalize the unforeseen photomodulatory effects of azo-NZ1 (**16**) [103]. This PCL was based on the benzodiazepine nitrazepam, which was conjugated with the azobenzene photoswitch and additionally a sulfonate group to improve solubility. Thus, the PCL design aimed at creating a photoswitchable positive allosteric modulator binding at the benzodiazepine binding site of GABA$_A$R, located at the $\alpha+\gamma$- interface in the ECD. Instead, electrophysiological data showed that *trans*-azo-NZ1 acts as a GABA$_A$R blocker, i.e., binds inside the TMD pore. Moreover, *trans*-azo-NZ1 inhibited GABA-mediated currents for some GABA$_A$R-ρ (or GABA$_C$R [104]) subtypes, as well as Gly-mediated currents for some GlyRs. This was completely unexpected, since GABA ρ subunits and GlyRs are insensitive to benzodiazepines. In order to understand the complete change of pharmacological activity of azo-NZ1 compared to the parent benzodiazepine, electrophysiological and mutagenesis experiments were combined with computational modeling. The homology model of $\alpha1\beta2\gamma2$ GABA$_A$R was taken from reference [105] (Figure 7a), whereas the homology models of GABA$_A$R-ρ were built following reference [106]. The main templates for these models are crystal structures of the homologous glutamate-gated chloride channel GluCl in the open state [107] and thus the resulting models are in the right functional state to study pore blockade upon GABA-triggered opening. Molecular docking (with flexible side chains for the pore-lining residues) showed that *trans*-azo-NZ1 has the right length to extend along the TMD ion pore (Figure 7b). The benzodiazepine core is positioned in the upper part of the TMD pore, forming hydrogen bonds with the 13′ residues (following the generalized numbering of the M2 helix residues for pLGICs) and hydrophobic interactions with the 6′ and 9′ residues, whereas the sulfonate group binds at the lower part (near the 2′ position). Interestingly, the negatively charged sulfonate overlaps with one of the chloride ion binding sites detected in cryo-EM structures [108]. Therefore, *trans*-azo-NZ1 hampers chloride conduction both sterically and electrostatically, consistent with the experimentally observed inhibitory effect. In contrast, the *cis* isomer binds in the middle region of the pore, which is wider, and thus blockade is less likely, in line with the inhibition relief upon UV irradiation. The docking calculations also provided molecular insights into the subtype selectivity of azo-NZ1 [103]. Electrophysiological experiments showed that both the heteropentameric $\alpha1\beta2\gamma2$ GABA$_A$R and the homopentameric GABA$_A$R-$\rho2$ were inhibited by *trans*-azo-NZ1, whereas GABA$_A$R-$\rho1$ was not. Moreover, sensitivity to azo-NZ1 depended on the residue at position 2': the S2'G GABA$_A$R-$\rho2$ mutation abolishes inhibition, whereas the P2'S GABA$_A$R-$\rho1$ mutation endows sensitivity to azo-NZ1. Based on the docking results on the three GABA$_A$Rs [103], it was proposed that *trans*-azo-NZ1-mediated inhibition requires either a hydrogen-bonding Ser or a residue of similar volume (Ala or Val) at position 2'.

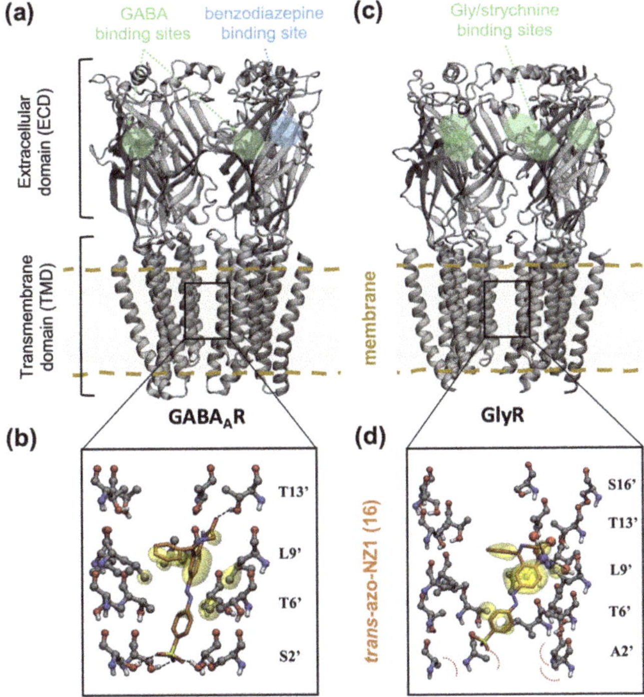

Figure 7. Computational models of the pore blocker *trans*-azo-NZ1 bound to γ-aminobutyric acid type A (GABA$_A$) and Gly receptors. (**a**) Computational model of $\alpha1\beta2\gamma2$ GABA$_A$R in complex with GABA [105]. The orthosteric GABA binding sites and the allosteric benzodiazepine binding site in the ECD are indicated with green and blue circles, respectively. For the sake of clarity, the front subunit of the heteropentamer is not shown. (**b**) Docking pose of *trans*-azo-NZ1 in the GABA$_A$R-$\rho2$ pore. Pore-lining residues are labeled according to the generalized numbering of the M2 helix residues for pLGICs. Hydrophobic interactions between azo-NZ1 and the receptor residues are marked as yellow transparent surfaces, hydrogen bond interactions and steric repulsion are represented with black and red dashed lines, respectively. Adapted from reference [103] with permission (CC-BY NC license) from the British Journal of Pharmacology, published by John Wiley and Sons (2019). (**c**) Cryo-EM structure of GlyR in complex with strychnine [109], a competitive antagonist that binds to the glycine neurotransmitter site, whose location is indicated with a green circle. For the sake of clarity, the front subunit is not shown. (**d**) Docking pose of *trans*-azo-NZ1 in the G2'A $\alpha1$ GlyR pore. Interactions are displayed following the same representation as in panel (**b**). Adapted from reference [110] with permission (CC-BY license) from eNeuro, published by the Society for Neuroscience (2020).

4.4. Glycine Receptors

Compared to $GABA_ARs$, the number of (photo)pharmacological agents for GlyRs is more limited [111,112]. Nonetheless, GlyRs are attracting growing attention as possible targets for painkillers [112,113]. Two PCLs based on azobenzene and a benzodiazepine core have been developed so far [110,114]. The aforementioned azo-NZ1 (**16** in Figure S3), as well as Glyght (**17**), exhibited light-modulated responses in an in vivo behavioral zebrafish assay. In vitro receptor screening using an heterologous expression system revealed that *trans*-azo-NZ1 acts as pore blocker for both $GABA_ARs$ and GlyRs, whereas *cis*-Glyght is a GlyR-selective negative allosteric modulator.

The *trans*-azo-NZ1-mediated pore blockade across several pLGICs resembles the inhibition mechanism of other pore blockers, such as picrotoxin [115,116]. Similar to $GABA_ARs$ [103], homology modeling and molecular docking were used to provide molecular insights into the binding of azo-NZ1 inside the GlyR ion pore, as well as to explain the GlyR subtype selectivity [110]. Capitalizing on the availability of cryo-EM structures of homomeric α1 GlyR [109] (Figure 7c), models of the homopentameric G2'A α1 GlyR mutant (as surrogate of α2 GlyR) and the heteropentameric α2/β GlyR were built. The subsequent docking calculations showed that the binding mode of azo-NZ1 inside the GlyR ion pore is very similar to the one of $GABA_ARs$, with the elongated *trans* isomer extending from the 13' to the 2' position (Figure 7d). The sulfonate group is still accommodated at the 2' position, despite the lack of a hydrogen-bonding residue, because the volume of the Ala 2' residue (present in GlyRs containing wild-type α2 and G2'A α1 mutant subunits) is similar to that of the Ser residue at the same position in ρ2 and γ2 $GABA_AR$ subunits. Instead, the Gly 2' residue in GlyRs containing wild-type α1 subunits is smaller; as a result, *trans*-azo-NZ1 is not able to completely block the pore, in line with the reduced inhibitory effect of the PCL for this GlyR subtype.

As mentioned above, the selective inhibitory effect of Glyght (**17**) on GlyRs when in *cis* form [114] was completely unexpected. The PCL was designed based on a nitrazepam/diazepam core and GlyRs do not contain a benzodiazepine binding site. Hence, a multilevel screening approach (Figure 8) was used to uncover the possible binding site of Glyght. A blind docking calculation was first run to identify putative binding pockets on the surface of the complete receptor. The results indicated that Glyght might potentially bind at the interface between ECD and TMD (Figure 8a,b). Therefore, an additional docking calculation was run, focused on this region. *Cis*-Glyght was found to bind in five symmetric sites located between two adjacent subunits in the homopentameric structure (Figure 8b,c). This region has been shown to participate in the allosteric coupling between neurotransmitter binding in the ECD and opening of the TMD ion pore [64]. Thus, it is likely that ligand binding at the ECD–TMD interface can thereby interfere in receptor activation. Lastly, the Glyght docking poses were further refined by using flexible docking centered on one out of the identified five symmetric intersubunit sites (Figure 8b,c). *Cis*-Glyght strengthens the interaction between M2–M3 and β8–β9 loops, which stabilizes the closed state. Such a "stapling" mechanism is in agreement with the stronger GlyR inhibition by *cis*-Glyght observed experimentally. Interestingly, the Glyght binding site in GlyR overlaps with the ECD/TMD interface region where the Pro molecular switch in 5-HT$_3$ receptors is located [76–78]. Thus, the light-modulated effect of Glyght on GlyR resembles the mechanism by which *trans-cis* Pro isomerization may mediate channel gating in the 5-HT$_3$ receptor [76–78]. This further supports the idea of the Pro molecular switch behaving as the endogenous counterpart of photoswitchable ligands.

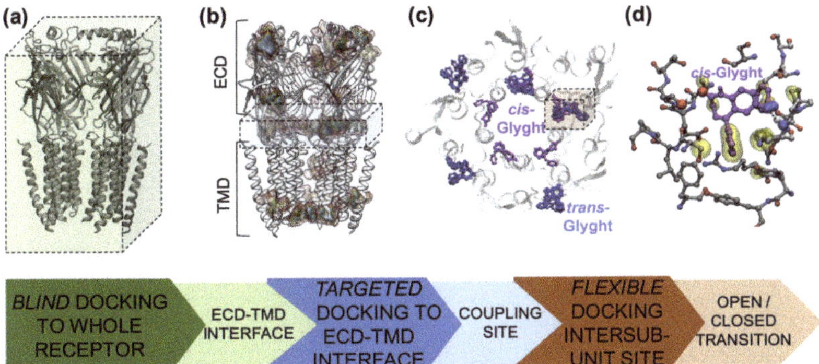

Figure 8. Multilevel docking screening to identify and characterize the binding site of Glyght in GlyR. (**a**) Blind docking of Glyght to the cryo-EM structure of α1 GlyR [109]. For the sake of clarity, the front subunit is not shown. The box size used for the blind docking calculation, encompassing the whole receptor, is shown as a green box. (**b**) Density map of the ligand poses of *cis*-Glyght obtained in the blind docking. Each contour line corresponds to a number density of 0.0006 particles/Å^3. Among the several high-density regions, the interface between ECD and TMD appears to be the most likely binding region for *cis*-Glyght [114], as it shows the largest differences compared to the *trans*-Glyght blind docking (data not shown). The box size for the subsequent targeted rigid docking is indicated as a blue box. (**c**) Docking poses obtained by rigid docking focused on the ECD–TMD interface. The five α1 GlyR subunits are represented in ribbons and colored in white, whereas the most populated ligand poses of *cis*- (violet) and *trans*-Glyght (blue) are shown as ball and sticks. The box size for the subsequent flexible docking is indicated as an orange box. (**d**) Detailed view of the *cis*-Glyght binding pose obtained by flexible docking centered at the ECD–TMD interface site. Hydrophobic contacts between *cis*-Glyght and the receptor residues are represented as a yellow surface, while the atoms involved in receptor–ligand hydrogen bonds are represented with larger spheres and a dashed line between them. Adapted from reference [114] with permission from Cell Chemical Biology, published by Elsevier (2020).

4.5. Ionotropic Glutamate Receptors

The design of the first photoswitchable ligands targeting ionotropic glutamate receptors [24,25,28] was based on inspection of X-ray structures of the LBD of the kainate receptor (GluK2, formerly known as GluR6) in complex with different ligands [23]. The LBD has a clamshell-like structure and the degree of closure upon ligand binding correlates with the degree of receptor activation. Moreover, these structures revealed an "exit tunnel" between the lips of the clamshell that could potentially accommodate the elongated *trans* form of the azobenzene photochrome. Based on these structural data, a PCL was designed consisting of the glutamate agonist and an azobenzene photoswitch (4-GluAzo, **18** in Figure S4), which acted as an agonist in the *trans* form but was inactive in the *cis* form [24], as intended with the original structure-guided design. In 2013, the crystal structure of GluK2 in complex with 4-GluAzo was solved [19] and confirmed the predicted binding mode of the *trans* isomer (Figure 2c). The glutamate group of 4-GluAzo interacts similarly to the endogenous glutamate agonist, while the *trans*-azobenzene is positioned between the lips of the clamshell. Complementarily, a manual docking calculation of the *cis* isomer on the same X-ray structure indicated that the distal phenyl ring in the *cis* configuration would clash with the lips of the clamshell and thus cause the opening of the LBD, explaining why 4-GluAzo is inactive upon irradiation. A follow-up computational study [117] used MD simulations together with umbrella sampling-based free energy calculations to further characterize the structural rearrangements occurring in the LBD upon PCL isomerization, as well as to estimate the change in binding affinity between *trans*- and *cis*-4-GluAzo.

An analysis of the available GluK2 X-ray structures [23] also allowed the design of a Cys mutagenesis screening to identify possible tethering sites of PTLs. Thereby, a light-modulated ionotropic glutamate receptor (LiGluR) was developed, containing a L439C mutation [25], whose activity was modulated by MAG compounds. Such PTLs consisted of maleimide, azobenzene and the agonist glutamate; the different modules were linked by one or more glycine units in order to vary the tether length. The first MAG compounds reported acted as photoswitchable LiGluR agonists when in the *cis* form [24,25,28]. Nonetheless, a subsequent study [118] reported other MAG compounds that were active in either the *cis* or *trans* forms, depending on the PTL tether length and the position where the tethering Cys was introduced (L439C or G486C). A model of the GluK2 LBD bound to the PTL MAG0 (**19**) was built based on the X-ray structure of the protein in complex with the parent compound glutamate [23]. Subsequent MD simulations and umbrella sampling-based free energy calculations [118] revealed that two factors contribute to determine which photoisomer is active: (i) the probability to properly orient the glutamate moiety inside the binding site of the open LBD and (ii) the degree of clamshell closure upon ligand binding.

A similar structure-based strategy was used to design photoswitchable ligands for other ionotropic glutamate receptors. In the case of NMDARs, light-modulated receptors (LiGluNs) were designed based on a structure-guided Cys mutagenesis screening to identify possible tethering sites for PTLs of the MAG series [119]. As described above for MAG compounds and LiGluRs, the attachment point and the length of the linker can yield either LiGluNs that are photoactivated (LiGluN2A-V173C or LiGluN2B-V714C) or photoinhibited (LiGluN1A-G172C and LiGluN1a-E406C) by MAGs. In the case of AMPAR, the crystal structure of the GluA2 LBD in complex with a benzyltetrazolyl-substituted AMPA (BnTetAMPA) [67] showed that the clamshell topology of the LBD is conserved among iGluRs. However, the degree of closure upon ligand binding is tighter for AMPARs compared to kainate receptors. As a result, the benzyl substituent of the AMPA analog is located in a cleft that opens into the solvent, suggesting that, upon replacement with azobenzene, the photochrome could still be accommodated in this cleft, but only if the azo group is added in the meta (or 3) position with respect to TetAMPA [120]. Hence, a new series of PCL compounds targeting AMPAR was designed called ATA; said compounds are composed of the azobenzene photoswitch, a tetrazolyl linker and the AMPA agonist [120]. The ATA PTLs were active in the *trans* form, as intended, and were selective for AMPARs over kainate receptors. A follow-up study [121] combined flexible docking, MD simulations and umbrella sampling-based free energy calculations to rationalize the differential effect of the two isomers of one of such ATA compounds, ATA-3 (**20**). Two ligand binding modes were found. One is similar to the crystallographic poses of other AMPAR agonists and corresponds to the most stable pose for the active *trans* isomer. The other could represent the position of the ligand immediately upon photoisomerization, in which a hydrogen bond is lost. From this second binding mode, the *cis* isomer could easily dissociate, unless it switches back to *trans*.

Photoswitchable ligands targeting ionotropic glutamate receptors are not limited to PCLs and PTLs binding to the LBD. A recent study [122] reported a photoswitchable pore blocker for the glutamate delta2 (GluD2) receptor. Glutamate delta receptors belong to the iGluR family due to their sequence and structure similarity to AMPA, NMDA and kainate receptors. However, they are considered to be orphan receptors because their LBD does not bind glutamate; instead, pore opening is regulated indirectly by metabotropic G protein-coupled glutamate receptors. Hence, Lemoine and coworkers [122] devised a PTL based on a known pore blocker (pentamidine), instead of an agonist binding to the LBD. MAGu (**21**) is composed of maleimide, azobenzene and a guanidinium head group that mimics the positively charged groups of pentamidine. The *trans* isomer is expected to reach the inside of the pore due to its elongated shape, whereas the shorter *cis* isomer will not. In order to identify possible positions to covalently attach MAGu on the pore-lining M3 helix, a homology model of GluD2 was built based on a crystal structure of the GluA2 receptor in the activated state. The thus-designed I677C mutant is blocked by *trans*-MAGu, but not by the *cis* form, and is denoted as the light-controllable

GluD2 (LiGluD2) channel. However, differently to pentamidine, the blockade by MAGu was not dependent on membrane voltage, suggesting that the two molecules may have different binding sites. This prediction was tested by performing covalent docking and ion pore calculations [122]. The positively charged group of MAGu does not seem to reach the inside of the pore, in line with the different blocking properties of MAGu and pentamidine observed experimentally. Instead, MAGu appears to alter the geometry and electrostatics of the pore, thus affecting ion conduction.

4.6. P2X Receptors

A recent review has compiled the photopharmacology applications for purinergic receptors reported until June 2021 [123]. Here, we showcase a study combining the use of a photoswitchable ligand and molecular dynamics simulations [124]. The used PTL (MAM, **22**) contains two maleimide groups separated by an azobenzene photoswitch, so that end-to-end distance of this Cys-Cys crosslinker changes with light. As previously carried out for other PTLs [125], the Cys screening of the transmembrane helices was designed using two homology models of the P2X2 receptor, based on experimental structures of the homologous P2X4 receptor in either closed (apo) or open (ATP-bound) states [126,127]. According to the experimental data for the P2X2 receptor, the two most promising Cys mutations were introduced into the equivalent positions (I336C and N353C) of the two aforementioned structures of the P2X4 receptor. Then, the MAM molecule was attached to these Cys mutants in either a "horizontal" (I336C/I336C mutant) or a "vertical" (I336C/N353C mutant) configuration and MD simulations were run for both P2X2 states. The simulations provided structural insights supporting the ideas that the P2X pore fluctuates among different open conformations and that receptor activation involves bending of the M2 helices at a conserved glycine residue that acts as a hinge for gating.

5. Conclusions and Perspectives

Light-controlled modulation of ion channels can be achieved by using photoswitchable ligands whose two isomers display different binding properties. Unraveling the molecular details of the complex between the target protein and the photoswitchable ligand in each of its two forms can help optimize this differential effect. Moreover, the addition of the photochromic group can modify the pharmacological properties of the parent bioactive molecule. Furthermore, the covalent attachment position and the length of the linker of photoswitchable tethered ligands can also result in different light-modulated effects. Given this complexity, a rational structure-based design approach is strongly recommended [2,6,28].

Indeed, the increasing availability of X-ray and cryo-EM structures has accelerated the development of both PCLs and PTLs for both VGICs [21,22] and LGICs [24,25]. Such experimental structural information has been complemented by computational modeling, in particular, homology modeling, molecular docking and molecular dynamics (Figure 3). In addition to providing structural models for ion channels without available experimental structures, these computational methods allow the modeling of the target protein bound to the photoswitchable ligand in either of its two forms. A comparison of the two counterpart models can be carried out either *a posteriori* (to explain the molecular basis of the observed light-modulation) or *a priori* (to predict whether the two photoswitch isomers will bind differently, as intended during the design). In the case of PTLs, covalent attachment can also be included in the model and thereby the fitness of the Cys mutant and the linker length can be explored for positioning the bioactive group in the correct binding pocket. Hence, nowadays, many photopharmacology studies integrate computational structural modeling together with the experimental data (Tables S1–S4). Based on the successful results obtained so far, we expect that computational methods will become instrumental in the field of ion channel photopharmacology in the coming years, especially when considering the ongoing methodological developments in the field.

Recent advancements in membrane protein structural biology, in particular cryo-EM, have substantially increased the number of available ion channel and receptor structures [128,129] that can be used as input to design photoswitchable ligands. In this regard, a recent study has reported crystal structures of the glutamate transporter homologue Glt_{Tk} in complex with a photoswitchable inhibitor in either *cis* or *trans* form [129]. In addition, these new experimental structures have broadened the range of templates available to build homology models for other ion channels without experimental structures. Furthermore, computational structural models can now be generated not only through homology modeling, but also using recently developed machine learning-based methods, such as AlphaFold [130] and RoseTTaFold [131].

Together with these computational protein structures, the accuracy of the predicted photoswitchable ligand binding modes is expected to improve thanks to the continuous development of molecular docking techniques and scoring functions [132]. Covalent docking methods [133,134] will be particularly useful for PTLs, whereas quantum mechanics (QM)-based docking approaches [135,136] may be applied to both PCLs and PTLs.

Complementarily, it is expected that the steady increase in computational resources, as well as the development of more efficient MD algorithms and enhanced sampling/free energy techniques [137–140], will also pave the way for further studies using classical simulations of photoswitchable ligands in complex with their target proteins. However, special care will be needed to develop accurate and transferable force field parameters for the two photoswitch forms [141]. Quantum mechanics/molecular mechanics (QM/MM) MD [142], in combination with excited state methods, will also allow us to study the photoisomerization of the azobenzene group within the PCL or PTL, either in solution or bound to the target protein. Nonetheless, going from the previous studies of azobenzene in solution [143,144] to photoswitchable ligands in complex with their target protein is likely to require adjustments in the theoretical treatment of the photochromic group excited states, as well as to address possible super-heating effects due to photoexcitation.

Hand in hand with these computational advancements, the toolkit of available photoswitchable ligands is rapidly expanding. The development of photoswitchable amino acids [82] and lipids [145] opens new avenues to investigate ion channel regulatory mechanisms. MD simulations are expected to be particularly useful to characterize at the molecular level the effect of these novel photoswitchable molecules [42,78,146,147]. Although the PCLs and PTLs mentioned in this review are based on azobenzene and its *trans-cis* isomerization upon light irradiation, other photochromic groups are increasingly being used, such as fulgimides [148], diarylethenes [149] and stilbenes [150], for which photoswitching involves bond formation. In these cases, the aforementioned QM- and QM/MM-based approaches could be combined with docking, MD and/or excited state calculations, to model the light-induced bond formation, providing an unprecedented atomistic picture of these exciting photoswitching processes.

Supplementary Materials: The following are available online at https://www.mdpi.com/1422-006 7/22/21/12072/s1.

Funding: CR and ANH are grateful for the financial support of the ERA SynBIO grant MODU-LIGHTOR (PCIN-2015-163-C02-01). MAP is funded by the Deutsche Forschungsgemeinschaft via the Research Unit FOR2518 "Functional Dynamics of Ion Channels and Transporters-DynIon", project P6. MAP also thankfully acknowledges the computer resources at MareNostrum III and IV and Mino-Tauro and the technical support provided by the Barcelona Supercomputing Center (RES activities BCV-2016-2-0002, BCV-2016-3-0005 and BCV-2017-2-0004). CM thanks the Engineering and Physical Sciences Research Council for computational support through the UKCP consortium (EPSRC grant P022472/1).

Acknowledgments: CR, ANH and MAP are greatly indebted to the MODULIGHTOR team (Pau Gorostiza, Piotr Begrestovski, Burkhard König and their lab members) for the collaborative projects and useful discussions regarding photopharmacology.

Conflicts of Interest: The authors declare no conflict of interest.

References and Notes

1. Velema, W.A.; Szymanski, W.; Feringa, B.L. Photopharmacology: Beyond Proof of Principle. *J. Am. Chem. Soc.* **2014**, *136*, 2178–2191. [CrossRef]
2. Broichhagen, J.; Frank, J.A.; Trauner, D. A Roadmap to Success in Photopharmacology. *Acc. Chem. Res.* **2015**, *48*, 1947–1960. [CrossRef] [PubMed]
3. Hüll, K.; Morstein, J.; Trauner, D. In Vivo Photopharmacology. *Chem. Rev.* **2018**, *118*, 10710–10747. [CrossRef] [PubMed]
4. Paoletti, P.; Ellis-Davies, G.C.R.; Mourot, A. Optical control of neuronal ion channels and receptors. *Nat. Rev. Neurosci.* **2019**, *20*, 514–532. [CrossRef]
5. Beharry, A.A.; Woolley, G.A. Azobenzene photoswitches for biomolecules. *Chem. Soc. Rev.* **2011**, *40*, 4422. [CrossRef]
6. Fehrentz, T.; Schönberger, M.; Trauner, D. Optochemical Genetics. *Angew. Chem. Int. Ed.* **2011**, *50*, 12156–12182. [CrossRef] [PubMed]
7. Ellis-Davies, G.C. Caged compounds: Photorelease technology for control of cellular chemistry and physiology. *Nat. Methods* **2007**, *4*, 619–628. [CrossRef]
8. Young, D.D.; Deiters, A. Photochemical control of biological processes. *Org. Biomol. Chem.* **2007**, *5*, 999–1005. [CrossRef] [PubMed]
9. Klán, P.; Solomek, T.; Bochet, C.G.; Blanc, A.; Givens, R.; Rubina, M.; Popik, V.; Kostikov, A.; Wirz, J. Photoremovable protecting groups in chemistry and biology: Reaction mechanisms and efficacy. *Chem. Rev.* **2013**, *113*, 119–191. [CrossRef]
10. Welleman, I.M.; Hoorens, M.W.H.; Feringa, B.L.; Boersma, H.H.; Szymański, W. Photoresponsive molecular tools for emerging applications of light in medicine. *Chem. Sci.* **2020**, *11*, 11672–11691. [CrossRef] [PubMed]
11. Deal, W.J.; Erlanger, B.F.; Nachmansohn, D. Photoregulation of biological activity by photochromic reagents, III. Photoregulation of bioelectricity by acetylcholine receptor inhibitors. *Proc. Natl. Acad. Sci. USA* **1969**, *64*, 1230–1234. [CrossRef]
12. Bartels, E.; Wassermann, N.H.; Erlanger, B.F. Photochromic Activators of the Acetylcholine Receptor. *Proc. Natl. Acad. Sci. USA* **1971**, *68*, 1820–1823. [CrossRef]
13. Silman, I.; Karlin, A. Acetylcholine Receptor: Covalent Attachment of Depolarizing Groups at the Active Site. *Science* **1969**, *164*, 1420–1421. [CrossRef] [PubMed]
14. Brejc, K.; van Dijk, W.J.; Klaassen, R.V.; Schuurmans, M.; van der Oost, J.; Smit, A.B.; Sixma, T.K. Crystal structure of an ACh-binding protein reveals the ligand-binding domain of nicotinic receptors. *Nature* **2001**, *411*, 269–276. [CrossRef]
15. Celie, P.H.; van Rossum-Fikkert, S.E.; van Dijk, W.J.; Brejc, K.; Smit, A.B.; Sixma, T.K. Nicotine and Carbamylcholine Binding to Nicotinic Acetylcholine Receptors as Studied in AChBP Crystal Structures. *Neuron* **2004**, *41*, 907–914. [CrossRef]
16. Bourne, Y.; Talley, T.T.; Hansen, S.B.; Taylor, P.; Marchot, P. Crystal structure of a Cbtx-AChBP complex reveals essential interactions between snake α-neurotoxins and nicotinic receptors. *EMBO J.* **2005**, *24*, 1512–1522. [CrossRef] [PubMed]
17. Hansen, S.B.; Sulzenbacher, G.; Huxford, T.; Marchot, P.; Taylor, P.; Bourne, Y. Structures of *Aplysia* AChBP Complexes Nicotinic Agon. Antagon. Reveal Distinctive Bind. Interfaces Conform. *EMBO J.* **2005**, *24*, 3635–3646. [CrossRef]
18. Lenaeus, M.J.; Vamvouka, M.; Focia, P.J.; Gross, A. Structural basis of TEA blockade in a model potassium channel. *Nat. Struct. Mol. Biol.* **2005**, *12*, 454–459. [CrossRef] [PubMed]
19. Reiter, A.; Skerra, A.; Trauner, D.; Schiefner, A. A Photoswitchable Neurotransmitter Analogue Bound to Its Receptor. *Biochemistry* **2013**, *52*, 8972–8974. [CrossRef] [PubMed]
20. Doyle, D.A. The Structure of the Potassium Channel: Molecular Basis of K^+ Conduction and Selectivity. *Science* **1998**, *280*, 69–77. [CrossRef]
21. Banghart, M.; Borges, K.; Isacoff, E.; Trauner, D.; Kramer, R.H. Light-activated ion channels for remote control of neuronal firing. *Nat. Neurosci.* **2004**, *7*, 1381–1386. [CrossRef]
22. Chambers, J.J.; Banghart, M.R.; Trauner, D.; Kramer, R.H. Light-Induced Depolarization of Neurons Using a Modified Shaker K^+ Channel and a Molecular Photoswitch. *J. Neurophysiol.* **2006**, *96*, 2792–2796. [CrossRef]
23. Mayer, M.L. Crystal Structures of the GluR5 and GluR6 Ligand Binding Cores: Molecular Mechanisms Underlying Kainate Receptor Selectivity. *Neuron* **2005**, *45*, 539–552. [CrossRef] [PubMed]
24. Volgraf, M.; Gorostiza, P.; Szobota, S.; Helix, M.R.; Isacoff, E.Y.; Trauner, D. Reversibly Caged Glutamate: A Photochromic Agonist of Ionotropic Glutamate Receptors. *J. Am. Chem. Soc.* **2006**, *129*, 260–261. [CrossRef]
25. Volgraf, M.; Gorostiza, P.; Numano, R.; Kramer, R.H.; Isacoff, E.Y.; Trauner, D. Allosteric control of an ionotropic glutamate receptor with an optical switch. *Nat. Chem. Biol.* **2005**, *2*, 47–52. [CrossRef] [PubMed]
26. Morstein, J.; Awale, M.; Reymond, J.L.; Trauner, D. Mapping the Azolog Space Enables the Optical Control of New Biological Targets. *ACS Cent. Sci.* **2019**, *5*, 607–618. [CrossRef]
27. Lee, E.H.; Hsin, J.; Sotomayor, M.; Comellas, G.; Schulten, K. Discovery through the computational microscope. *Structure* **2009**, *17*, 1295–1306. [CrossRef]
28. Gorostiza, P.; Isacoff, E. Optical switches and triggers for the manipulation of ion channels and pores. *Mol. Biosyst.* **2007**, *3*, 686. [CrossRef] [PubMed]
29. Bautista-Barrufet, A.; Izquierdo-Serra, M.; Gorostiza, P. Photoswitchable ion channels and receptors. In *Novel Approaches for Single Molecule Activation and Detection*; Benfenati, F., Di Fabrizio, E., Torre, V., Eds.; Springer: Berlin/Heidelberg, Germany, 2014; pp. 169–188.
30. Lerch, M.M.; Hansen, M.J.; van Dam, G.M.; Szymanski, W.; Feringa, B.L. Emerging Targets in Photopharmacology. *Angew. Chem. Int. Ed.* **2016**, *55*, 10978–10999. [CrossRef] [PubMed]

31. Kienzler, M.A.; Isacoff, E.Y. Precise modulation of neuronal activity with synthetic photoswitchable ligands. *Curr. Opin. Neurobiol.* **2017**, *45*, 202–209. [CrossRef]

32. Bregestovski, P.; Maleeva, G.; Gorostiza, P. Light-induced regulation of ligand-gated channel activity. *Br. J. Pharmacol.* **2017**, *175*, 1892–1902. [CrossRef] [PubMed]

33. Bregestovski, P.D.; Maleeva, G.V. Photopharmacology: A Brief Review Using the Control of Potassium Channels as an Example. *Neurosci. Behav. Physiol.* **2019**, *49*, 184–191. [CrossRef]

34. Bregestovski, P.D.; Ponomareva, D.N. Photochromic Modulation of Cys-loop Ligand-gated Ion Channels. *J. Evol. Biochem. Physiol.* **2021**, *57*, 354–371. [CrossRef]

35. Olivella, M.; Gonzalez, A.; Pardo, L.; Deupi, X. Relation between sequence and structure in membrane proteins. *Bioinformatics* **2013**, *29*, 1589–1592. [CrossRef]

36. Piccoli, S.; Suku, E.; Garonzi, M.; Giorgetti, A. Genome-wide Membrane Protein Structure Prediction. *Curr. Genom.* **2013**, *14*, 324–329. [CrossRef]

37. Sousa, S.; Ribeiro, A.; Coimbra, J.; Neves, R.; Martins, S.; Moorthy, N.; Fernandes, P.; Ramos, M. Protein-Ligand Docking in the New Millennium—A Retrospective of 10 Years in the Field. *Curr. Med. Chem.* **2013**, *20*, 2296–2314. [CrossRef] [PubMed]

38. Bucher, D.; Guidoni, L.; Carloni, P.; Rothlisberger, U. Coordination Numbers of K^+ and Na^+ Ions Inside the Selectivity Filter of the KcsA Potassium Channel: Insights from First Principles Molecular Dynamics. *Biophys. J.* **2010**, *98*, L47–L49. [CrossRef]

39. Salari, R.; Murlidaran, S.; Brannigan, G. Pentameric ligand-gated ion channels: Insights from computation. *Mol. Simul.* **2014**, *40*, 821–829. [CrossRef]

40. Cournia, Z.; Allen, T.W.; Andricioaei, I.; Antonny, B.; Baum, D.; Brannigan, G.; Buchete, N.V.; Deckman, J.T.; Delemotte, L.; del Val, C.; et al. Membrane Protein Structure, Function, and Dynamics: A Perspective from Experiments and Theory. *J. Membr. Biol.* **2015**, *248*, 611–640. [CrossRef]

41. Comitani, F.; Melis, C.; Molteni, C. Elucidating ligand binding and channel gating mechanisms in pentameric ligand-gated ion channels by atomistic simulations. *Biochem. Soc. Trans.* **2015**, *43*, 151–156. [CrossRef]

42. Crnjar, A.; Comitani, F.; Melis, C.; Molteni, C. Mutagenesis computer experiments in pentameric ligand-gated ion channels: The role of simulation tools with different resolution. *Interface Focus* **2019**, *9*, 20180067. [CrossRef] [PubMed]

43. Howard, R.J.; Carnevale, V.; Delemotte, L.; Hellmich, U.A.; Rothberg, B.S. Permeating disciplines: Overcoming barriers between molecular simulations and classical structure-function approaches in biological ion transport. *Biochim. Biophys. Acta Biomembr.* **2018**, *1860*, 927–942. [CrossRef]

44. Carnevale, V.; Delemotte, L.; Howard, R.J. Molecular Dynamics Simulations of Ion Channels. *Trends Biochem. Sci.* **2021**, *46*, 621–622. [CrossRef] [PubMed]

45. Yu, F.H.; Catterall, W.A. The VGL-Chanome: A Protein Superfamily Specialized for Electrical Signaling and Ionic Homeostasis. *Sci. Signal.* **2004**, *2004*, re15. [CrossRef] [PubMed]

46. Moreau, A.; Gosselin-Badaroudine, P.; Chahine, M. Biophysics, pathophysiology, and pharmacology of ion channel gating pores. *Front. Pharmacol.* **2014**, *5*. [CrossRef]

47. Alexander, S.P.H.; Mathie, A.; Peters, J.A.; Veale, E.L.; Striessnig, J.; Kelly, E.; Armstrong, J.F.; Faccenda, E.; Harding, S.D.; Pawson, A.J.; et al. The Concise Guide to Pharmacology 2019/20: Ion channels. *Br. J. Pharmacol.* **2019**, *176*. [CrossRef] [PubMed]

48. Shen, H.; Liu, D.; Wu, K.; Lei, J.; Yan, N. Structures of human Nav1.7 channel in complex with auxiliary subunits and animal toxins. *Science* **2019**, *363*, 1303–1308. [CrossRef]

49. Mourot, A.; Herold, C.; Kienzler, M.A.; Kramer, R.H. Understanding and improving photo-control of ion channels in nociceptors with azobenzene photo-switches. *Br. J. Pharmacol.* **2017**, *175*, 2296–2311. [CrossRef]

50. Fehrentz, T.; Huber, F.M.E.; Hartrampf, N.; Bruegmann, T.; Frank, J.A.; Fine, N.H.F.; Malan, D.; Danzl, J.G.; Tikhonov, D.B.; Sumser, M.; et al. Optical control of L-type Ca^{2+} channels using a diltiazem photoswitch. *Nat. Chem. Biol.* **2018**, *14*, 764–767. [CrossRef]

51. Palmisano, V.F.; Gómez-Rodellar, C.; Pollak, H.; Cárdenas, G.; Corry, B.; Faraji, S.; Nogueira, J.J. Binding of azobenzene and P-Diaminoazobenzene Hum. Volt.-Gated Sodium Channel Nav1.4. *Phys. Chem. Chem. Phys.* **2021**, *23*, 3552–3564. [CrossRef]

52. Trads, J.B.; Hüll, K.; Matsuura, B.S.; Laprell, L.; Fehrentz, T.; Görldt, N.; Kozek, K.A.; Weaver, C.D.; Klöcker, N.; Barber, D.M.; et al. Sign Inversion in Photopharmacology: Incorporation of Cyclic Azobenzenes in Photoswitchable Potassium Channel Blockers and Openers. *Angew. Chem. Int. Ed.* **2019**, *58*, 15421–15428. [CrossRef]

53. Stein, M.; Breit, A.; Fehrentz, T.; Gudermann, T.; Trauner, D. Optical Control of TRPV1 Channels. *Angew. Chem. Int. Ed.* **2013**, *52*, 9845–9848. [CrossRef]

54. Kokel, D.; Cheung, C.Y.J.; Mills, R.; Coutinho-Budd, J.; Huang, L.; Setola, V.; Sprague, J.; Jin, S.; Jin, Y.N.; Huang, X.P.; et al. Photochemical activation of TRPA1 channels in neurons and animals. *Nat. Chem. Biol.* **2013**, *9*, 257–263. [CrossRef] [PubMed]

55. Cunha, M.R.; Bhardwaj, R.; Lindinger, S.; Butorac, C.; Romanin, C.; Hediger, M.A.; Reymond, J.L. Photoswitchable Inhibitor of the Calcium Channel TRPV6. *ACS Med. Chem. Lett.* **2019**, *10*, 1341–1345. [CrossRef] [PubMed]

56. Curcic, S.; Tiapko, O.; Groschner, K. Photopharmacology and opto-chemogenetics of TRPC channels-some therapeutic visions. *Pharmacol. R.* **2019**, *200*, 13–26. [CrossRef]

57. Lichtenegger, M.; Tiapko, O.; Svobodova, B.; Stockner, T.; Glasnov, T.N.; Schreibmayer, W.; Platzer, D.; de la Cruz, G.G.; Krenn, S.; Schober, R.; et al. An optically controlled probe identifies lipid-gating fenestrations within the TRPC3 channel. *Nat. Chem. Biol.* **2018**, *14*, 396–404. [CrossRef] [PubMed]

58. Jorgensen, C.; Domene, C. Location and Character of Volatile General Anesthetics Binding Sites in the Transmembrane Domain of TRPV1. *Mol. Pharm.* **2018**, *15*, 3920–3930. [CrossRef]

59. Oakes, V.; Domene, C. Combining Structural Data with Computational Methodologies to Investigate Structure–Function Relationships in TRP Channels. In *TRP Channels*; Springer: New York, NY, USA, 2019; Volume 1987, pp. 65–82. [CrossRef]

60. Koleva, M.N.; Fernandez-Ballester, G. In Silico Approaches for TRP Channel Modulation. In *TRP Channels*; Springer: New York, NY, USA, 2019; Volume 1987, pp. 187–206. [CrossRef]

61. Chernov-Rogan, T.; Gianti, E.; Liu, C.; Villemure, E.; Cridland, A.P.; Hu, X.; Ballini, E.; Lange, W.; Deisemann, H.; Li, T.; et al. TRPA1 modulation by piperidine carboxamides suggests an evolutionarily conserved binding site and gating mechanism. *Proc. Natl. Acad. Sci. USA* **2019**, *116*, 26008–26019. [CrossRef]

62. Yazici, A.T.; Gianti, E.; Kasimova, M.A.; Lee, B.H.; Carnevale, V.; Rohacs, T. Dual regulation of TRPV1 channels by phosphatidylinositol via functionally distinct binding sites. *J. Biol. Chem.* **2021**, *296*, 100573. [CrossRef]

63. Collingridge, G.L.; Olsen, R.W.; Peters, J.; Spedding, M. A nomenclature for ligand-gated ion channels. *Neuropharmacology* **2009**, *56*, 2–5. [CrossRef]

64. Jaiteh, M.; Taly, A.; Hénin, J. Evolution of Pentameric Ligand-Gated Ion Channels: Pro-Loop Receptors. *PLoS ONE* **2016**, *11*, e0151934. [CrossRef]

65. Although eukaryotic pLGICs have been traditionally denoted as Cys-loop receptors, the lack of such disulfide bond-containing loop in the prokaryotic members of this family makes the term pLGIC preferable, according to the latest guidelines of the NC-IUPHAR; see reference [47]. Alternatively, the name Pro-loop receptors has been proposed in reference [64].

66. Zhao, Y.; Liu, S.; Zhou, Y.; Zhang, M.; Chen, H.; Xu, H.E.; Sun, D.; Liu, L.; Tian, C. Structural basis of human α7 nicotinic acetylcholine receptor activation. *Cell Res.* **2021**, *31*, 713–716. [CrossRef]

67. Sobolevsky, A.I.; Rosconi, M.P.; Gouaux, E. X-ray structure, symmetry and mechanism of an AMPA-subtype glutamate receptor. *Nature* **2009**, *462*, 745–756. [CrossRef]

68. Mansoor, S.E.; Lü, W.; Oosterheert, W.; Shekhar, M.; Tajkhorshid, E.; Gouaux, E. X-ray structures define human P2X3 receptor gating cycle and antagonist action. *Nature* **2016**, *538*, 66–71. [CrossRef]

69. Tochitsky, I.; Banghart, M.R.; Mourot, A.; Yao, J.Z.; Gaub, B.; Kramer, R.H.; Trauner, D. Optochemical control of genetically engineered neuronal nicotinic acetylcholine receptors. *Nat. Chem.* **2012**, *4*, 105–111. [CrossRef] [PubMed]

70. Law, R.J.; Henchman, R.H.; McCammon, J.A. A gating mechanism proposed from a simulation of a human α7 nicotinic acetylcholine receptor. *Proc. Natl. Acad. Sci. USA* **2005**, *102*, 6813–6818. [CrossRef] [PubMed]

71. Xu, Z.; Shi, L.; Jiang, D.; Cheng, J.; Shao, X.; Li, Z. Azobenzene Modified Imidacloprid Derivatives as Photoswitchable Insecticides: Steering Molecular Activity in a Controllable Manner. *Sci. Rep.* **2015**, *5*, 13962. [CrossRef] [PubMed]

72. Zhang, C.; Xu, Q.; Xu, Z.; Wang, L.; Liu, Z.; Li, Z.; Shao, X. Optical Control of Invertebrate nAChR and Behaviors with Dithienylethene-Imidacloprid. *bioRxiv* **2021**. [CrossRef]

73. Kesters, D.; Thompson, A.J.; Brams, M.; van Elk, R.; Spurny, R.; Geitmann, M.; Villalgordo, J.M.; Guskov, A.; Danielson, U.H.; Lummis, S.C.R.; et al. Structural basis of ligand recognition in 5-HT$_3$ receptors. *EMBO Rep.* **2013**, *14*, 49–56. [CrossRef]

74. Rustler, K.; Maleeva, G.; Bregestovski, P.; König, B. Azologization of serotonin 5-HT$_3$ receptor antagonists. *Beilstein J. Org. Chem.* **2019**, *15*, 780–788. [CrossRef] [PubMed]

75. Hassaine, G.; Deluz, C.; Grasso, L.; Wyss, R.; Tol, M.B.; Hovius, R.; Graff, A.; Stahlberg, H.; Tomizaki, T.; Desmyter, A.; et al. X-ray structure of the mouse serotonin 5-HT$_3$ receptor. *Nature* **2014**, *512*, 276–281. [CrossRef]

76. Lummis, S.C.R.; Beene, D.L.; Lee, L.W.; Lester, H.A.; Broadhurst, R.W.; Dougherty, D.A. *Cis-trans* isomerization at a proline opens the pore of a neurotransmitter-gated ion channel. *Nature* **2005**, *438*, 248–252. [CrossRef]

77. Melis, C.; Bussi, G.; Lummis, S.C.R.; Molteni, C. *Trans-Cis* Switching Mechanisms in Proline Analogs and Their Relevance for the Gating of the 5-HT$_3$ Receptor. *J. Phys. Chem. B* **2009**, *113*, 12148–12153. [CrossRef] [PubMed]

78. Crnjar, A.; Comitani, F.; Hester, W.; Molteni, C. *Trans-Cis* Proline Switch. A Pentameric Ligand-Gated Ion Channel: How They Are Affect. How They Affect Biomol. Environ. *J. Phys. Chem. Lett.* **2019**, *10*, 694–700. [CrossRef] [PubMed]

79. Schmidpeter, P.A.M.; Rheinberger, J.; Nimigean, C.M. Prolyl isomerization controls activation kinetics of a cyclic nucleotide-gated ion channel. *Nat. Commun.* **2020**, *11*. [CrossRef] [PubMed]

80. Leone, V.; Lattanzi, G.; Molteni, C.; Carloni, P. Mechanism of Action of Cyclophilin A Explored by Metadynamics Simulations. *PLoS Comput. Biol.* **2009**, *5*, e1000309. [CrossRef]

81. Maschio, M.C.; Fregoni, J.; Molteni, C.; Corni, S. Proline isomerization effects in the amyloidogenic protein β_2-microglobulin. *Phys. Chem. Chem. Phys.* **2021**, *23*, 356–367. [CrossRef]

82. Klippenstein, V.; Mony, L.; Paoletti, P. Probing Ion Channel Structure and Function Using Light-Sensitive Amino Acids. *Trends Biochem. Sci.* **2018**, *43*, 436–451. [CrossRef]

83. Sieghart, W.; Savić, M.M. International Union of Basic and Clinical Pharmacology. CVI: GABA$_A$ Receptor Subtype- and Function-selective Ligands: Key Issues in Translation to Humans. *Pharmacol. Rev.* **2018**, *70*, 836–878. [CrossRef]

84. Olsen, R.W.; Sieghart, W. GABA$_A$ receptors: Subtypes provide diversity of function and pharmacology. *Neuropharmacology* **2009**, *56*, 141–148. [CrossRef]

85. Belelli, D.; Hales, T.G.; Lambert, J.J.; Luscher, B.; Olsen, R.; Peters, J.A.; Rudolph, U.; Sieghart, W. GABA$_A$ receptors (version 2019.4) in the IUPHAR/BPS Guide to Pharmacology Database. *IUPHAR/BPS Guide Pharmacol. CITE* **2019**, *2019*, 4. [CrossRef]

86. Puthenkalam, R.; Hieckel, M.; Simeone, X.; Suwattanasophon, C.; Feldbauer, R.V.; Ecker, G.F.; Ernst, M. Structural Studies of GABA$_A$ Receptor Binding Sites: Which Experimental Structure Tells us What? *Front. Mol. Neurosci.* **2016**, *9*. [CrossRef]

87. Scott, S.; Aricescu, A.R. A structural perspective on GABA$_A$ receptor pharmacology. *Curr. Opin. Struct. Biol.* **2019**, *54*, 189–197. [CrossRef] [PubMed]

88. Kim, J.J.; Hibbs, R.E. Direct Structural Insights into GABA$_A$ Receptor Pharmacology. *Trends Biochem. Sci.* **2021**, *46*, 502–517. [CrossRef] [PubMed]

89. Lin, W.C.; Davenport, C.M.; Mourot, A.; Vytla, D.; Smith, C.M.; Medeiros, K.A.; Chambers, J.J.; Kramer, R.H. Engineering a Light-Regulated GABA$_A$ Receptor for Optical Control of Neural Inhibition. *ACS Chem. Biol.* **2014**, *9*, 1414–1419. [CrossRef]

90. Lin, W.C.; Tsai, M.C.; Davenport, C.M.; Smith, C.M.; Veit, J.; Wilson, N.M.; Adesnik, H.; Kramer, R.H. A Comprehensive Optogenetic Pharmacology Toolkit for *In Vivo* Control of GABA$_A$ Receptors and Synaptic Inhibition. *Neuron* **2015**, *88*, 879–891. [CrossRef] [PubMed]

91. Lin, W.C.; Tsai, M.C.; Rajappa, R.; Kramer, R.H. Design of a Highly Bistable Photoswitchable Tethered Ligand for Rapid and Sustained Manipulation of Neurotransmission. *J. Am. Chem. Soc* **2018**, *140*, 7445–7448. [CrossRef]

92. Yue, L.; Pawlowski, M.; Dellal, S.S.; Xie, A.; Feng, F.; Otis, T.S.; Bruzik, K.S.; Qian, H.; Pepperberg, D.R. Robust photoregulation of GABA$_A$ receptors by allosteric modulation with a propofol analogue. *Nat. Commun.* **2012**, *3*. [CrossRef]

93. Stein, M.; Middendorp, S.J.; Carta, V.; Pejo, E.; Raines, D.E.; Forman, S.A.; Sigel, E.; Trauner, D. Azo-Propofols: Photochromic Potentiators of GABA$_A$ Receptors. *Angew. Chem. Int. Ed.* **2012**, *51*, 10500–10504. [CrossRef]

94. Rustler, K.; Maleeva, G.; Gomila, A.M.J.; Gorostiza, P.; Bregestovski, P.; König, B. Optical Control of GABA$_A$ Receptors with a Fulgimide-Based Potentiator. *Chem. Eur. J.* **2020**, *26*, 12722–12727. [CrossRef]

95. O'Mara, M.; Cromer, B.; Parker, M.; Chung, S.H. Homology Model of the GABA$_A$ Receptor Examined Using Brownian Dynamics. *Biophys. J.* **2005**, *88*, 3286–3299. [CrossRef]

96. Mortensen, M.; Huckvale, R.; Pandurangan, A.P.; Baker, J.R.; Smart, T.G. Optopharmacology reveals a differential contribution of native GABA$_A$ receptors to dendritic and somatic inhibition using azogabazine. *Neuropharmacology* **2020**, *176*, 108135. [CrossRef]

97. Laverty, D.; Desai, R.; Uchański, T.; Masiulis, S.; Stec, W.J.; Malinauskas, T.; Zivanov, J.; Pardon, E.; Steyaert, J.; Miller, K.W.; et al. Cryo-EM structure of the human α1β3γ2 GABA$_A$ receptor in a lipid bilayer. *Nature* **2019**, *565*, 516–520. [CrossRef] [PubMed]

98. Borghese, C.M.; Wang, H.Y.L.; McHardy, S.F.; Messing, R.O.; Trudell, J.R.; Harris, R.A.; Bertaccini, E.J. Modulation of α1β3γ2 GABA$_A$ receptors expressed in *X. Laevis* Oocytes Using A Propofol Photoswitch Tethered Transmembrane Helix. *Proc. Natl. Acad. Sci. USA* **2021**, *118*, e2008178118. [CrossRef]

99. Bertaccini, E.J.; Yoluk, O.; Lindahl, E.R.; Trudell, J.R. Assessment of Homology Templates and an Anesthetic Binding Site within the γ-Aminobutyric Acid Receptor. *Anesthesiology* **2013**, *119*, 1087–1095. [CrossRef]

100. Cayla, N.S.; Dagne, B.A.; Wu, Y.; Lu, Y.; Rodriguez, L.; Davies, D.L.; Gross, E.R.; Heifets, B.D.; Davies, M.F.; MacIver, M.B.; et al. A newly developed anesthetic based on a unique chemical core. *Proc. Natl. Acad. Sci. USA* **2019**, *116*, 15706–15715. [CrossRef] [PubMed]

101. GABA receptors composed by α, β and γ subunits are assembled in a β-α-γ-β-α heteropentameric arrangement. In order to distinguish the different subunit interfaces, a principal (+) and a complementary (−) sides are defined, so that the heteropentamer contains the following interfaces: β+α-, α+γ-, γ+β- and α+β-. The two GABA binding sites are then located at the two β+α- interfaces in the extracellular domain.

102. Kim, J.J.; Gharpure, A.; Teng, J.; Zhuang, Y.; Howard, R.J.; Zhu, S.; Noviello, C.M.; Walsh, R.M.; Lindahl, E.; Hibbs, R.E. Shared structural mechanisms of general anaesthetics and benzodiazepines. *Nature* **2020**, *585*, 303–308. [CrossRef]

103. Maleeva, G.; Wutz, D.; Rustler, K.; Nin-Hill, A.; Rovira, C.; Petukhova, E.; Bautista-Barrufet, A.; Gomila-Juaneda, A.; Scholze, P.; Peiretti, F.; et al. A photoswitchable GABA receptor channel blocker. *Br. J. Pharmacol.* **2019**, *176*, 2661–2677. [CrossRef]

104. Although GABA receptors formed by ρ-subunits have been traditionally refer to as GABA$_C$ receptors, nowadays the NC-IUPHAR recommends their classification as GABA$_A$ receptors, based on structural and functional criteria [85].

105. Bergmann, R.; Kongsbak, K.; Sørensen, P.L.; Sander, T.; Balle, T. A Unified Model of the GABA$_A$ Receptor Comprising Agonist and Benzodiazepine Binding Sites. *PLoS ONE* **2013**, *8*, e52323. [CrossRef] [PubMed]

106. Naffaa, M.M.; Chebib, M.; Hibbs, D.E.; Hanrahan, J.R. Comparison of templates for homology model of ρ1 GABA$_C$ receptors: More insights to the orthosteric binding site's structure and functionality. *J. Mol. Graph. Model.* **2015**, *62*, 43–55. [CrossRef]

107. Hibbs, R.E.; Eric, G. Principles of activation and permeation in an anion-selective Cys-loop receptor. *Nature* **2011**, *474*, 54–60. [CrossRef]

108. Laverty, D.; Thomas, P.; Field, M.; Andersen, O.J.; Gold, M.G.; Biggin, P.C.; Gielen, M.; Smart, T.G. Crystal structures of a GABA$_A$-receptor chimera reveal new endogenous neurosteroid-binding sites. *Nat. Struct. Mol. Biol.* **2017**, *24*, 977–985. [CrossRef]

109. Du, J.; Lü, W.; Wu, S.; Cheng, Y.; Gouaux, E. Glycine receptor mechanism elucidated by electron cryo-microscopy. *Nature* **2015**, *526*, 224–229. [CrossRef] [PubMed]

110. Maleeva, G.; Nin-Hill, A.; Rustler, K.; Petukhova, E.; Ponomareva, D.; Mukhametova, E.; Gomila, A.M.; Wutz, D.; Alfonso-Prieto, M.; König, B.; et al. Subunit-Specific Photocontrol of Glycine Receptors by Azobenzene-Nitrazepam Photoswitcher. *eNeuro* **2020**, *8*. [CrossRef]

111. Lynch, J.W.; Zhang, Y.; Talwar, S.; Estrada-Mondragon, A. Glycine Receptor Drug Discovery. In *Advances in Pharmacology*; Geraghty, D.P.; Rash, L.D., Eds.; Elsevier: Amsterdam, The Netherlands, 2017; Volume 79, pp. 225–253. [CrossRef]

112. Zeilhofer, H.U.; Acuña, M.A.; Gingras, J.; Yévenes, G.E. Glycine receptors and glycine transporters: Targets for novel analgesics? *Cell. Mol. Life Sci.* **2017**, *75*, 447–465. [CrossRef] [PubMed]
113. Huang, X.; Shaffer, P.L.; Ayube, S.; Bregman, H.; Chen, H.; Lehto, S.G.; Luther, J.A.; Matson, D.J.; McDonough, S.I.; Michelsen, K.; et al. Crystal structures of human glycine receptor α3 bound to a novel class of analgesic potentiators. *Nat. Struct. Mol. Biol.* **2016**, *24*, 108–113. [CrossRef]
114. Gomila, A.M.; Rustler, K.; Maleeva, G.; Nin-Hill, A.; Wutz, D.; Bautista-Barrufet, A.; Rovira, X.; Bosch, M.; Mukhametova, E.; Petukhova, E.; et al. Photocontrol of Endogenous Glycine Receptors *In Vivo*. *Cell Chem. Biol.* **2020**, *27*, 1425–1433.e7. [CrossRef]
115. Zhorov, B.S.; Bregestovski, P.D. Chloride Channels of Glycine and GABA Receptors with Blockers: Monte Carlo Minimization and Structure-Activity Relationships. *Biophys. J.* **2000**, *78*, 1786–1803. [CrossRef]
116. Gielen, M.; Corringer, P.J. The dual-gate model for pentameric ligand-gated ion channels activation and desensitization. *J. Physiol.* **2018**, *596*, 1873–1902. [CrossRef]
117. Guo, Y.; Wolter, T.; Kubař, T.; Sumser, M.; Trauner, D.; Elstner, M. Molecular Dynamics Investigation of gluazo, a Photo-Switchable Ligand for the Glutamate Receptor GluK2. *PLoS ONE* **2015**, *10*, e0135399. [CrossRef]
118. Numano, R.; Szobota, S.; Lau, A.Y.; Gorostiza, P.; Volgraf, M.; Roux, B.; Trauner, D.; Isacoff, E.Y. Nanosculpting reversed wavelength sensitivity into a photoswitchable iGluR. *Proc. Natl. Acad. Sci. USA* **2009**, *106*, 6814–6819. [CrossRef] [PubMed]
119. Berlin, S.; Szobota, S.; Reiner, A.; Carroll, E.C.; Kienzler, M.A.; Guyon, A.; Xiao, T.; Trauner, D.; Isacoff, E.Y. A family of photoswitchable NMDA receptors. *eLife* **2016**, *5*, e12040. [CrossRef]
120. Stawski, P.; Sumser, M.; Trauner, D. A Photochromic Agonist of AMPA Receptors. *Angew. Chem. Int. Ed.* **2012**, *51*, 5748–5751. [CrossRef] [PubMed]
121. Wolter, T.; Steinbrecher, T.; Trauner, D.; Elstner, M. Ligand Photo-Isomerization Triggers Conformational Changes in iGluR2 Ligand Binding Domain. *PLoS ONE* **2014**, *9*, e92716. [CrossRef] [PubMed]
122. Lemoine, D.; Mondoloni, S.; Tange, J.; Lambolez, B.; Faure, P.; Taly, A.; Tricoire, L.; Mourot, A. Probing the ionotropic activity of glutamate GluD2 receptor in HEK cells with genetically-engineered photopharmacology. *eLife* **2020**, *9*, e59026. [CrossRef]
123. Wang, T.; Ulrich, H.; Semyanov, A.; Illes, P.; Tang, Y. Optical control of purinergic signaling. *Purinergic Signal.* **2021**, *17*, 385–392. [CrossRef]
124. Habermacher, C.; Martz, A.; Calimet, N.; Lemoine, D.; Peverini, L.; Specht, A.; Cecchini, M.; Grutter, T. Photo-switchable tweezers illuminate pore-opening motions of an ATP-gated P2X ion channel. *eLife* **2016**, *5*, e11050. [CrossRef]
125. Lemoine, D.; Habermacher, C.; Martz, A.; Mery, P.F.; Bouquier, N.; Diverchy, F.; Taly, A.; Rassendren, F.; Specht, A.; Grutter, T. Optical control of an ion channel gate. *Proc. Natl. Acad. Sci. USA* **2013**, *110*, 20813–20818. [CrossRef]
126. Kawate, T.; Carlisle Michel, J.; Birdsong, W.T.; Eric, G. Crystal structure of the ATP-gated P2X4 ion channel in the closed state. *Nature* **2009**, *460*, 592–598. [CrossRef]
127. Hattori, M.; Eric, G. Molecular mechanism of ATP binding and ion channel activation in P2X receptors. *Nature* **2012**, *485*, 207–212. [CrossRef]
128. Li, F.; Egea, P.F.; Vecchio, A.J.; Asial, I.; Gupta, M.; Paulino, J.; Bajaj, R.; Dickinson, M.S.; Ferguson-Miller, S.; Monk, B.C.; et al. Highlighting membrane protein structure and function: A celebration of the Protein Data Bank. *J. Biol. Chem.* **2021**, *296*, 100557. [CrossRef]
129. Arkhipova, V.; Fu, H.; Hoorens, M.W.H.; Trinco, G.; Lameijer, L.N.; Marin, E.; Feringa, B.L.; Poelarends, G.J.; Szymanski, W.; Slotboom, D.J.; et al. Structural Aspects of Photopharmacology: Insight into the Binding of Photoswitchable and Photocaged Inhibitors to the Glutamate Transporter Homologue. *J. Am. Chem. Soc.* **2021**, *143*, 1513–1520. [CrossRef] [PubMed]
130. Jumper, J.; Evans, R.; Pritzel, A.; Green, T.; Figurnov, M.; Ronneberger, O.; Tunyasuvunakool, K.; Bates, R.; Žídek, A.; Potapenko, A.; et al. Highly accurate protein structure prediction with AlphaFold. *Nature* **2021**, *596*, 583–589. [CrossRef]
131. Baek, M.; DiMaio, F.; Anishchenko, I.; Dauparas, J.; Ovchinnikov, S.; Lee, G.R.; Wang, J.; Cong, Q.; Kinch, L.N.; Schaeffer, R.D.; et al. Accurate prediction of protein structures and interactions using a three-track neural network. *Science* **2021**, *373*, 871–876. [CrossRef]
132. Sabe, V.T.; Ntombela, T.; Jhamba, L.A.; Maguire, G.E.; Govender, T.; Naicker, T.; Kruger, H.G. Current trends in computer aided drug design and a highlight of drugs discovered via computational techniques: A review. *Eur. J. Med. Chem.* **2021**, *224*, 113705. [CrossRef]
133. Bianco, G.; Goodsell, D.S.; Forli, S. Selective and Effective: Current Progress in Computational Structure-Based Drug Discovery of Targeted Covalent Inhibitors. *Trends Pharmacol. Sci.* **2020**, *41*, 1038–1049. [CrossRef] [PubMed]
134. Scarpino, A.; Ferenczy, G.G.; Keserű, G.M. Binding Mode Prediction and Virtual Screening Applications by Covalent Docking. In *Protein-Ligand Interactions and Drug Design*; Ballante, F., Ed.; Springer: New York, NY, USA, 2021; Volume 2266, pp. 73–88. [CrossRef]
135. Bryce, R.A. What Next for Quantum Mechanics in Structure-Based Drug Discovery? In *Quantum Mechanics in Drug Discovery*; Heifetz, A., Ed.; Springer: New York, NY, USA, 2020; Volume 2114, pp. 339–353. [CrossRef]
136. Aucar, M.G.; Cavasotto, C.N. Molecular Docking Using Quantum Mechanical-Based Methods. In *Quantum Mechanics in Drug Discovery*; Heifetz, A., Ed.; Springer: New York, NY, USA, 2020; Volume 2114, pp. 269–284. [CrossRef]
137. Vivo, M.D.; Masetti, M.; Bottegoni, G.; Cavalli, A. Role of Molecular Dynamics and Related Methods in Drug Discovery. *J. Med. Chem.* **2016**, *59*, 4035–4061. [CrossRef] [PubMed]

138. Gioia, D.; Bertazzo, M.; Recanatini, M.; Masetti, M.; Cavalli, A. Dynamic Docking: A Paradigm Shift in Computational Drug Discovery. *Molecules* **2017**, *22*, 2029. [CrossRef]
139. Śledź, P.; Caflisch, A. Protein structure-based drug design: From docking to molecular dynamics. *Curr. Opin. Struct. Biol.* **2018**, *48*, 93–102. [CrossRef]
140. Lazim, R.; Suh, D.; Choi, S. Advances in Molecular Dynamics Simulations and Enhanced Sampling Methods for the Study of Protein Systems. *Int. J. Mol. Sci.* **2020**, *21*, 6339. [CrossRef]
141. Klaja, O.; Frank, J.A.; Trauner, D.; Bondar, A.N. Potential energy function for a photo-switchable lipid molecule. *J. Comput. Chem.* **2020**, *41*, 2336–2351. [CrossRef] [PubMed]
142. Kotev, M.; Sarrat, L.; Gonzalez, C.D. User-Friendly Quantum Mechanics: Applications for Drug Discovery. In *Quantum Mechanics in Drug Discovery*; Heifetz, A., Ed.; Springer: New York, NY, USA, 2020; Volume 2114, pp. 231–255. [CrossRef]
143. Cembran, A.; Bernardi, F.; Garavelli, M.; Gagliardi, L.; Orlandi, G. On the Mechanism of the *Cis-Trans* Isomerization Lowest Electron. States Azobenzene: S0, S1, T1. *J. Am. Chem. Soc* **2004**, *126*, 3234–3243. [CrossRef] [PubMed]
144. Quick, M.; Dobryakov, A.L.; Gerecke, M.; Richter, C.; Berndt, F.; Ioffe, I.N.; Granovsky, A.A.; Mahrwald, R.; Ernsting, N.P.; Kovalenko, S.A. Photoisomerization Dynamics and Pathways of *Trans-cis*-Azobenzene in Solution from Broadband Femtosecond Spectroscopies and Calculations. *J. Phys. Chem. B* **2014**, *118*, 8756–8771. [CrossRef] [PubMed]
145. Morstein, J.; Impastato, A.C.; Trauner, D. Photoswitchable Lipids. *ChemBioChem* **2020**, *22*, 73–83. [CrossRef]
146. Dämgen, M.A.; Biggin, P.C. Computational methods to examine conformational changes and ligand-binding properties: Examples in neurobiology. *Neurosci. Lett.* **2019**, *700*, 9–16. [CrossRef] [PubMed]
147. Duncan, A.L.; Song, W.; Sansom, M.S. Lipid-Dependent Regulation of Ion Channels and G Protein–Coupled Receptors: Insights from Structures and Simulations. *Annu. Rev. Pharmacol. Toxicol.* **2020**, *60*, 31–50. [CrossRef]
148. Lachmann, D.; Lahmy, R.; König, B. Fulgimides as Light-Activated Tools in Biological Investigations. *Eur. J. Org. Chem.* **2019**, *2019*, 5018–5024. [CrossRef]
149. Komarov, I.V.; Afonin, S.; Babii, O.; Schober, T.; Ulrich, A.S. Efficiently Photocontrollable or Not? Biological Activity of Photoisomerizable Diarylethenes. *Chem. Eur. J.* **2018**, *24*, 11245–11254. [CrossRef]
150. Villarón, D.; Wezenberg, S.J. Stiff-Stilbene Photoswitches: From Fundamental Studies to Emergent Applications. *Angew. Chem. Int. Ed.* **2020**, *59*, 13192–13202. [CrossRef]

MDPI

St. Alban-Anlage 66

4052 Basel

Switzerland

Tel. +41 61 683 77 34

Fax +41 61 302 89 18

www.mdpi.com

International Journal of Molecular Sciences Editorial Office

E-mail: ijms@mdpi.com

www.mdpi.com/journal/ijms